MW00910914

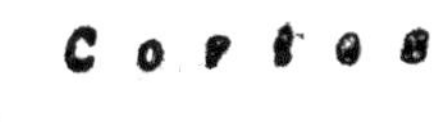
C o r t e s

# SPATIAL ANALYSIS
## A Guide for Ecologists

The spatial and temporal dimensions of ecological phenomena have always been inherent in the conceptual framework of ecology, but it is only recently that they have been incorporated explicitly into ecological theory, sampling design, experimental design and models. The number and variety of statistical techniques for spatial analysis of ecological data are burgeoning and many ecologists are unfamiliar with what is available and how the techniques should be used correctly. This book gives an overview of the wide range of spatial statistics available to analyse ecological data, and provides advice and guidance for graduate students and practising researchers who are either about to embark on spatial analysis in ecological studies or who have started but are unsure how to proceed. Only a basic understanding of statistics is assumed and many schematic illustrations are given to complement or replace mathematical technicalities, making the book accessible to ecologists wishing to enter this important and fast-growing field for the first time.

MARIE-JOSÉE FORTIN is an associate professor in the Department of Zoology at the University of Toronto. Her research focuses on investigating the spatial dynamic processes responsible for creating and maintaining landscape heterogeneity, which in turn facilitates the persistence of species and their conservation. She has active research projects in landscape and conservation ecology, spatial ecology, spatial statistics and forest ecology.

MARK DALE is a professor in the Department of Biological Science at the University of Alberta, and Dean of the university's Faculty of Graduate Studies and Research. His area of research is statistical plant ecology, most recently focusing on the development of spatial pattern in plant communities, much of which is summarized in his book *Spatial Pattern Analysis in Plant Ecology* (Cambridge University Press, 1999). More generally, he works on the development and evaluation of statistical methods with which to test ecological hypotheses, and on their application in answering ecological questions.

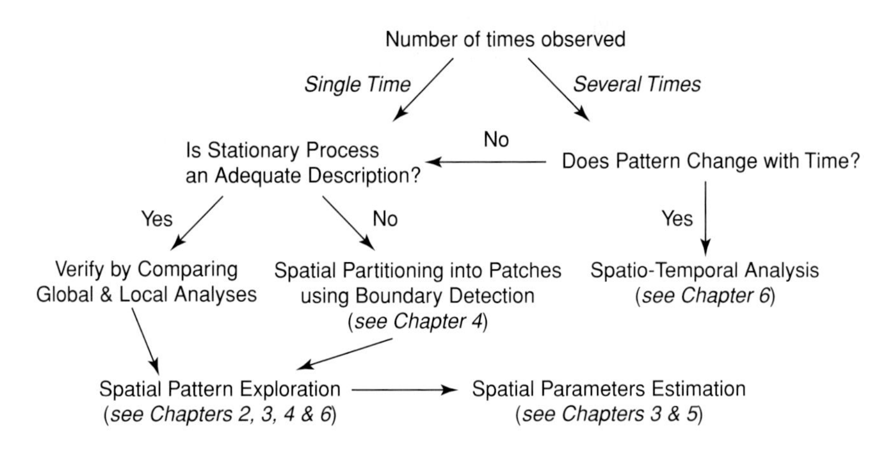


Decision-Tree of the Spatial Analyses Presented in the Book

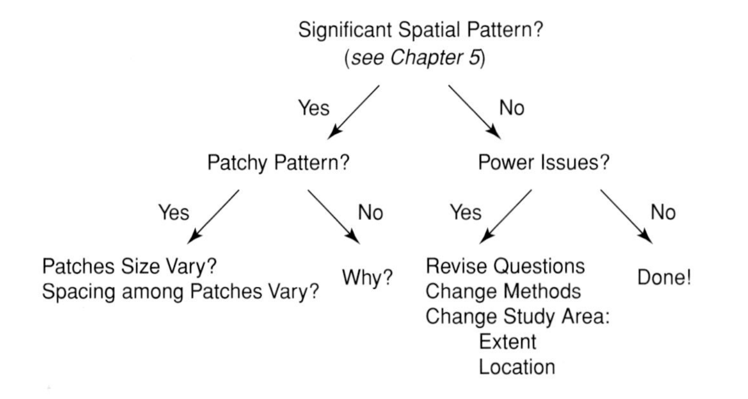


Key Questions to Ask while Analysing Spatial Ecological Data

# SPATIAL ANALYSIS

## A Guide for Ecologists

MARIE-JOSÉE FORTIN
MARK R. T. DALE

CAMBRIDGE UNIVERSITY PRESS
Cambridge, New York, Melbourne, Madrid, Cape Town, Singapore, São Paulo

Cambridge University Press
The Edinburgh Building, Cambridge CB2 2RU, UK

www.cambridge.org
Information on this title: www.cambridge.org/9780521804345



First published 2005

Printed in the United Kingdom at the University Press, Cambridge

*A catalogue record for this book is available from the British Library*

*Library of Congress Cataloguing in Publication data*
Fortin, Marie-Josée, 1958–
Spatial analysis: a guide for ecologists / Marie-Josée Fortin, Mark R. T. Dale.
p. cm.
Includes bibliographical references and index.
ISBN 0 521 80434 5 (alk. paper) ISBN 0 521 00973 1 (alk. paper)
1. Ecology – Statistical methods. 2. Spatial analysis (Statistics) I. Dale,
Mark R. T. (Mark Randall Thomas), 1951– II. Title.
QH541.15.S72F66 2005
577′.015′195 – dc22 2004048884

ISBN-13 978-0-521-80434-5 hardback
ISBN-10 0-521-80434-5 hardback

ISBN-13 978-0-521-00973-1 paperback
ISBN-10 0-521-00973-1 paperback

À Ferko et à Ian
To Phyllis, John and Martha

# Contents

# Preface

Spatial analysis has become the most rapidly growing field in ecology. This popularity is directly related to at least three factors: (1) a growing awareness among ecologists that it is important to include spatial structure in ecological thinking; (2) the alteration of landscapes around us at an increasing rate, which requires a constant re-evaluation of their spatial heterogeneity; and (3) the availability of software designed specifically to perform spatial analyses. One major problem with spatial statistics software is that they are often not used correctly. Incorrect application arises because: (1) ecologists have not been properly trained about issues of scale; and (2) ecologists do not realize fully the implications of the fact that spatially autocorrelated data are not independent, and thus violate the assumptions of the familiar parametric statistics. The purpose of this book is to fill the gap between the current need for spatial analysis and the uncertainty of many ecologists on how to perform these kinds of analysis correctly.

The motivation for this book is as the title suggests; it is intended as a guide for ecologists through the large array of methods available for spatial analysis. Given that the scope of this book is quite broad, it is not as specialized as Dale (1999), which concentrates on the analysis of static spatial pattern. It is crafted as a reference book that could be used as a text in a course introducing ecology students to spatial analysis. The intent is that the book will be a useful guide to help both those who do not know how to start dealing with spatial analysis in ecological studies and those who have started but are unsure how to proceed. Each chapter is more or less self-contained but there are several treads that link them together, including the application of methods and their usefulness in addressing ecological questions. Our goal is to provide a broad overview, as much as possible, of the various well-established spatial methods. Hence, we do not provide much of the theoretical background or mathematical derivations (which are both available elsewhere, in more specialized texts such as Cressie 1993); but we hope that we provide sufficient detail for ecologists to apply and understand the methods. We do not cover all the methods that have ever appeared in print; we have been selective, but we have tried to go beyond what is readily available in the ecological literature, and to include references from fields such as geography, geology and epidemiology, where appropriate.

Most ecological questions are aimed at a better understanding of the complexity of nature and how it works, by testing hypotheses about ecological processes and their interactions. This knowledge-building is based on observation, pattern detection, experimentation and modelling. Hence for ecologists, pattern recognition is only one step in a series to disentangle the complexity of natural systems. Thus, the ecological motivation for performing spatial analysis is to detect pattern, but that is only the beginning of **answering a bigger question**. Ecologists then want to understand the process that generates the pattern. Geographers are probably like ecologists in that the description is of interest, but not the final goal. Epidemiology is essentially applied ecology: looking for pattern to find the process. The classic example is John Snow's study in the 1850s that used the spatial pattern of disease incidence to determine that the Broad Street pump in London was the source of a cholera outbreak (cf. Haining 2003). Identifying the pattern leads to an understanding of the system that gave rise to it. In ecology, however, many of the puzzles are of much greater complexity than tracing the source of disease. Consider the complexities of the processes that give rise to the spatial arrangement of 20 species of tree in a temperate forest . . . and then those for a tropical forest with hundreds of tree species . . . and then all the insects in the tropical forest . . .

This book stems from years of teaching by both authors in their respective universities. Also, it results from career-long learning and from collaborating with our mentors and colleagues: Barry Boots, Ferko Csillag, Geoffrey M. Jacquez, Pierre Legendre, Neal Oden, Chris Pielou, Robert Sokal, Tony Yarranton, the NCEAS working group on 'Integrating the Statistical Modeling of Spatial Data in Ecology', and many others.

We were fortunate to have several people helping along the way with all the details. We thank those who helped: creating the figures: Gillian Forbes, Patrick James, Stephanie Melles and Agnes Wong; editing the various versions of the text: Gillian Forbes, Stephanie Melles and Rebecca Torretti; carrying out the spatial analyses of the data: Patrick James, Yuanyuan Liang, Stephanie Melles and Agnes Wong; with the field work: Ilka Bauer, Vernon Peters, Steve Kembel, Michael Simpson and Agnes Wong. Also, we were privileged to have access to excellent software packages, thanks to Mike Rosenberg (PASSAGE) and Geoffrey Jacquez (BoundarySeer and ClusterSeer by TerraSeer 2001).

For their comments and help on earlier versions of the chapters, we are grateful to: Ferko Csillag, Stewart Fotheringham, Norm Kenkel, Charles Krebs, Pierre Legendre, Stephanie Melles, Evie Merrill, Joe Perry and Mike Rosenberg. We need to thank Joe Perry also for discussion on animal movement analysis. Ferko Csillag provided indispensable technical support for the wavelet analysis example: many thanks. Furthermore, one of us (M.-J. F.) benefited from a constant source of spatial statistics clarification and stimulating discussion, as well as moral support, in the person of Ferko Csillag; it was immensely appreciated.

Finally, we acknowledge the financial support that made possible the research that contributed to the material in the book from the Natural Sciences and Engineering Research Council of Canada and from the University of Alberta. Thanks also to the University of Toronto and the University of Alberta (and our sympathetic immediate 'bosses', James Thomson and Carl Amrhein) for the time to complete this project.

# 1
# Introduction

## Introduction

Processes in natural systems and the patterns that result from them occur in ecological space and time. To study natural structure and to understand the functional processes we need to identify the relevant spatial and temporal scales at which these all occur. While the spatial and temporal dimensions of ecological phenomena have always been an inherent part of the conceptual framework of ecology, it is only recently that they have been incorporated explicitly into ecological theory, sampling design, experimental design and models (Levin 1992). For example, McIntosh (1985), in describing the concepts and theory that form the background of ecological studies, included very little discussion about the spatial aspects of ecological processes. In recent years, however, a growing number of texts have addressed spatial questions by providing both a spatial framework perspective and spatial statistics to perform spatial analyses, for example Cliff & Ord (1973, 1981), Getis & Boots (1978), Ripley (1981), Upton & Fingleton (1985), Anselin (1988), Haining (1990, 2003), Cressie (1991, 1993), Bailey & Gatrell (1995), Manly (1997), Legendre & Legendre (1998), Dale (1999), Fotheringham *et al.* (2000) and O'Sullivan & Unwin (2003).

In this book, we will concentrate on the spatial aspects of ecological data analysis to provide some advice and guidance to practising ecologists. Because all phenomena of ecological interest have both a spatial location, which can be designated by geographic coordinates, and other characteristics, such as measured attributes, we can have different perspectives on how to analyse them:

- their spatial locations can be included explicitly for the purpose of understanding spatial structure and pattern;
- other characteristics of these phenomena can be analysed separately by ignoring, or controlling for, their relative positions (e.g. their topology defined by neighbours) or absolute spatial locations ($x$–$y$ coordinates); or

- spatial locations can be incorporated directly into the evaluation of those other characteristics.

Currently, several advanced spatial statistical books are available that cover the formal mathematics of these methods (e.g. Ripley 1981; Cressie 1993), but they may not be readily accessible to most ecologists and may not provide immediate application in the ecological context. In fact, there is a potential for the misapplication of some of these techniques to lead to incorrect inferences. It is our intention to present the concepts needed to perform valid spatial analyses and interpretation. To enhance the presentation (we hope), we use various data sets to illustrate the behaviour of the methods, and the relationships among them.

There is a large number of spatial statistics and new methods are constantly being developed, and so our presentation will not cover all possible approaches but will concentrate on those that we think are key for ecological analysis. We acknowledge that we are omitting several important fields of research and schools of thought. For example, we are not attempting to cover spatial issues related to diversity, information theory and spatio-temporal modelling since these topics would deserve and require a whole book each.

This book aims to guide ecologists through the broad field of spatial analysis by providing the essential basics to perform spatial studies. The intended audience is graduate students and other practising researchers. The structure of the book is straightforward. We begin by introducing important terms and concepts, taking the opportunity to clarify how they will be used in subsequent discussion. There are then five main chapters that present spatial methods based on their objectives: population (fully censused) data methods, methods for sampled data, boundary detection, methods dealing with spatial autocorrelation and spatio-temporal analysis. Each chapter includes a description of methods, some examples, an evaluation of the methods' characteristics, and concluding remarks including our advice on the choice of method. The last chapter asks and tries to answer the question 'Where do we go from here?', describing what we see as the direction for future development in this field and the areas where we perceive the need for more work. We also summarize our thoughts on the themes and threads that run through the book and unify it, and we provide some advice to students of ecology on the kinds of skills that we think they need in their future work.

## 1.1 Process and pattern

In ecological studies, explicit considerations of spatial structure have come to play an important role in efforts to understand and to manage ecological processes. Therefore, in our quest to comprehend the complexity of nature, the description and quantification of ecological patterns, both spatial and temporal, are important

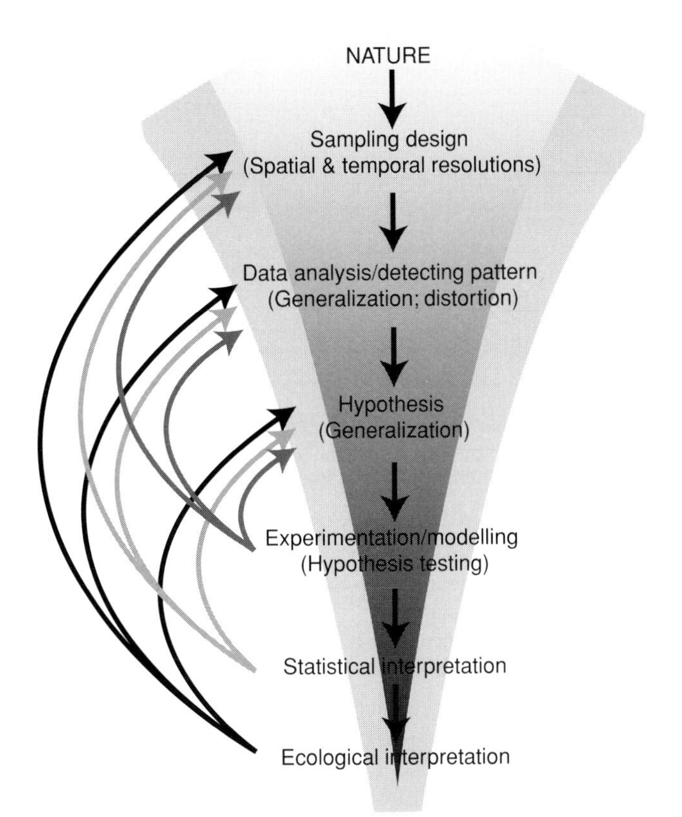


Figure 1.1 Flow of the steps involved in the study of nature and its complexity. As nature acts at several temporal and spatial scales, the selected sampling design narrows down the temporal and spatial limits of the domain under study (as indicated by the funnel effect illustrated in grey). By imposing arbitrary and potentially inappropriate scales by means of the sampling design, the identified spatial patterns can be distorted. From these spatial patterns, generalizations and hypotheses can be drawn about the ecological processes. Then specific experiments or models can be used to test the newly defined hypotheses. And finally, some statistical interpretations and ecological understanding can be reached. At each step, the spatial and temporal domains of inference of the findings diminished.

first steps. Description is not usually an end in itself, but rather the beginning of a process that leads to insight into natural complexity, and which in turn generates new ecological hypotheses that can be tested either by experiments or by modelling (Figure 1.1). Therefore, ecological research is an iterative process that, at each iteration, provides some insights about the underlying ecological processes through the quantification of ecological patterns. Unfortunately, the match between pattern and process is far from perfect, because changes in process intensity can create different patterns, and because several different processes can generate the same

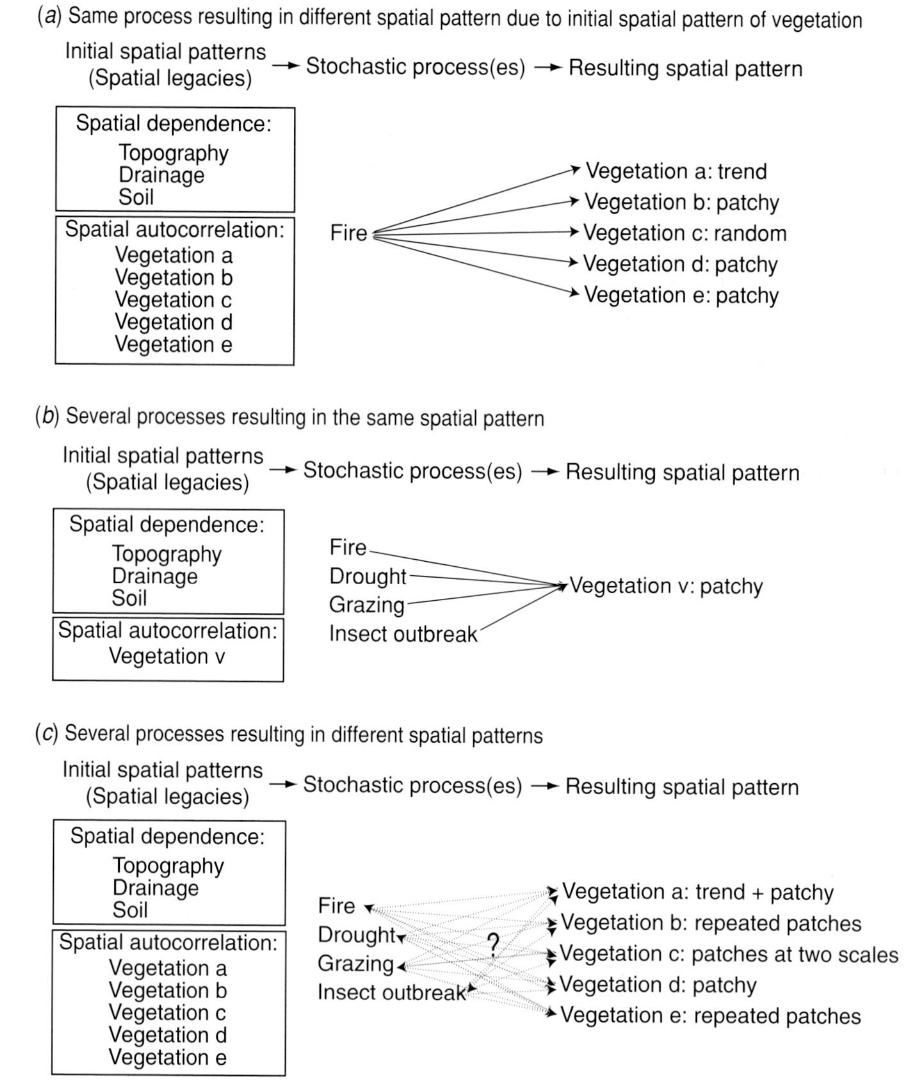


Figure 1.2 Relations between pattern and process. (*a*) Given the initial conditions of the environmental factors and the legacy of the landscape spatial structures, the same intensity of a process can result in different spatial patterns. (*b*) For a given spatial legacy, several processes can generate a given spatial pattern. (*c*) Most of the time there are several spatial legacies nested within each other, which are affected by several processes resulting in several distinct spatial patterns.

pattern signature (Figure 1.2). Furthermore, the processes may create a mosaic of intermingled and confounded spatial patterns, and the spatial legacy of this heterogeneity affects the intensity and types of ecological processes that act on them through time. These feedback effects between processes and patterns are difficult to distinguish (Figure 1.2*c*). Prior knowledge of the scope of these processes can help to guide the scale chosen for the investigation of spatial patterns. The term 'scale' is used by ecologists to refer to any of several concepts including the extent of the processes and the spatial and temporal resolution of the data. For a more detailed presentation and discussion of the concepts of scale, we refer readers to Csillag *et al.* (2000) and Dungan *et al.* (2002). When the scale at which the processes are realized is unknown, analysing spatial patterns using different approaches and scales of observation can provide an overall consensus that contributes to our understanding of ecological complexity.

To clarify this discussion we need to define 'pattern' exactly and to circumscribe the analytic limits of detecting it accurately. One definition of 'a pattern' is 'a distinctive form' (Webster 1989), implying that a pattern can be detected and described. Another definition (Fowler & Fowler 1976) is 'regular form or order' and hence the term 'pattern' is sometimes used as the opposite of 'random'. Either definition can then be qualified according to whether one is interested in spatial or temporal component of a pattern. These definitions lack the implication that pattern in ecological systems is dynamic, evolving and changing. Indeed, a spatial pattern is usually 'a single realization' or 'snapshot' of a process or of a combination of processes at one given time (Fortin *et al.* 2003). This is why spatial pattern is so important in ecology and why we emphasize its analysis. Furthermore, our perception of the spatial structure of an area is directly related, and limited, to both the study area or 'extent' and sampling unit size or 'grain' at which we analyse it (Wiens 1989). Thus, depending on the spatial scale of observations, an area may be considered homogeneous when the extent is small (e.g. one forest stand), or heterogeneous when the extent is large (e.g. a mosaic of forest stands). The physical distances that determine local vs. global can be very different depending on the system studied; just as 'landscape' is a level of organization with the distance encompassed determined by the organism of interest: beetle vs. coyote.

Ecological data usually include several kinds of spatial pattern, which are confounded (Figure 1.3): (1) trends at larger scales, (2) patchiness at intermediate and local scales, and (3) random fluctuations or noise at the smallest scale. Therefore, ecological data are the result of embedded and confounded processes and, hence, as ecologists, we are trying to disentangle the spatial scales of these processes using spatial analysis. The components that affect our ability to identify spatial patterns and their underlying processes accurately are numerous, but they can be organized into three main categories (Figure 1.4; Dungan *et al.* 2002): (1) the extent of spatial expression of the processes themselves; (2) the sampling design used to measure ecological data (sample vs. population data; local vs. global level); and (3) the statistical tools used to characterize the spatial pattern of either the entire sampling area (i.e. global spatial statistics) or each sampling location (i.e. local spatial statistics). In the following sections we will present the implications of the three components and how they are tied together through the notion of 'scale'. Indeed,

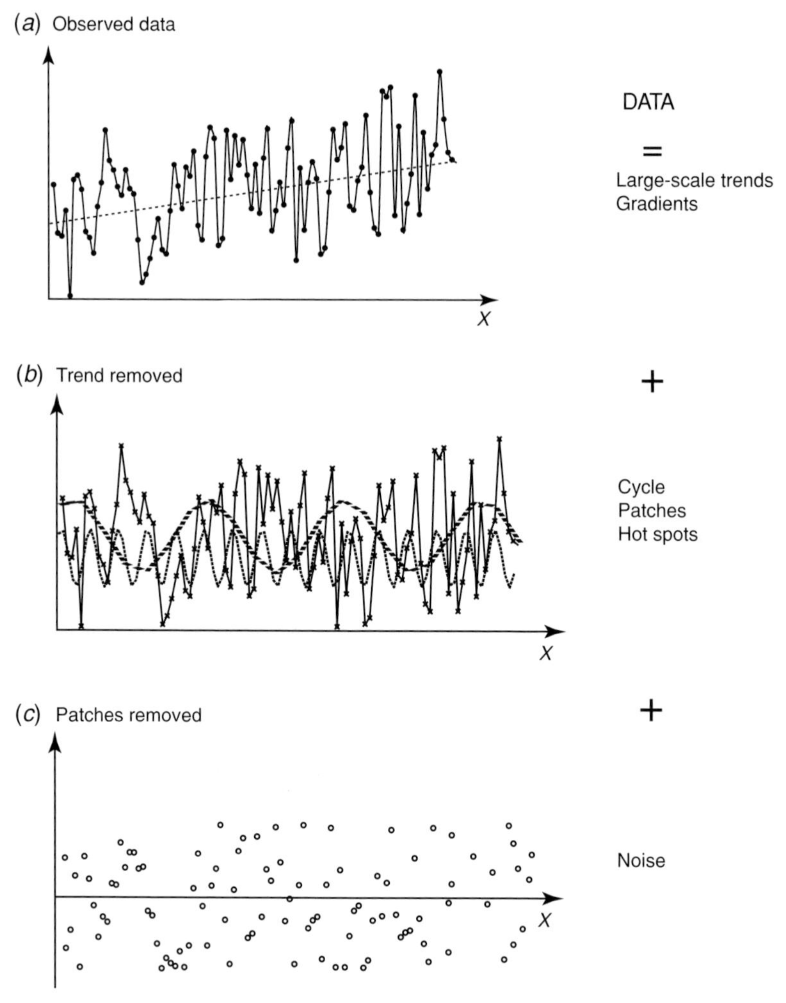


Figure 1.3 Nested spatial patterns (signals) imbedded in ecological data: (*a*) if the data are gathered along a temperature gradient, tree height can increase in a linear fashion at large scale; (*b*) both topography and spatial dispersal processes can generate patchy patterns at intermediate, landscape, scale; and (*c*) there is only random noise at the micro, local, scale.

the identification of ecological processes through the detection of the spatial and temporal patterns that they create vary according to the different aspects of the scale of observation arbitrarily imposed by the sampling design and analytic tools used (Bradshaw & Fortin 2000; Csillag *et al.* 2000; Dungan *et al.* 2002).

## 1.2 Spatial pattern: spatial dependence versus spatial autocorrelation

Spatial structures and patterns can take several forms: (1) trend, gradient (Figure 1.5*a*); (2) aggregation, clumping, patchy (Figure 1.5*b*); (3) random (Figure 1.5*c*);

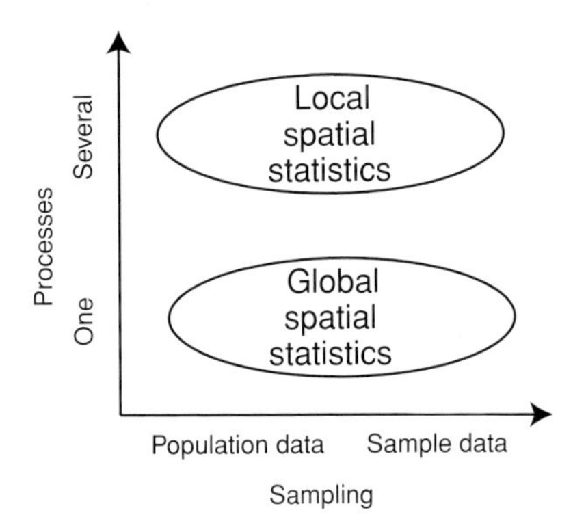


Figure 1.4 Three main components that interact and affect our ability to identify and characterize spatial patterns accurately: the scale of expression of processes, the sampling design being used at the plot level or landscape level, and the spatial statistics characterizing either the spatial structure of each sampling location (local spatial statistics) or the entire study area (global spatial statistics).

and (4) for point patterns, uniform, regular or overdispersed patterns. Either exogenous or endogenous processes can generate these patterns. In the case of exogenous pattern generation, the identified spatial pattern is generated by factors independent of the variable or characteristics of interest. Several factors can act at the same time, interacting either additively when the factors are linear, or multiplicatively or otherwise when the factors are non-linear. Several spatial patterns can be identified when the variables of interest, such as species abundances, respond to an exogenous process such as a disturbance or to underlying environmental conditions, such as a moisture gradient on a slope for plants or the spatial configuration of habitats for animals. For example, soil patchiness can result in patches of plants within which the locations of the individual plants are randomly arranged or even overdispersed. In these cases, the values associated with the plants are likely to be similar, not due to internal processes, but rather because the species are responding to external processes which have their own spatial structure; for example, these plants may grow only on a specific type of soil that is itself patchy in its distribution. Hence the spatial structure of plant species is due to the spatial structure of the environmental variables only. On the other hand, when there are endogenous ecological processes involved, such as dispersal, spatial competition or spatial inhibition, plant patchiness is an inherent property of the variable of interest.

In fact, most ecological data have at least some degree of spatial structure, often described by what is known as the first law of geography: 'Everything is related to everything else, but near things are more related than distant things' (Tobler 1970). These spatial patterns usually result from a mixture of both exogenous ('induced') and endogenous ('inherent') processes acting on species spatial structure resulting in spatial dependence among individual organisms (e.g. plants).

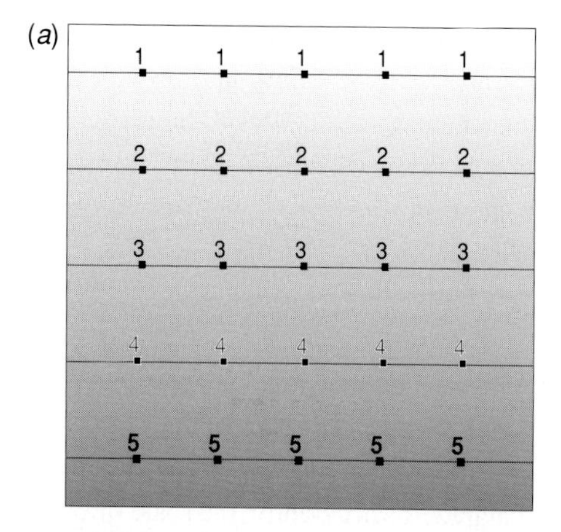


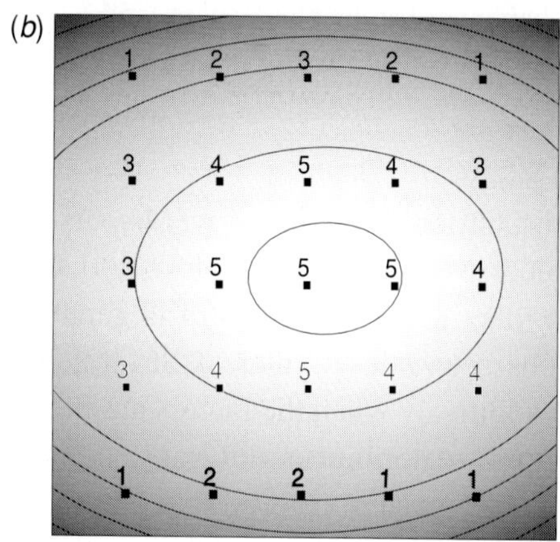


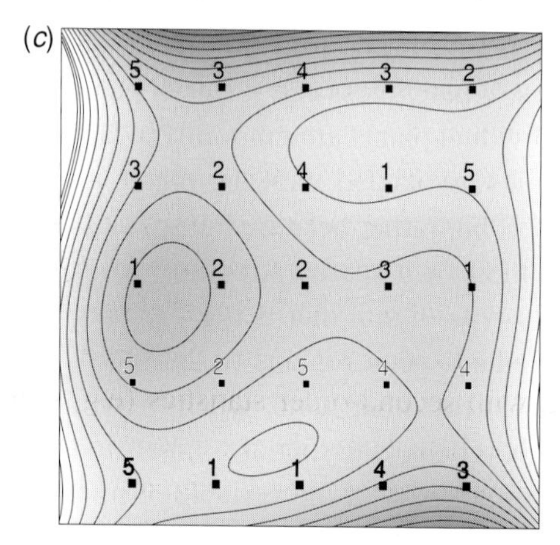


Figure 1.5 Spatial patterns: (*a*) gradient, (*b*) single patch and (*c*) random (although the isolines seem to suggest a patchy pattern). Note that each panel has the same number of sampling locations ($5 \times 5 = 25$), as well as the same frequency distribution of the count of individuals (5 ones; 5 twos; 5 threes; 5 fours and 5 fives).

Here, the term 'spatial dependence' is used broadly to include a mix of both the species' response to the underlying (exogenous) processes and the species' spatial autocorrelation due to endogenous processes. The term 'autocorrelation' refers to the degree of correlation of a variable and itself ('auto'). By adding the adjective 'spatial', it implies that the relationship among the values of a given variable is a function of the spatial distances between them or their locations in space. Hence, the notion of spatial dependence implies that there is a lack of independence among data from nearby locations. This definition of spatial dependence is the most widely used by spatial statisticians and geographers (Cressie 1993; Haining 2003). Bailey & Gatrell (1995, p. 32) defined spatial dependence using an analogy to first- and second-order moments: a first-order effect is due to variation in the mean value of a process over the study area, corresponding to the large global trend illustrated in Figure 1.3*a*, and second-order effects are due to spatial autocorrelation of the process, implying that deviations of process values from the mean are more alike at neighbouring sampling locations, and hence are equated with localized trends and small-scale patchiness (Figure 1.3*b*).

Therefore, although Legendre (1993) used the term 'false' spatial autocorrelation to refer to species' response to the spatial structure of exogenous processes, we will not use this terminology in this book for clarity and for compatibility with other textbooks on spatial analysis in other fields. Instead, we will refer to this phenomenon as 'induced spatial dependence', which is a more general term that includes species response to spatially structured environmental processes at more than one spatial scale.

In describing spatial dependence of plants, where exogenous processes predominate, we would say that the spatial dependence is 'induced' by the underlying variable that is itself spatially autocorrelated. Such spatial patterns can be modelled by means of regression where the independent variables are themselves spatially structured (Legendre & Legendre 1998). For cases of endogenous processes, individuals of a species are more likely to be spatially adjacent in a patchy fashion, related to what is referred to as 'true' (Legendre 1993; Legendre & Legendre 1998) or 'inherent' spatial autocorrelation. This means that nearby values of a variable are more likely to be similar than they would be by chance. The spatial structure can therefore be modelled with second-order statistics (e.g. spatial covariance rather than just mean value) that characterize the local spatial variability of the variable.

The degree of spatial dependence can be estimated by comparing the value at one location with those at given distances apart (termed spatial lag or distance interval), say at 1 m, 2 m, and so on. In Figure 1.6, we present a situation where spatial autocorrelation occurs only due to seed dispersal from a tree. Due to the dispersal process, we expect to find fewer and fewer seeds as the distance from the source increases (Figure 1.6). The degree of spatial autocorrelation in space

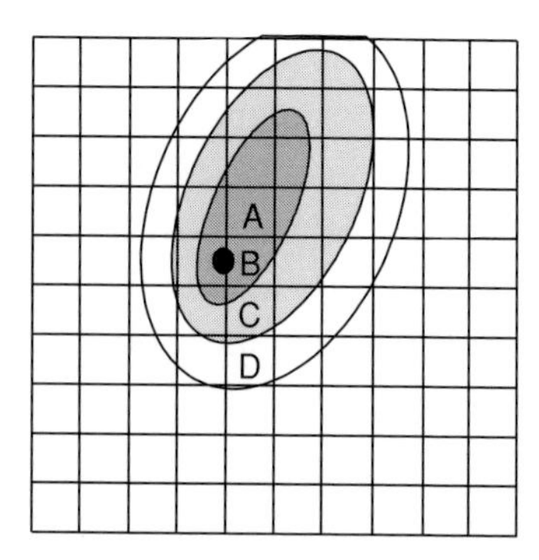


Figure 1.6 Seed abundance from a tree source. The filled circle indicates the location of the tree source from which seeds are dispersed by wind. As the distance from the tree increases, the amount of seeds decreases (as indicated by the grey-shaded gradient: dark grey for high abundance; light grey for low abundance; white for no seeds). Positive spatial autocorrelation exists between adjacent sampling units A and B; no significant spatial autocorrelation exits between A and C; and negative spatial autocorrelation exists between A and D.

will also decrease as the spatial distance increases, for example from locations A to D in Figure 1.6. At short distances from the tree releasing seeds, values of seed abundance should be similar within patches or at nearby locations along a gradient, giving positive autocorrelation, and as the distance at which the comparison is made increases, the values are less likely to be similar. They can become either independent, with no spatial autocorrelation, or dissimilar, with negative autocorrelation. Over large areas, plants can have a patchy pattern that repeats itself to create spatial structure at two scales: (1) a within-patch scale of plants and (2) a between-patch scale of patches in their landscape.

The magnitude of the ecological process usually has a direct effect on the degree of spatial autocorrelation in the variable that it influences (e.g. its intensity, spatial range or sign). The intensity of spatial autocorrelation can vary according to direction (Figures 1.6 and 1.7). In the previous example of seed abundance, with the presence of strong directional wind, because seeds are more likely to be dispersed downwind (say northeast–southwest), an elongated, elliptical, patch of seeds results (Figure 1.6). This kind of spatial pattern is said to be 'anisotropic', because the intensity and range of spatial autocorrelation vary with the orientation or direction; the opposite is 'isotropic' where spatial autocorrelation intensity varies similarly with distance in all directions (Figure 1.7). Various types of internal and external processes can create anisotropic pattern: topography, gradients (e.g. brousse tigrée; see Chapter 2; Lejeune & Tlidi 1999; Wu *et al.* 2000), stream and riparian strips, etc. Anisotropic spatial patterns can appear as artefacts of the shape of the sampling units used to collect the data as will be discussed in more detail in Section 1.4.2 (cf. Fortin 1999a).

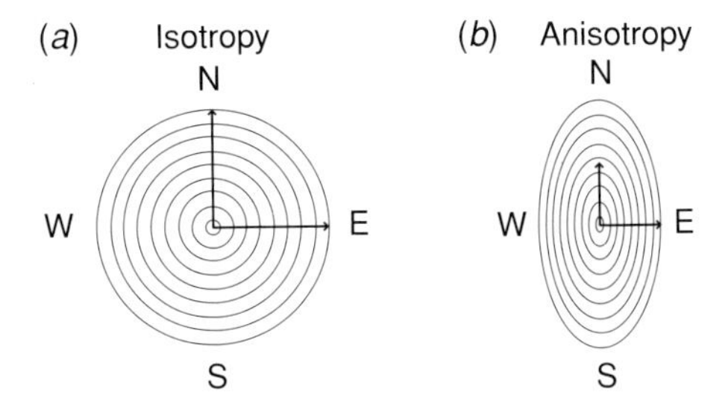


Figure 1.7 Pattern directionality. (*a*) Isotropic and (*b*) anisotropic spatial patterns. Each isoline indicates the same value of the variable decreasing from the highest value at the centre to the lowest value at the periphery.

## 1.3 The concept of stationarity

The spatial pattern in a given area is a synthesis of dynamic processes operating at various spatial and temporal scales (Figure 1.3). Hence the spatial structure at any given time can be viewed as one realization among several potential outcomes of the interactions among these processes (see Figure 1.2). To make meaningful ecological interpretations of the spatial pattern, we need to make some assumptions about the underlying processes (Figure 1.4). Similarly, spatial statistics are usually based on the assumption that the process being studied is stationary (Figure 1.8). In a spatial context, a process, or the model of a process, is stationary (or homogeneous) if its properties are independent of the absolute location and direction in space (Haining 1990; Burrough 1987). In other words, the parameters of the process, such as the mean and the variance, should be the same in all parts of the study area and in all directions. This assumption of stationarity is equivalent to the assumption of the independence of the observations for parametric statistics, implying that the data follow a known distribution and are homoscedastic.

The property of stationarity is required for making inferences from a model that characterizes the process of the spatial structure of data at locations that are not sampled. This implies that the assumption of stationarity is related to a property of the model or the process, and not to a property of the data. Note that a stationary process can generate a spatial pattern that looks non-stationary, perhaps because of a trend. In such cases, the data can be detrended. Furthermore, the property of stationarity is scale dependent, as illustrated in Figure 1.8. When data values vary from place to place, resulting in spatial heterogeneous patterns due to changes in both mean and variance, the assumption of stationarity required for spatial statistics is not fulfilled. The implications are that the identified spatial pattern can be distorted and inaccurate (Boots 2002; Fortin *et al.* 2003), so that any subsequent spatial inferences will be invalid (see Chapter 5; Legendre 1993; Dale & Fortin 2002). Thus, it is important that spatial statistics are calculated over areas (or subregions of

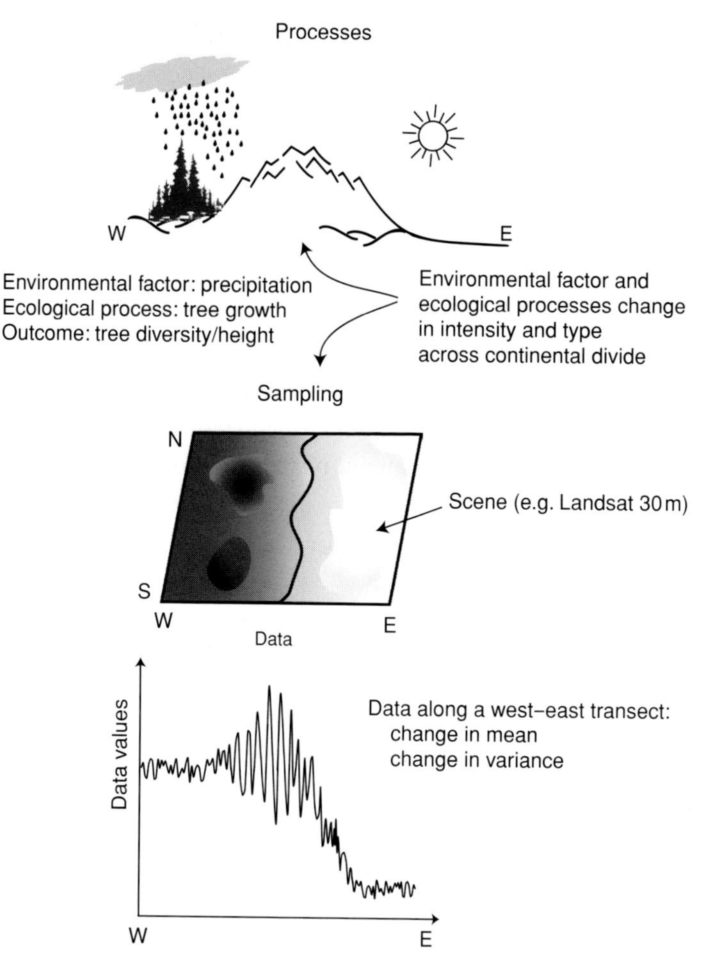


Figure 1.8 The concept of stationarity: a process level issue. From the west to the east of the continental divide, the amount of precipitation varies, which affects tree growth and diversity. While sampling this region using remotely sensed imagery (a scene) to estimate tree net primary productivity, the extent of the scene includes both sides of the continental divide, which are not under the same process regime. The mean and variance of net primary productivity change along a transect from west to east.

them) for which stationarity can be safely assumed (Figures 1.8 and 1.9). Because stationarity is a property of the process, it cannot be tested directly, but we can determine whether or not the landscape is homogeneous by computing the mean and the variance of the data using sliding windows of varying sizes. When processes are obviously not stationary (as illustrated in Figures 1.8 and 1.9*b*, *c*), we need first to identify homogeneous subregions by means of spatial partitioning methods (as presented in Chapter 4).

(*a*) Homogeneous (*b*) Heterogeneous (*c*) Locally homogeneous Globally heterogeneous

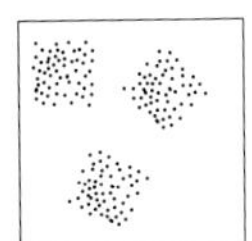

Figure 1.9 Homogeneous (*a*) versus heterogeneous landscapes (*b*, *c*). In (*b*), the study area includes subregions where plants cannot grow, such as a lake, on rock or in inappropriate soil types. In (*c*) the plants can grow only in subregions where the soil type is appropriate.

In studies where the scale of observation is at the landscape level, the spatial pattern of the data may change but the assumption of stationarity of the process is still required at the analysis stage. Data cannot usually be collected as intensively at the landscape level as at the stand or plot level. Hence, at the landscape level, the data are usually obtained by using remote sensing or as vegetation inventory maps created from air photo interpretation. Such information is mostly in broad categories, such as forest vs. water, vs. urban, or mature spruce forest vs. peatland. So, while at the plot level, spatial statistics are used to characterize spatial pattern from quantitative data (among others, Haining 1990, Cressie 1993), at the landscape scale, landscape indices (O'Neill *et al.* 1988; Baker & Cai 1992; McGarigal & Marks 1995; Gustafson 1998) are commonly used to summarize the spatial configuration of categorical data.

From the plot to the landscape level, the inherent stationarity of the process needs to be ensured to avoid distorting spatial pattern identification and incorrect ecological interpretations. Spatial statistics, summarizing the spatial pattern for the entire study area with a single number, should not be applied to a study area with non-stationary processes. Instead, 'local spatial statistics' should be used to estimate the spatial pattern for each sampling location (Boots 2002; Table 1.1). Only when stationary processes prevail in all the study area, can spatial statistics be used to characterize the study area with a single value (see Tables 1.2 and 1.3; Appendices 1 and 2). In such cases, they are referred to as 'global spatial statistics' to stress the fact that the statistics are considered valid for the entire area (Table 1.1). In this book, global spatial statistics will be referred to loosely as 'spatial statistics'; local statistics will be designated explicitly.

## 1.4 Sampling

Natural complexity occurs at several spatial and temporal scales. Thus, as mentioned above, several processes may be acting in combination, either additively or multiplicatively, where linear (e.g. trend) and non-linear (e.g. threshold, sigmoid) spatial

Table 1.1 *Implications of process properties on sampling design, spatial and statistical analyses*

| Stationarity processes | Non-stationarity processes |
| --- | --- |
| Random sampling design | Stratified sampling design |
| Global spatial statistics | Local spatial statistics |
| Parametric statistics | Restricted randomization tests |

patterns are confounded, making it hard or impossible to disentangle them. Any sampling design for studying ecological processes imposes an arbitrary template, or filter, with its own specific temporal and spatial units (Bradshaw & Fortin 2000). To be efficient, a sampling design needs to be thought out and crafted carefully by a series of steps to obtain meaningful insights about the ecological processes (see Figures 1.1 and 1.2; summarized in Table 1.2 and presented in Chapters 2 and 3): (1) define explicitly the spatial and temporal domains of expression of the process(es) under study; (2) determine that the spatial and temporal resolution of the sampling design is able to capture the process under study; and (3) ensure that the spatial and statistical analyses are appropriate for the data type. These three steps interact with each other and must be considered as a whole before going out to do any sampling (Dungan *et al.* 2002). Any ecological study requires, therefore, a comprehensive overview of all the components and the steps involved (Figure 1.1). Indeed, once the data are gathered, one cannot obtain more information from the data set than it actually contains. Therefore, determination of the appropriate spatial and temporal domains of the study is one of the most important steps in data analysis from which all subsequent statistical and ecological interpretations will be either meaningful or meaningless.

### 1.4.1 Ecological data

Various kinds of measurements can be considered as ecological data, ranging from qualitative records (e.g. taxonomic species), semi-quantitative (e.g. non-additive values such as pH, ranked data with uneven interval classes), to quantitative measures (e.g. abundance data, height, weight). These measurements can be made for individuals (point data: e.g. discrete objects, organisms), along a line (transect data), over an area (surface data: e.g. within a sampling unit) or in a volume (e.g. phytoplankton productivity in a water column with $x$, $y$ and $z$ coordinates); see Figure 1.10. When sampling units are used, these can either be spatially adjacent, contiguous to one another or separated by a constant or variable distance (Figure 1.10, Table 1.2). In either case, the measurements are subject to several

Table 1.2 *Classification of spatial statistics according to objectives*

| | Data[a] | | |
|---|---|---|---|
| | Population | | Sample |
| Objectives | Point ($x$–$y$) | Lattice ($x$–$y$, $v$) | Sparse ($x$–$y$, $v$) |
| Topology (linking sampling locations) | Networks (nearest-neighbour, relative neighbourhood graph, minimum spanning tree, Gabriel, Delaunay) | Networks (rook, bishop, queen) | Networks (nearest-neighbour, minimum spanning tree, Gabriel, Delaunay) |
| Exploration (detection of spatial structure) | Aggregation indices<br>*k*-nearest-neighbours [i, r, t]<br>Join count [i, r, t]<br>Ripley's *K* (uni-, bi-multivariate [i, r, p]<br>Dixon's method [i, p]<br>Circumcircle [r, p]<br>Fractal dimension [i, r] | Join count [i, r, t]<br>Global Moran's *I*, Geary's *c* [i, r, d, t]<br>Semi-variance [i, r, d, p]<br>Local Moran's *I*, Getis' *G**, Ord's *O* [i, r, t]<br>Mantel correlogram [i, r, t]<br>Mantel and partial Mantel tests [i, t]<br>Fractal dimension [i, r]<br>Lacunarity [i, r]<br>Block variance [i, r]<br>Mark correlation [i, r]<br>Circumcircle [i, r, p]<br>Spectral analysis [i, r]<br>Wavelets [i, r] | Global Moran's *I*, Geary's *c* [i, r, d, t]<br>Semi-variance [i, r, d, p]<br>Local Moran's *I*, Getis' *G**, Ord's *O* [i, r, t]<br>Mantel and partial Mantel tests [i, t] |
| Inference (parameters estimation, testing hypotheses) | | Mantel and partial Mantel tests [i, t]<br>Semi-variance [i, r, d, p]<br>Autoregressive models<br>Conditional annealing | Mantel and partial Mantel tests [i, t]<br>Semi-variance [i, r, d, p]<br>Autoregressive models<br>Conditional annealing |
| Interpolation (mapping) | Voronoi polygons | Trend surface analysis<br>Kriging | Trend surface analysis<br>Thin-spline<br>Kriging |
| Partition (segmentation) | | Spatial clustering<br>Boundary detection [p]<br>Wavelets | Spatial clustering<br>Boundary detection [p] |

[a] ($v$: 'value' of a given variable (either qualitative or quantitative); i: the method can estimate the intensity of the spatial structure; r: the method can estimate the spatial range (zone of influence) of the spatial pattern; d: the method can estimate the intensity of spatial pattern according to the orientation/directionality (so it can differentiate isotropic from anisotropic patterns); t: significance tests (either analytic or randomization tests); p: significance tests based on randomization tests).

Table 1.3 *Requirements, assumptions and rules of thumb for spatial statistics*

| Methods | Requirements | | |
|---|---|---|---|
| | Objectives | Assumption | Rules of thumb/limits |
| Networks | Linking locations | Exhaustive mapping of all events | Planar graphs are more useful because they reduce the number of links in a network |
| Aggregation indices | Testing for spatial structure | Stationarity | Cannot differentiate among spatial structures |
| Block variance methods | Computing the intensity and range of spatial structure | Contiguous sampling units | Computed up to 1/10 of the length of the transect |
| Ripley's $K$ | Computing and testing the intensity and range of spatial structure | Exhaustive mapping of all events | Extents having rectangular shape are favoured for computation<br>Edge effect needs to be corrected for<br>Statistic computed up to 1/3 to 1/2 of the shortest distance of the extent |
| Moran's $I$, Geary's $c$ | Description of spatial structure; testing for the presence of spatial autocorrelation | Stationarity | Edge effect needs to be corrected for<br>Statistic computed up to 1/3 to 1/2 of the shortest distance of the extent<br>Minimum of 20–30 sampling locations |
| Semi-variance | Description of the spatial structure; estimation of spatial parameters | Pseudo-stationarity | Edge effect needs to be corrected for<br>Statistic computed up to 1/3 to 1/2 of the shortest distance of the extent<br>Minimum of 50 sampling locations<br>Search neighbourhood of either the range distance or 12–25 neighbouring sampling locations |
| Mantel and partial Mantel tests | Correlation between spatially autocorrelated data | Stationarity | Overall, synthetic, value of the linear relationship between distance matrices, not the raw data |

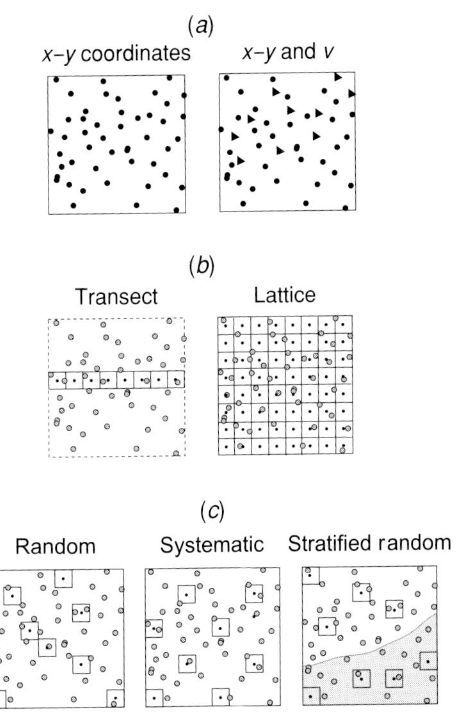


Figure 1.10 Spatial sampling strategies to collect ecological data: (*a*) point data methods: exhaustive survey of the geographic *x–y* coordinates of all the individuals of a species (left panel) or of more species (right panel; here two species, where *v* indicates the attribute of each individual – in this case, the species' name); (*b*) contiguous sampling units: transect (left panel) and lattice (right panel); and (*c*) sparse sampling units (random, systematic, stratified random). See text for more details.

kinds of precision and accuracy issues: (1) for quantitative measurements, their quality is a function of the precision and accuracy of the instrument, of an observer to count species abundance or to estimate per cent cover with the same accuracy over time; (2) for qualitative data, the accuracy with which the attribute is identified is directly related to the ability of the observer to identify species correctly; (3) positional accuracy of the coordinates of either the individuals or sampling units depends on the precision and accuracy of the instrument used (GPS, telemetry, laser, tape measure, etc.); (4) precision in data gathering and transfer to digital form (accuracy of transcription); and (5) appropriate match between the sampling unit size and the variable measured (Fortin 1999a; Bradshaw and Fortin 2000). All these accuracy levels and types of errors will affect the identification and quantification of spatial patterns (Burrough & McDonnell 1998; Hunsaker *et al.* 2001; Appendix 1). All these accuracy problems cannot be eliminated but they can be minimized or at least acknowledged while analysing and interpreting spatial structure.

### 1.4.2 Sampling design

In view of the accumulated evidence (among others, Fortin *et al.* 1989; Jelinski & Wu 1996; Qi & Wu 1996; Fortin 1999a; Dungan *et al.* 2002), all the steps and decisions involved in sampling data will affect the identified spatial pattern:

(1) the sample size (the number of observations '*n*');
(2) the size of the study area (the extent);
(3) the size (the grain) and shape of the sampling or observational units (Figure 1.11);
(4) the sampling strategy or spatial layout of the sampling units used to collect the data (e.g. transect, lattice, random, systematic, stratified, . . . , see Figure 1.10); and,
(5) the spatial lag, spatial distance, among sampling units: (a) non-contiguous sampling units where the spatial lag is measured either from edge to edge ($L_e$) of the sampling units or from centroid to centroid ($L_c$) of the sampling units (Figures 1.10 and 1.11); (b) contiguous sampling units having no edge-to-edge spatial lag but a centroid-to-centroid spatial lag equivalent to the length of the sampling unit.

In any ecological study, the choice of the sample size '*n*' is one of the most important decisions that confront ecologists. In the context of spatial analysis, this choice needs to be guided by the minimum requirement for subsequent spatial statistics and analysis (Fortin *et al.* 1989; Legendre & Fortin 1989; Fortin 1999a; Table 1.3). For example, a minimum of 30 sampling locations is recommended to detect significant spatial autocorrelation (Legendre & Fortin 1989; Table 1.3). In cases where the spatial pattern is very strong, it may be detected with as few as 20 sampling locations, but this would be exceptional. Reliable estimation of spatial structure and spatial model parameters may require 100 or more sampling locations.

The detection of spatial pattern is also directly related to the spatial scale at which ecological data are measured (Figures 1.2 and 1.11). Spatial scale has at least two aspects: the size of the study area, or 'extent', and the size of the sampling unit used to collect the data, the 'sampling grain'. The extent is the total area under consideration and aims to capture the domain of the ecological process under study, while grain refers to the minimum spatial resolution at which information is measured. Several studies have showed that spatial statistics are very sensitive to both of these (Jelinski & Wu 1996; Qi & Wu 1996; Fortin 1999a; Dungan *et al.* 2002). As a guideline in determining the extent of the study area, O'Neill *et al.* (1996, 1999) suggested that it should be at least two to five times larger than the spatial extent of the largest process under study. If the study area is too small in relation to the ecological process studied, not enough of the pattern is included, and because it is not picked up by the data, it may not be identified. If the extent is too large, several processes may be included and non-stationarity may become a problem with different processes affecting the subregions of the study area differently (as illustrated in Figures 1.8, 1.9 and 1.11*a*). This is especially true

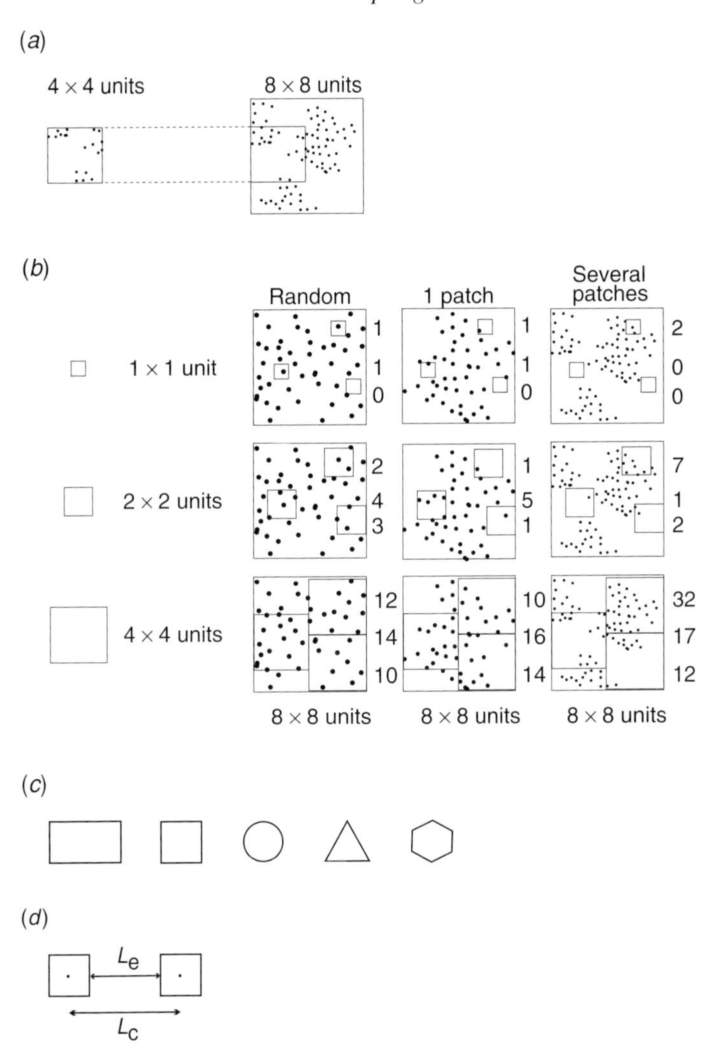


Figure 1.11 Sampling design: (*a*) spatial extent; (*b*) sampling unit size, where the numbers to the right of each study plot are the count of individuals per sampling unit; (*c*) sampling unit shape; and (*d*) spatial lag ($L_e$ indicates the distance 'edge-to-edge' between the sampling units, $L_c$ indicates the distance 'centroid-to-centroid' between the sampling units). See text for more details.

for remotely sensed images because it is unlikely that the study area corresponds exactly with naturally defined homogeneous areas. Unfortunately, the smaller the extent, the more likely it is that a high proportion of patches will be truncated by the limits of the study area (Figure 1.11*a*).

The size of the sampling unit also needs to be considered (Figure 1.11*b*); it sets the smallest spatial resolution at which data are measured and at which spatial structure can be characterized. When a landscape is analysed using a remotely sensed image, the sampling unit size is the pixel resolution. In such cases, it is unlikely that the pixel resolution matches the ecological process of interest, which affects the detected spatial pattern (Bradshaw & Fortin 2000). Again, O'Neill *et al.* (1996, 1999) suggested that the sampling unit size should be two to five times smaller than the patch or other features of interest. The sampling unit should be, however, large enough to contain more than one individual (Figure 1.11*b*), but not so big that there is too much within-unit variability, or that the smallest scale cannot be detected (Figures 1.3 and 1.11*b*). For randomly distributed objects, the size and shape of the sampling unit do not affect our ability to determine the absence of spatial pattern (Figure 1.11*b*). However, when the data are not randomly distributed (Figure 1.11*b*), a sampling unit that is too small (e.g. $1 \times 1$ in the figure) will increase the variance and a sampling unit that is too large (e.g. $4 \times 4$) will reduce the variability (unless the extent includes a non-stationary process). Here, the optimal sampling unit size is $2 \times 2$. In general, if there is a choice of sampling unit size we suggest that a smaller sampling unit should be favoured because small units can be aggregated into larger ones without the loss of information, but the reverse is not true.

When we use a sampling unit that is more-or-less isotropic, such as a square, a circle or a hexagon, we may be assuming implicitly that the spatial pattern is also isotropic. Ecologists often use rectangular sampling units (and very rarely triangular ones) to reduce within-sampling unit variability along a gradient (Fortin 1999a). Such an anisotropic shape can alter the spatial pattern detected by artificially generating the appearance of an anisotropic spatial pattern (Fortin 1999a). When it is not known in advance whether the ecological data are isotropic or anisotropic, we recommend using small isotropic sampling units so that the spatial pattern can be characterized better (more details will be presented about this issue in Chapter 3). In short, both the sampling unit size and shape affect the accuracy of spatial pattern detection and using an inappropriate sampling unit to study a process at a given scale may result in detecting less spatial structure than is actually present (Fortin 1999a; Bradshaw & Fortin 2000; Chapter 3).

Once both the extent and the grain are defined for a study, there are still important decisions on the spatial arrangement of sampling to be made: should we use contiguous or spaced units (Figure 1.10)? Using contiguous sampling units in a transect or a lattice allows a finer description of the spatial pattern because there is no information missing due to unsampled space. In such cases, the extent is exhaustively sampled and the resulting data represent the entire population of sampling units within the extent. This does not guarantee that the data are representative of the entire ecological process being studied, 'population of inference' in the terms used in parametric statistics, but rather of the spatial representation of the extent. On the other hand, when we use a different sampling strategy, such as random, systematic or stratified samples, the sampling units are not spatially contiguous; the extent is not completely surveyed and information is missing about the spatial pattern. We

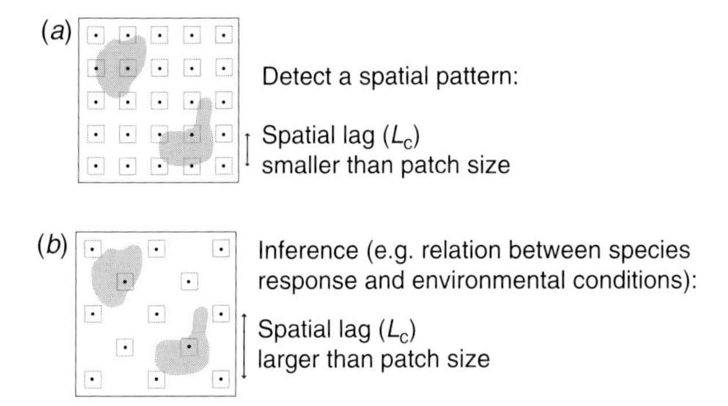


Figure 1.12 Spatial sampling design. (*a*) Spatial lag must be set according to the objective of the study: to detect a spatial pattern, the spatial lag should be smaller than the patch size; (*b*) to perform inference statistics, the spatial lag should exceed the patch size.

will refer to such spatial layout of the sampling units as sparse. Spatial lag between the sampling units is directly related to the previous decisions about sample size ($n$), the extent and the grain, and of the shape of the sampling unit (Dungan *et al.* 2002): as the sample size increases, the spatial lag decreases; as the extent size increases, the spatial lag increases; and, as the sampling unit size increases, the spatial lag decreases.

The goal of the study should also guide the choice of the spatial lag between the sampling units. There are two main reasons to collect ecological data in this context (Figure 1.12):

(1) To detect, characterize and quantify any spatial pattern in the data in order to obtain insights about ecological processes through the spatial signature of the ecological data (as illustrated in Figure 1.5). This implies that first the aim is to test for significant spatial dependence. In order to detect the spatial pattern, the spatial lag among sampling units needs to be smaller than the size of the patch or the process structure that we want to characterize (Figure 1.12). In this way, there will be several sampling locations within each patch.
(2) To establish the relationship between two or more kinds of ecological data. Here, we are not so interested in the spatial structure of the ecological data but rather in, for example, the species' response to various environmental conditions, once the spatial structure is accounted for. As mentioned above, spatial dependence and autocorrelation are considered a nuisance when using inferential tests that require that the observations are independent (Legendre & Legendre 1998). As Fortin *et al.* (1989) showed, random sampling designs ensure only that each sampling unit is drawn independently from the others and will be representative of the population. It does not guarantee, however, that there is no spatial autocorrelation in the data. In fact, in the presence of spatial autocorrelation, it is almost impossible to obtain spatially independent data (see Chapter 5).

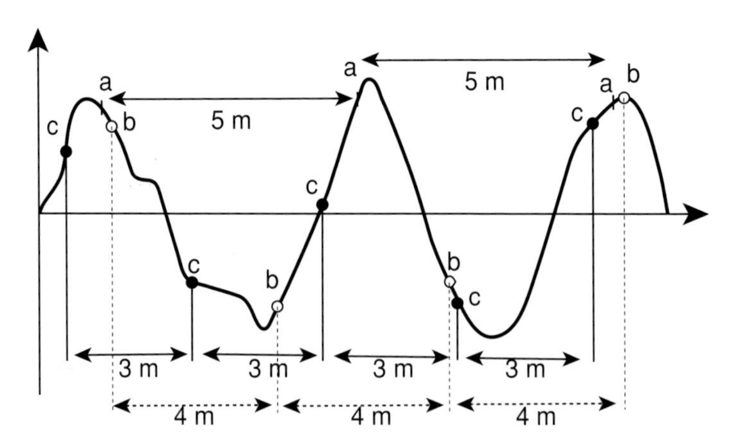


Figure 1.13 Importance of spatial lag while using a systematic sampling design: when the spatial lag is 5 m ('a' sampling locations) the spatial structure identified is uniform, flat; when the spatial lag is 4 m ('b' sampling locations) high and low periodicity is detected; and when the spatial lag is 3 m ('c' sampling locations), patchiness can be characterized.

Furthermore, the choice of the spatial lag is crucial while using systematic sampling design (Fortin *et al.* 1989): when the spatial lag (say 5 m) matches the spatial pattern of the data peak as indicated by sampling locations 'a' (Figure 1.13), the identified pattern of the data will be flat (uniform); when the spatial lag is 4 m (indicated by sampling locations 'b'), the periodicity of high and low values of the variable starts to be detected; and when the spatial lag in smaller, say 3 m (indicated by sampling locations 'c'), the spatial pattern detected can better describe the patchiness and periodicity of the data. When the spatial scale of the process is unknown, nested sampling designs with several spatial lag distances are preferred, so that the spatial pattern can be identified (Fortin *et al.* 1989; Webster & Oliver 2001).

Finally, interaction between the extent and the sampling unit size generates a more-or-less pronounced edge effect, which in turn biases the estimation of spatial pattern at small and large spatial distances (Figure 1.14; Haining 1990, 2003; Cressie 1993). Several of the spatial statistics as presented in Chapters 2 and 3 estimate spatial patterns at several of the increasing distances among the sampling locations (for example, between A and B, A and C, A and D in Figure 1.6). In Figure 1.14, sampling units along the border of the study area (in grey) have fewer neighbours at short distances than at intermediate ones (e.g. at 1.5 units apart there are 200 pairs of sampling units, while close to 600 pairs at 4.5 units apart) and almost none at the largest distance (two pairs at 10.5 units apart). The appropriate edge effect correction procedure should be selected according to the data type and the spatial statistics used (Cressie 1993, Haase 1995). Some generic rules of thumb can be

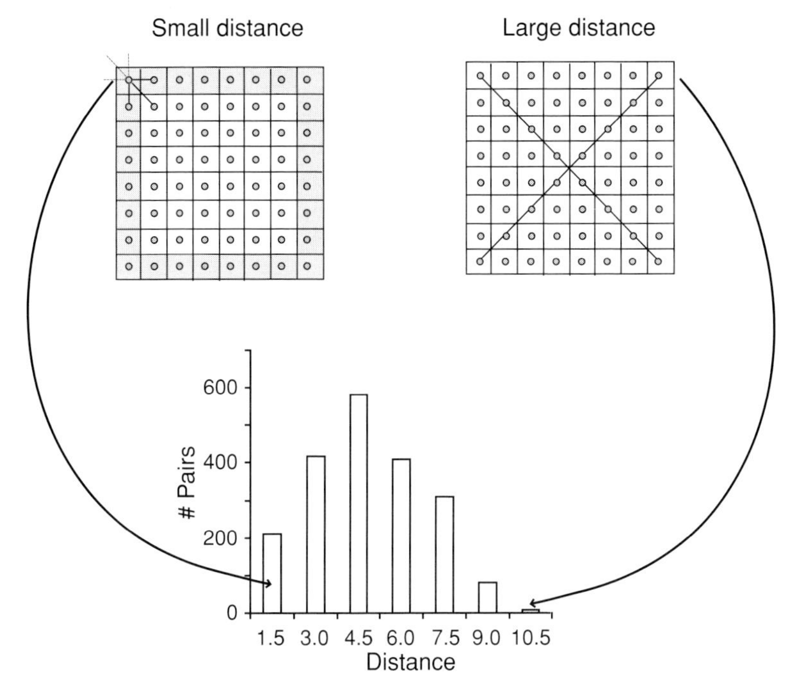


Figure 1.14 Edge effect affecting the sampling units along the border of the study area (the grey sampling units): at small distance (left panel; 1.5 units apart) and large distance (right panel; say 10.5 units apart), where the number of pairs of sampling locations used to estimate the spatial pattern is lower than for intermediate distances (e.g. 4.5 units apart).

used to minimize the edge effect either at the sampling design phase or during the analysis phase (Figure 1.15, Table 1.3). For example, during the sampling phase, a buffer zone can be sampled around the study area (Figure 1.15*a*). By doing so, estimation of the spatial pattern at small distances is based on the sampling locations at the border, including sampling locations inside (filled circles) and outside (open circles) the extent of the study area (Figure 1.15*a*). When the required extra resources are not available, or when the surroundings of the study area are not homogeneous, calculation of spatial statistics should be limited to the centre of the extent (filled circles), using the sampling locations around the border (open circles) only as neighbours at small distances (Figure 1.15*a*). Another technique that can be used during the analysis, assuming that the extent is a homogeneous area, is the computation of 'torus distances' (Figure 1.15*b*). This can be achieved by wrapping together opposite borders, the north and south borders as well as the west and the east, to create a doughnut-shaped structure called a torus. Then torus distances are computed among all the sampling locations, i.e. the sampling

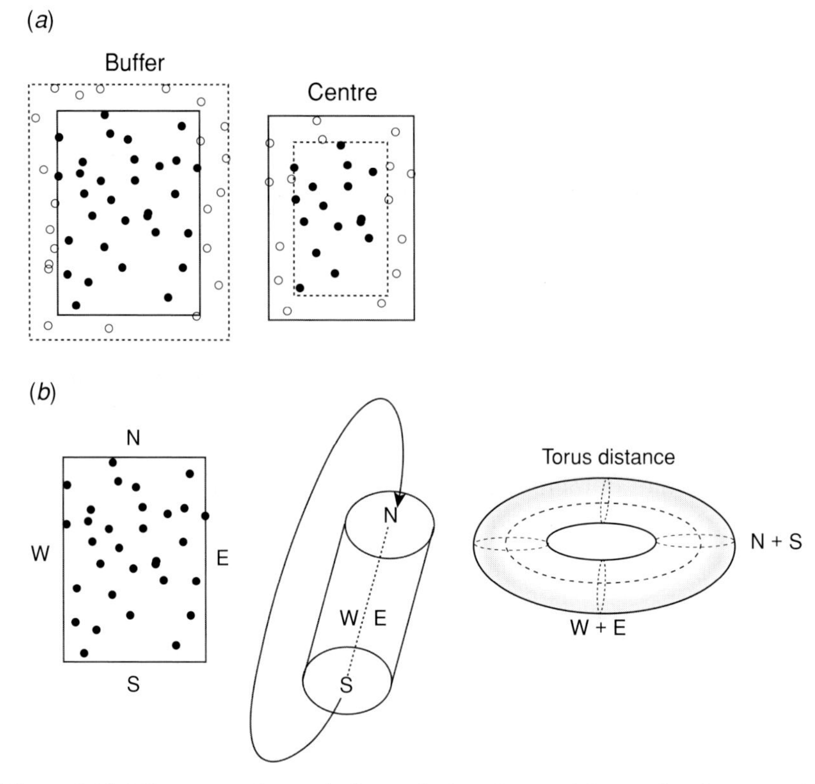


Figure 1.15 Edge correction techniques during: (*a*) spatial sampling by surveying a buffer area (open circles) around the extent (filled circles) or by analysing only the centre region (filled circles) of the extent; and (*b*) analysis where torus distances are used instead of Euclidean distances. See text for more details.

locations at the northern border are used as neighbours, at small distances, for the sampling locations at the southern border. The distribution of pairs of sampling locations will be more uniform, minimizing the edge effect. The torus correction should only be used when the entire study area is under a stationary process. Indeed if, for example, at the northern edge of the study plot there is a patch that does not occur at the southern edge, then the torus distances will artificially mix two different spatial patterns, resulting in a distorted spatial structure. (This technique is closely related to a restricted randomization technique known as the 'toroidal shift', which will be described later in this chapter.)

In conclusion, design of an optimal spatial sampling scheme requires a careful balance between sampling locations that are too close to one another, thus not providing enough new information (data highly autocorrelated), and sampling locations that are too sparse, so that processes at other spatial scales introduce too much variability (Haining 1990).

## 1.5 Spatial statistics

Because natural spatial heterogeneity affects most systems, it is not surprising that many kinds of spatial statistics have emerged from a variety of disciplines (plant ecology, human geography, mining engineering, etc.), trying to determine how spatial patterns are generated by one or several processes (Figure 1.2). Over the last half-century or more, a series of 'spatial statistics' (a generic term that includes all statistical methods) have been developed, either building on earlier methods or in parallel by different disciplines to describe spatial patterns, or to estimate and predict spatial processes. Not all spatial statistics have exactly the same goal (Table 1.2) or assume the same kind of underlying spatial processes and data type (Appendix 1) or were developed using the same underlying mathematical approach (Appendix 2). Therefore, each of these spatial statistics has its own requirements, assumptions and rules of thumb for its application (Table 1.3).

Spatial exploration methods aim to identify and describe spatial patterns. This can be done using:

- First-order statistics, such as aggregation indices based on the species abundance data. These can detect trends in the data over the entire study area (i.e. the mean value). These statistics can indicate whether there is a spatial pattern or not, but not its intensity. For example, the variance–mean ratio, a concept based on the Poisson distribution, may distinguish between only three types of spatial pattern: random, when the mean and the variance of species abundance per sampling unit are equal; patchiness, when the variance is greater than the mean; and uniform or regular, when the variance is smaller than the mean (see Chapter 2 for more detail). As shown in Dale (1999, p. 226), there are some problems with the logic of this approach to point pattern evaluation.
- Second-order statistics that measure local spatial pattern (i.e. the spatial intensity) in the data by computing the deviations of the values at neighbouring locations from the mean (i.e. the spatial variance).

In this book, we will not emphasize first-order statistics but we refer the readers to other textbooks (among others, Krebs 2002). Instead, we will concentrate on the many second-order statistics commonly used by ecologists (Table 1.2).

Spatial statistics can also be classified according to type of ecological process to which they can be applied (Appendix 1). Some processes act on the actual locations of individual organisms and it is the spatial pattern of these locations that is of interest. These are called point pattern processes (see Figure 1.10*a*). Other processes affect the quantitative values of variables and the spatial pattern is continuous in space. Several spatial statistics can be used to study these continuous processes such as spatial autocorrelation coefficients and spatial variance estimators (see Table 1.2 and Appendices 1 and 2). These spatially continuous processes can be sampled using either contiguous sampling units (Figure 1.10*b*) or spatially separated units

(Figure 1.10*c*). Then, some processes involve qualitative changes within an area and surface pattern methods are used to analyse them (e.g. join count statistics). Such categorical processes usually require contiguous sampling units. There are, however, some 'grey zones' in the way these three families of processes can be sampled and analysed.

Once the spatial pattern is identified and described, we may be interested in either estimating spatial parameters to model the process for prediction or interpolation purposes (see Chapter 3), or we may wish to test the relationships among ecological data (see Chapter 6). In the cases either of estimating parameters or of testing the relation among variables, the goal of the study and prior knowledge about the ecological data should be used as a guide to decide whether the spatial structure should be explicitly modelled or detrended before modelling or analysis (see Chapters 3 and 6). For example, we could be interested in the relationship, and potential causality, between soil moisture and plant growth. As illustrated in Figure 1.16*a*, both variables have a spatial structure based on multiple regression using the $x$ and $y$ coordinates of the sampling locations as independent variables. The relationship between the two variables (Figure 1.16*b*) is, however, only strong and significant when the raw data are used (left panel) but not when the spatially detrended data (residual data where the spatial structures are modelled by multiple regression) are used (right panel). This exercise clearly shows that before claiming causality between two ecological variables, we should test whether they have significant spatial structure. In the present case, the spatial structure is due to the spatial dependence of both variables on the slope from which the samples were obtained. This may be an example of spurious correlation. If the residuals from the multiple regressions still have some degree of spatial pattern, the ecological variables could have been both spatially dependent at one spatial scale and spatially autocorrelated at another. Thus, it is important to define the spatial scale of the question asked adequately so that the appropriate spatial statistics can be used adequately.

### *1.5.1 Significance testing of ecological data*

The significance of an observed measure is evaluated based on the assumption that the statistic computed using the observed data follows a reference distribution. When this reference distribution is known, and can be derived analytically, parametric tests can be used. Otherwise, randomization procedures can be used to generate the reference distribution from the data (Good 1993; Edgington 1995; Manly 1997; Figure 1.17*a*). When inference to the population level is required, bootstrap procedures and Monte Carlo simulations can be used (Figure 1.17*b*; Efron & Tibshirani 1993; Manly 1997).

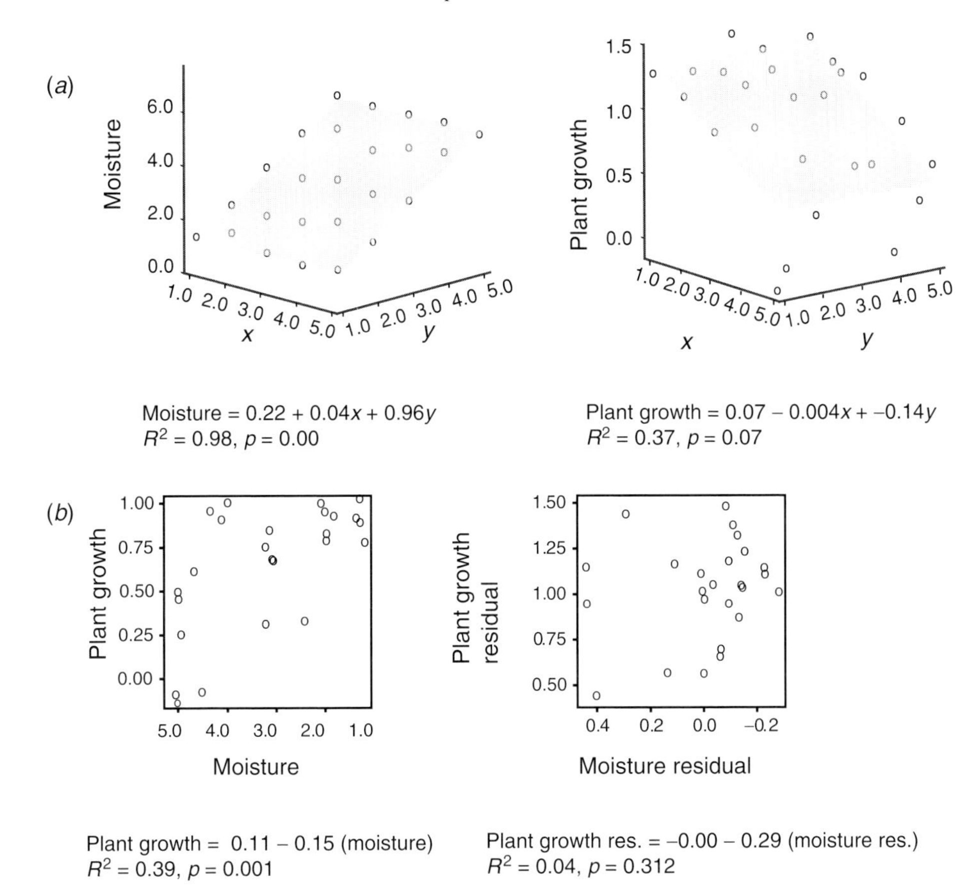


Figure 1.16 Example of the relationship between soil moisture and plant growth (artificial data). (*a*) Multiple regressions, using the $x$ and $y$ coordinates of the sampling locations, as independent variables (soil moisture: left panel; plant growth: right panel). (*b*) Relationship (linear regression) between the two variables using the raw data (left panel) and residuals from the multiple regressions (right panel). Note that the relationship changes from significant to non-significant as the data are detrended for the spatial structure.

The basic procedure to generate the null reference distribution, to which the observed statistic can be compared, is:

- to re-allocate, using a complete simple randomization procedure, the sampled values of a variable over the sampling locations;
- to re-compute the statistic; and,
- to repeat these two steps many times (thousands).

The probability at which the statistical decision of accepting or rejecting the null statistical hypothesis is made is proportional to the number of randomizations generated. When the number of observations is small, all permutations can be examined

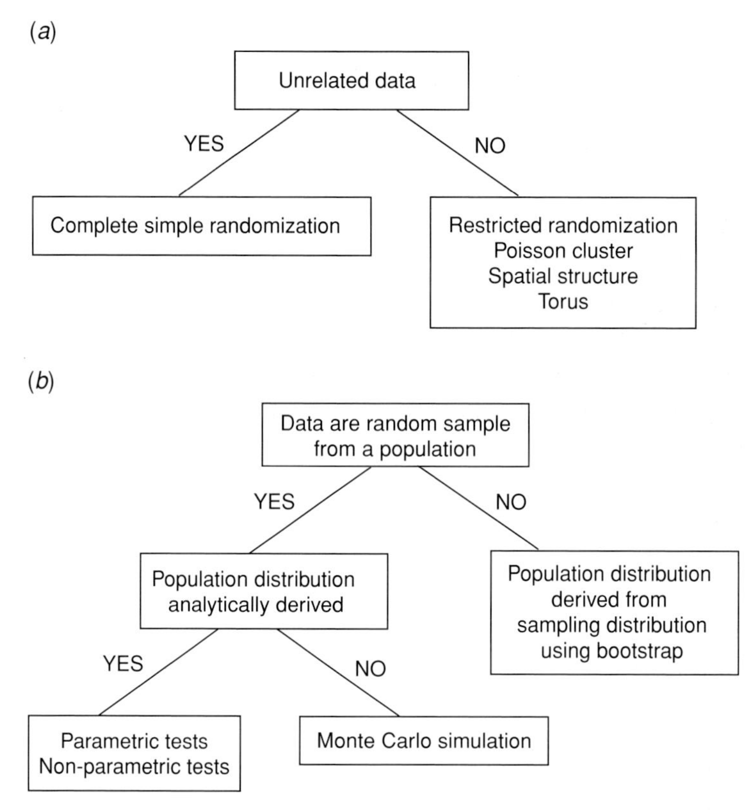


Figure 1.17 Decision trees to select appropriate significance tests at (*a*) the sample level and (*b*) the population level.

(Good 1993). When the number of observations is large, only a subsample of all possible permutations of the data can be computed and it is recommended to perform 10,000 or more randomizations (Manly 1997). Then, this reference distribution is used to assess the probability of the observed data where the precision of the probability depends on the number of randomizations. As an example, 1,000 values (999 randomized statistics and one observed) imply that the smallest probability can be 0.001 (i.e. 1/1,000), but the accuracy of the estimate of the probability will not be good; 10,000 or more is the recommended number of iterations.

The presence of spatial dependence, however, can impair the use of both parametric and randomization tests (Legendre & Legendre 1998). Parametric tests require that the errors are independent, so that each observation or data point brings a full degree of freedom. The presence of positive spatial dependence makes nearby sampling units more alike and therefore the value of a variable at location A (Figure 1.6) is a good predictor of the value at location B. As a consequence, a spatially autocorrelated datum does not bring a full degree of freedom, but rather a fraction of it, which is inversely proportional to the degree of autocorrelation in the data

(Legendre 1993; Dale & Fortin 2002). For example, the values at locations A and B are more positively autocorrelated than the values between locations A and D (Figure 1.6). Several techniques at the sampling design stage (see Section 1.4) and at the stage of statistical analysis (Legendre & Legendre 1998; Dale & Fortin 2002) can handle this issue by either correcting for it, so that parametric tests can be used, or by explicitly incorporating the spatial pattern into the analysis. This issue, of not having independent errors, is at the core of the analysis of ecological data and will be discussed at length in Chapter 5.

Some randomization tests are also invalidated by data that are spatially dependent. Randomization tests assume that the values of a variable are exchangeable, so that any arrangement that might arise by shuffling them is equally likely (Manly 1997). Under a complete randomization, all arrangements of the observations are equiprobable and this implies complete spatial randomness (CSR). In many instances, such a null hypothesis is inappropriate when analysing spatially dependent data, and thus other forms of randomness which incorporate some degree of spatial structure can be used (Cressie 1993; Venables & Ripley 2002; Figure 1.18). Restricted randomization procedures can also be used (Cressie 1993; Manly 1997). There are several different ways to restrict the randomization procedure:

- Randomize only within subsets of the sampling locations, i.e. subregions, where the data are considered spatially independent within the subregion but where there is some spatial structure when all the subregions are considered.
- Retain the sequential spatial order of the data. For spatial data, this is achieved using a two-dimensional torus constructed by connecting the map margins and then sliding one variable map over the other (the 'toroidal shift'). This procedure maintains most of the spatial structure of the data. Hence the relationship between two variables can be tested using this sequential restricted randomization by sliding the values of one variable over the other, one spatial lag at a time, and re-computing the statistic of interest. Depending on the number of restricted randomizations performed, i.e. sliding, given the shape and size of the study area, this test can be too liberal (Fortin *et al.* 1996).
- Generate several realizations of a stochastic spatial process based on the same spatial structure as the data. This can be achieved by first estimating the structure of spatial autocorrelation in the observed data and then using the estimation of the parameters that will generate simulated data with a similar spatial structure (but not the same spatial pattern). Techniques called 'the conditional annealing simulation algorithm' (Journel & Huijbregts 1978; Isaaks & Srivastava 1989; Cressie 1993) and 'conditional autoregressive modelling' (Getis & Boots 1978; Cliff & Ord 1981; Haining 1990; Cressie 1993; Fotheringham *et al.* 2000) are available to do so (Fortin *et al.* 2003).

Note that the type of restriction appropriate for the randomization technique is related to the null hypothesis under consideration, and we will revisit this issue in Chapter 7.

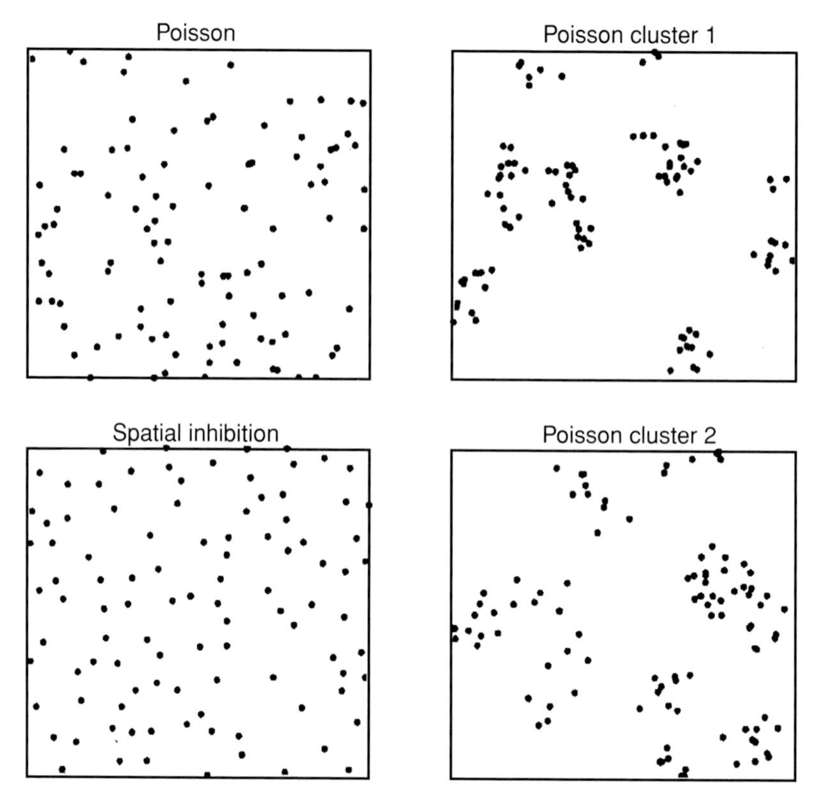


Figure 1.18 Different point pattern processes generating different types of random patterns: Poisson process (i.e. complete simple randomness); Poisson clusters (1 and 2 using different parameter values for spacing among clusters); spatial inhibition.

## 1.6 Concluding remarks

The goal of this chapter was to provide the background and the concepts needed to understand the spatial analysis of ecological processes and ecological data. The key purpose in performing spatial analysis is to determine whether the data lack independence: and if so, what is the nature of the spatial dependence. We rely on independence to allow us to make trustworthy interpretations and predictions. There are several ways in which the detection of significant spatial dependence provides meaningful insights about ecological data and their underlying process(es) (Griffith 1992):

- The presence of spatial autocorrelation in residuals can be used as a diagnostic tool indicating whether one or more processes are not included in the model or were not parameterized adequately.
- Under some circumstances, the degree of spatial autocorrelation can be used as a surrogate for unmeasured variables that are too expensive or too difficult to measure.

- The presence of spatial autocorrelation is a nuisance for parametric and randomization tests that require independent errors.
- Sometimes, the presence of negative spatial autocorrelation can indicate that the sampling unit size or shape is inappropriate to capture the scale of the process adequately.
- The finding of weak or absent significant spatial autocorrelation at small distances can be an indication that the sampling unit size and shape, the spatial lag among the sampling locations, and the spatial distance class used to estimate the degree of spatial autocorrelation during analysis are inappropriate to capture the scale of the process.
- Spatial pattern in the variable of interest may be due to the variable's response to other spatially structured variables. Hence this is a case of induced spatial autocorrelation. Such spatial structure should be referred to as spatial dependence.

This list of insights should convince the reader of the importance and usefulness of detecting, testing and predicting spatial structure in ecological data. It also reinforces the motivation for this book, which is to guide ecologists in determining and interpreting correctly the spatial structures of natural systems and their origins. The following chapters will explore and provide greater detail on all these challenging topics.

# 2

# Spatial analysis of population data

## Introduction

In this chapter, we describe methods for analysing completely censused 'population' data. Such data can be in one of two formats. The first we will discuss is the situation where there is a map of the locations of all 'events', such as individual organisms of a particular species, in an area, with or without accompanying information about each event, such as tree height or condition (see Figure 1.10*a*). This is a complete census for the entire extent of the study area, not a sample (although we will comment on the analysis of sample data in Chapter 3), and in most ecological examples the map is in two dimensions (with $x$–$y$ coordinates), but we can consider one- or three-dimensional maps, as well (with only $x$ or with $x$–$y$–$z$ coordinates). The methods for analysing event location data can be based on determining the neighbours of each event and making calculations based on the distances to them, or may involve counting the events in circles of a given range of sizes, centred on the events or on randomly placed points. In studying the spatial structure of a one-species population of mature trees, the analysis is univariate. When two kinds of events are of interest, for example mature trees of a species and its seedlings, the analysis is bivariate. Multivariate analysis is applied when there are several kinds of events, as in a study of a multispecies forest community. If the events have a quantitative variable associated with them (such as stem diameter), rather than categorical (e.g. the species to which they belong), versions of what is referred to as 'marked process' analysis can be used.

The second form of data would be a map, in one, two or three dimensions, consisting of information collected in each unit of an array of contiguous sample units such as quadrats (see Figure 1.10*b*). The information for each unit of the array might be a measure of density, a count of events, or an estimate of cover of all plant species in a quadrat. Again, the data do not represent a sample of the study array; there is information for all units. The size of the units determines the minimum size of the 'grain' that can be studied, like the pixels of a satellite image (as discussed in Chapter 1). The methods of analysis may be based on combining the information from units into larger 'blocks', or by calculations based on all units a particular set of distances apart. Depending on the data, as described above, analysis can be univariate, bivariate or multivariate.

All the methods described in this chapter are unified by the concept of analysing the data using a window function or template, with which the data are selected or compared (Dale *et al.* 2002). At the end of the chapter we provide a summary table that illustrates this interpretation of the material we present.

## 2.1 Mapped point data in two dimensions

The first set of methods are those used to analyse maps of the positions (e.g. $x$–$y$ coordinates) of all points (events) of a particular type (tree stems) in a study plot. In many of these methods, a test statistic is calculated from the data and then compared to the expected value of the statistic under the null hypothesis of complete spatial randomness (CSR). One theme in the analysis of a spatial point pattern is the distinction between patterns that are random (complete spatial randomness) and those that are underdispersed (clumped or aggregated) or overdispersed (spaced or regular) by comparison. It is important to acknowledge that the appearance and interpretation of pattern can change with the scale of study; for example, if the events occur as clumps, and only one clump is examined, the events appear to be overdispersed (see Figure 1.10 of Dale 1999, and our Figure 2.1). The classification of spatial point patterns into the three categories is probably an oversimplification and more sophisticated methods of analysis will address that issue, as we will illustrate and discuss below.

### *2.1.1 Distance to neighbours methods*

One basic approach would be to measure directly, or to calculate from a map, the distances between neighbouring events and determine whether the average distance is greater than or less than that expected from CSR. If the average distance is significantly less than expected, the conclusion would be that the events are clumped; if greater, that they are overdispersed.

There is, however, a number of different ways of defining which events are the neighbours of a particular event (see Section 2.3). A simple definition is to determine each event's nearest neighbour. In some cases, pairs of events will be each other's nearest neighbour, but not always. Given a map of the positions of all events in a study area, we can choose to use only a sample of them, or we can choose to use them all.

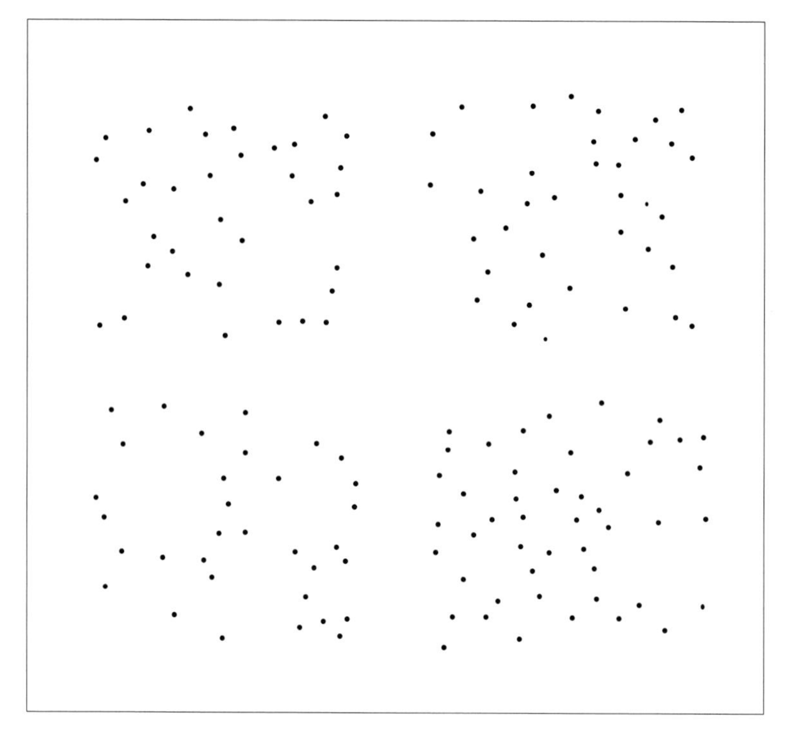

Figure 2.1 Artificial point pattern data where events only occur in the centres of the four quadrants. The events are overdispersed at small scales but underdispersed (clumped) at large scales.

Although the focus of this chapter is complete census data, we will begin the discussion of nearest neighbour distance statistics by considering a sample of events first. In using distances to study spatial pattern, it is often more convenient to use the square of the measured distances for calculating a test statistic. For example, let $W_{i1}$ be the distance between any event $i$ and its first nearest neighbour, and let $\lambda$ be the density of events per unit area; then for a sample of the population of events, the test statistic

$$Q = \pi\lambda \left( \sum_{i=1}^{n} W_{i1}^{2} \right) \Big/ \; n \tag{2.1}$$

can be compared to the normal distribution with a mean of 1 and a variance of $1/n$, often designated $N(1, 1/n)$ (Pielou 1959). This is just one of a very large number of statistics that have been proposed, some of which compare the average distance of an event to the nearest neighbour event with the distance between randomly placed points and their nearest events, $X_{i1}$. Upton & Fingleton (1985) provide a summary chart of these in their Table 1.10. More sophisticated measures look at the distances not just to the first nearest neighbour but to the first nearest, second nearest and so on, such as $W_{i3}$ or $X_{i4}$. Liu (2001) provides a comparison of methods using the first to fifth nearest neighbours ($j = 1, 2, 3, 4, 5$). While different statistics have different strengths and weaknesses, we can summarize his findings as recommending the modification of Pollard's (1971) statistic. It is

$$P(j) = 12j^2n\left[n\ln\left(\sum_{i=1}^{n} X_{ij}^2/n\right) - \sum_{i=1}^{n}\ln\left(X_{ij}^2\right)\right] \Big/ [(6jn + n + 1)(n - 1)]. \tag{2.2}$$

$P(j)$ takes the value 1 for CSR; values less than one indicate overdispersion and values greater than one indicate aggregation. The significance of departures from unity is tested by comparing $(n - 1)P(j)$ to the $\chi^2$ distribution with $n - 1$ degrees of freedom.

For example, Figure 2.1 shows artificial data in which the events occur only in the centres of the four quadrants of the study area. This pattern was evaluated using 80 randomly placed points and Pollard's index. In this case, $P(1)$ is 1.11, which suggests clumping but is not significant, but $P(2)$ is 1.31 and $P(3)$ is 1.29, both of which are significant either by comparison with the $\chi^2$ distribution with 79 degrees of freedom or by a Monte Carlo test with 1,000 realizations of CSR. In fact, Liu (2001) suggests that $P(3)$, $P(4)$ and $P(5)$ are to be preferred.

In using all the mapped events, rather than just a sample of them, it is tempting to proceed with any one of the many methods available for sampled events. That would be wrong, of course, because if we use all of the events, the sources of information we are using are no longer independent. In addition, we need to be concerned that we are now using all the events that are close to the edge of the study plot, and there should be some consideration of edge effects (see Chapter 1). These concerns lead to a somewhat different approach to the analysis, referred to by Diggle (1979) as 'refined' nearest neighbour analysis. Given a complete census of the locations of all events, we could use just a sample of them for analysis, as described above, provided we were willing to discard the information from those not included in the sample. Under most situations, it is preferable to adjust the method of analysis so that the full census can be used, in order to take advantage of all the information available.

### 2.1.2 Refined nearest neighbour analysis

Refined nearest neighbour analysis is a Monte Carlo procedure because it compares a value or set of values calculated from the data, with the same values calculated from a number of realizations of CSR using the same plot size and shape, and the same number of events. Manly (1997) illustrates one such procedure using $W_{ij}$, the average distance between events and their $j$th nearest neighbours for $j = 1$ to

$j = 10$, and 499 realizations of CSR. In that example, the events are pine seedlings, and only $W_1$ and $W_2$ are significantly greater than in CSR, indicating the effects of competitive inhibition at short distances.

The approach suggested by Diggle (1979) is slightly more complicated. For any given distance, $w$, calculate the proportion of those events that are further than $w$ from any boundary for which the distance to their nearest neighbour is less than $w$. Call that proportion $G(w)$; it is an estimate of the cumulative probability distribution of the distance from any event to its nearest neighbour event. It is sensitive to local clustering or inhibition of neighbours. If the events are randomly arranged in the plot, the expected value of $G(w)$, $E(G(w))$, is:

$$E[G(w)] = 1 - e^{-\lambda \pi w^2}. \tag{2.3}$$

One possible test statistic is the largest difference between $G(w)$ and $E(G(w))$ over the range of values of $w$:

$$d_w = \max |E[G(w)] - G(w)|. \tag{2.4a}$$

The observed test statistic is then compared with the values found in a 1,000 or more random configurations of $n$ events in an area the same as that of the study.

A test based on the distance between random points and their nearest neighbour event, call it $u$, rather than on $w$, can be calculated in almost exactly the same way, $d_u$:

$$d_u = \max |E[F(u)] - F(u)|, \tag{2.4b}$$

where $F$ is defined for $u$ as $G$ was for $w$. $F(u)$ is an estimate of the cumulative probability distribution of the distance from a randomly chosen point to the nearest event. Because it is sensitive to gaps in the pattern of events, it is sometimes referred to as the 'empty space function'. Further insight into the spatial pattern of the events can be gained by plotting $G(w) - E(G(w))$ or $F(u) - E(F(u))$ over the range of values of $w$ or $u$ (see Upton & Fingleton 1985). Diggle (1979) also suggests the use of a test statistic based on the differences between the two kinds of distance functions, event–event vs. point–event:

$$S_d = \sum_{i=1}^{j} [F(z_i) - G(z_i)], \tag{2.5}$$

where $z_i$ represents a series of distances.

Again, evaluation of the test statistic would be based on comparison with the range of values found in a large number of CSR realizations. One advantage of this approach is that the test can be modified to use a distribution other than that based on CSR as the null hypothesis.

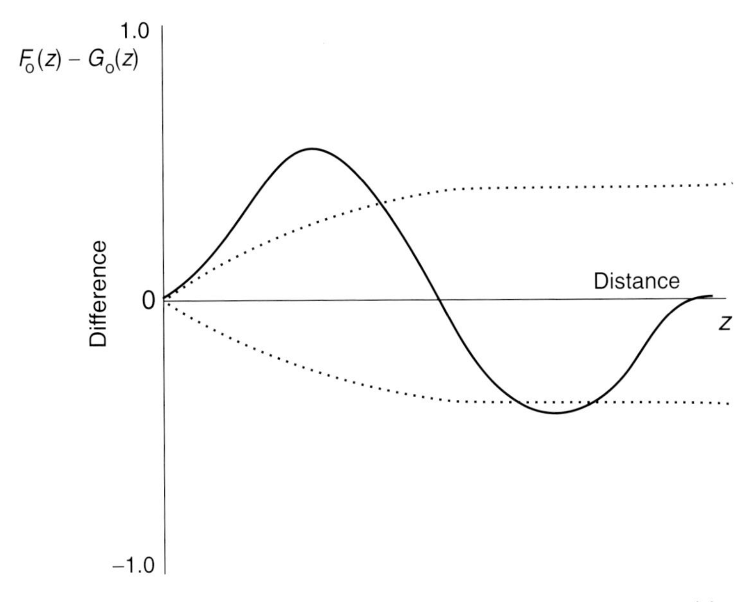


Figure 2.2 Artificial example of Diggle's difference statistic where positive and negative differences may cancel each other out.

Our own experience suggests that a better choice of statistic is:

$$S_a = \sum_{i=1}^{j} |F(z_i) - G(z_i)|, \tag{2.6}$$

where $z_i$ represents a series of distances.

The reason for this suggestion is that it is possible for the positive and negative differences to cancel each other out, even if they exceed, in places, the randomization envelopes (Figure 2.2). Figure 2.3 shows the example of living lodgepole pine (*Pinus contorta* Loudon) stems at the Fort Assiniboine site with the difference between $F$ and $G$ plotted as a function of $z$. It illustrates the large-scale clumping of these stems, but does not detect significant overdispersion at smaller scales.

### 2.1.3 Second-order point pattern analysis

The next set of methods is used to analyse the mapped positions of events in the plane, such as the stems of trees, and assume a complete census of the objects of interest in the area under study. One of the most commonly used methods is called Ripley's $K$ (Ripley 1976). The approach is based on the concept that, if $\lambda$ is the density of events per unit area, the expected number of points in a circle radius $t$ centred on a randomly chosen point is $\lambda K(t)$, where $K(t)$ is some function of $t$ that

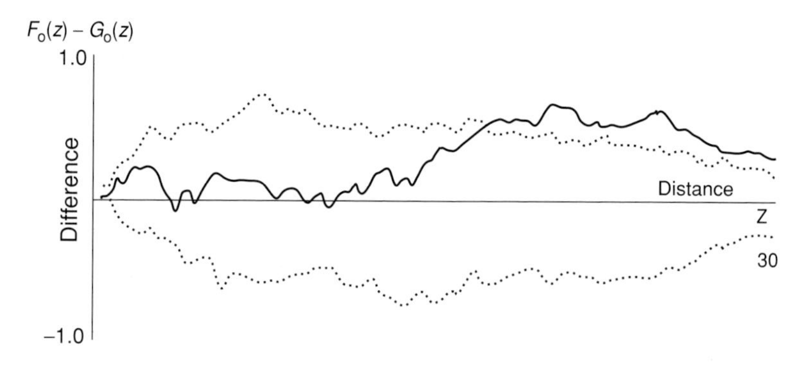


Figure 2.3 Refined neighbour analysis on living lodgepole pine trees at Fort Assiniboine (data in Figure 2.5*a*), where $F_0(z) - G_0(z)$ is the difference between event–event and point–event functions for a suitable chosen distance $z$.

depends on the pattern of the points. For example, if the points are overdispersed, $K(t)$ will be close to 0 for small radii and increase for larger distances.

The calculation of the statistic for a given radius, $t$, is based on counting all pairs of points separated by distance less than $t$. The statistic $\hat{K}(t)$ is an estimate of $K(t)$:

$$\hat{K}(t) = A \sum_{\substack{i=1 \\ i \neq j}}^{n} \sum_{\substack{j=1 \\ j \neq i}}^{n} w_{ij} I_t(i, j)/n^2, \tag{2.7a}$$

where $A$ is the area of the plot, with $d_{ij}$ being the distance between points $i$ and $j$, $I_t(i, j)$ an indicator function, taking the value 1 if $d_{ij} \leq t$ and 0 otherwise, and $w_{ij}$ is a weight that corrects for edge effects. If the circle centred on $i$ with radius $d_{ij}$ is completely within the study plot, $w_{ij} = 1$; otherwise it is the reciprocal of the proportion of that circle's circumference within the plot (Diggle 1983). A number of authors have provided explicit formulae for the edge correction based on geometric arguments (Haase 1995; Goreaud & Pélissier 1999). These can be complicated and an alternative is to use a 'quick and dirty' numerical estimation by dividing the circumference into many sectors, say 120, and counting how many of them are within the boundaries of the study area.

Our own investigations suggest that the weight $w_{ij}$ can be replaced with a weight that depends on $i$ and $t$, rather than on $i$ and $j$: weight $h_i(t) = 1$ if the circle centred on $i$ with radius $t$ is completely within the study plot, otherwise the reciprocal of the proportion of that circle's area within the plot:

$$\hat{K}(t) = A \sum_{\substack{i=1 \\ i \neq j}}^{n} \sum_{\substack{j=1 \\ j \neq i}}^{n} h_i(t) I_t(i, j)/n^2. \tag{2.7b}$$

This approach reduces the number of calculations that need to be made: one weight for each event and each radius, rather than one for each pair of points and

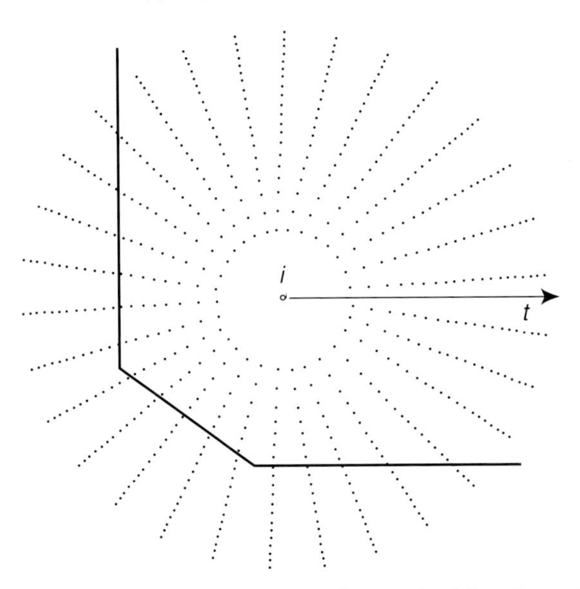


Figure 2.4 An illustration of the edge correction method for plots with irregular boundaries; the number of radial sectors within the plot is calculated.

each radius. Again, because of the complexities of a strictly geometric calculation, a numerical estimation can be achieved by dividing the area of the circle into a large number of radial sectors, say 600 (where we suggested 120 for the circumference), and counting the number that fall within the boundary (Figure 2.4). This approach makes it possible to deal with boundaries with irregular shapes. For further discussion of edge correction techniques see Ripley (1988), Cressie (1993), Haase (1995) and Gignoux *et al.* (1999).

If the events follow CSR, the number of points in a circle follows a Poisson distribution and the expected number of events in a circle of radius $t$ is $n\pi t^2/A$. $\hat{K}(t)$ is compared with this expected value by subtracting the observed from the expected:

$$\hat{L}(t) = t - \sqrt{\hat{K}(t)/\pi}. \tag{2.8}$$

In some versions, the expected is subtracted from the observed (e.g. Bailey & Gatrell 1995) and in interpreting results, the reader needs to be careful to determine which version is being used! $\hat{L}(t)$, as given in (2.8), is plotted as a function of $t$, with negative values indicating clumping and positive values indicating overdispersion. (As a mnemonic, remember that under the line means underdispersed; over the line means overdispersed.) For example, Figure 2.5*b* shows the analysis of the data in Figure 2.5*a*, a map of the living lodgepole pine trees in the Fort Assiniboine example; the features of the data are clear in the analysis: overdispersion at scales less than a metre and clumping at all larger scales.

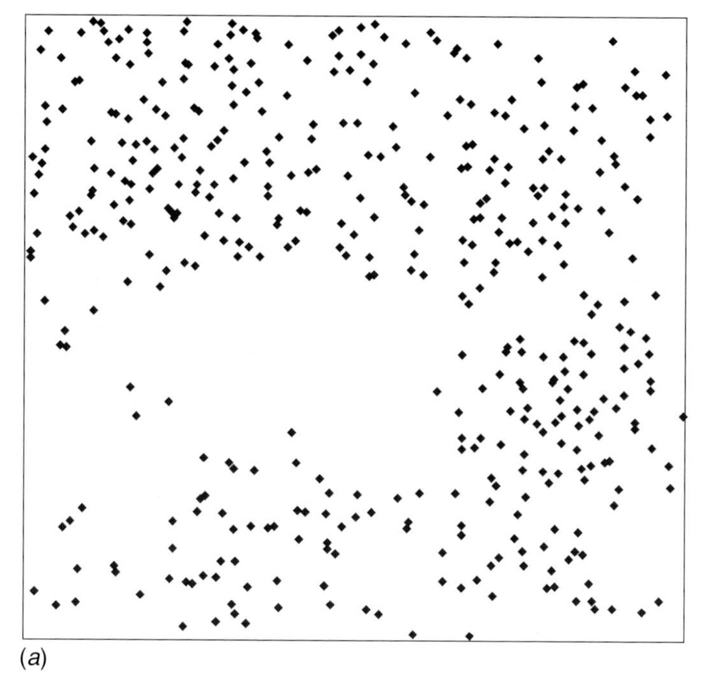


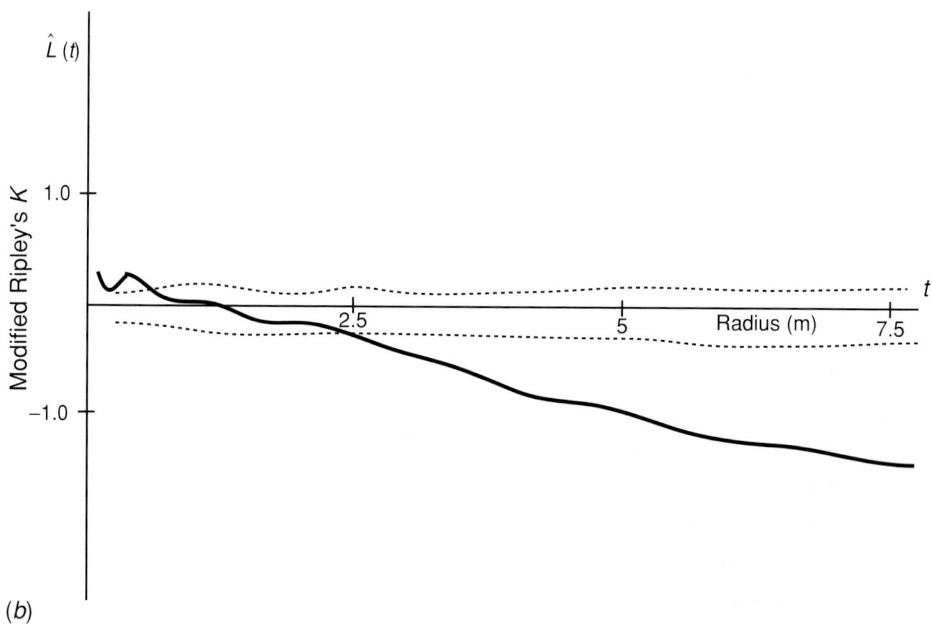


Figure 2.5 (*a*) Map of living lodgepole pine at Fort Assiniboine, Alberta. The plot is 50 m on each side. (*b*) Standardized univariate Ripley's *K*-function analysis of living lodgepole pine at Fort Assiniboine, Alberta, where $\hat{L}$ is Ripley's standardized statistic and *t* is distance.

The results of the analysis can be assessed using the approximate confidence intervals for $\hat{L}(t) = 0$, provided by Ripley (1976) of $\pm 1.42(A^{1/2})/n$ for $\alpha = 5\%$ or $\pm 1.68(A^{1/2})/n$ for $\alpha = 1\%$. Using that criterion, in Figure 2.5*b*, the value at 0.375 m is significantly large, and values at distances greater than 2 m are significantly less than 0. In many published accounts, authors have used Monte Carlo techniques (cf. Manly 1997) to evaluate the significance of the results, thus avoiding problems of distribution theory (Andersen 1992; Haase 1995). This is the approach we recommend, and Figure 2.5*b* shows the 99% envelope generated from 100 realizations of CSR. As in many such situations, use of the Monte Carlo approach makes it possible, in theory at least, to use a dispersion pattern other than CSR as the null hypothesis. For example, several authors, including Reich *et al.* (1997), have used as a null model the Neyman–Scott process, which is based on completely random *clusters* of events (not random *events* as in CSR), in each of which the number of events follows a Poisson distribution. (Pielou 1977, p. 119, refers to this as a Neyman Type A or Poisson–Poisson distribution.) Kenkel (1993) found that populations of the clonal plant *Aralia nudicaulis* were well described by a Markov model with an inhibition distance of 18 cm, corresponding to the radius of the plant's shoots.

In Chapter 1, we made the distinction between global analysis, which summarizes the characteristics of the spatial pattern over the whole study area (perhaps with the implicit assumption of stationarity), and local analysis, which makes explicit the differences in the pattern observed among parts of the study area. The discussion of second-order statistics for point patterns has, so far, been global, but the method can be adapted to produce a spatially explicit result (cf. Getis & Franklin 1987). For each event, $i$, and distance, $t$, a score can be assigned to the position of the event that compares the observed count of events in the circle radius $t$ centred on event $i$ with the expected count based on the area of the circle and CSR. Contour maps of those scores can then be drawn, for particular ranges of radii, to make interpretation easy. Getis & Franklin (1987) suggest including the scores for randomly or regularly placed points, as well as those of events, so that the regions where events are sparse can be covered as well. This parallels, for $K$-function analysis, the inclusion of the 'empty space' function in refined nearest neighbour analysis. They provide examples of this technique, which give us the option of spatially explicit results to Ripley's $K$-function approach. Figure 2.6 illustrates this approach using artificial data in which the events are overdispersed in the bottom left corner (hard core repulsion of 5 m) but clumped or random elsewhere in the $100 \times 100$ m square. Figure 2.6 *a–c* show the $L$-transform scores of Ripley's $K$-function (Eqn (2.8)) for 3, 10, and 20 m. These spatially explicit results clearly show the non-stationarity in the pattern of events, which could be very useful in the exploration of the characteristics of real data sets.

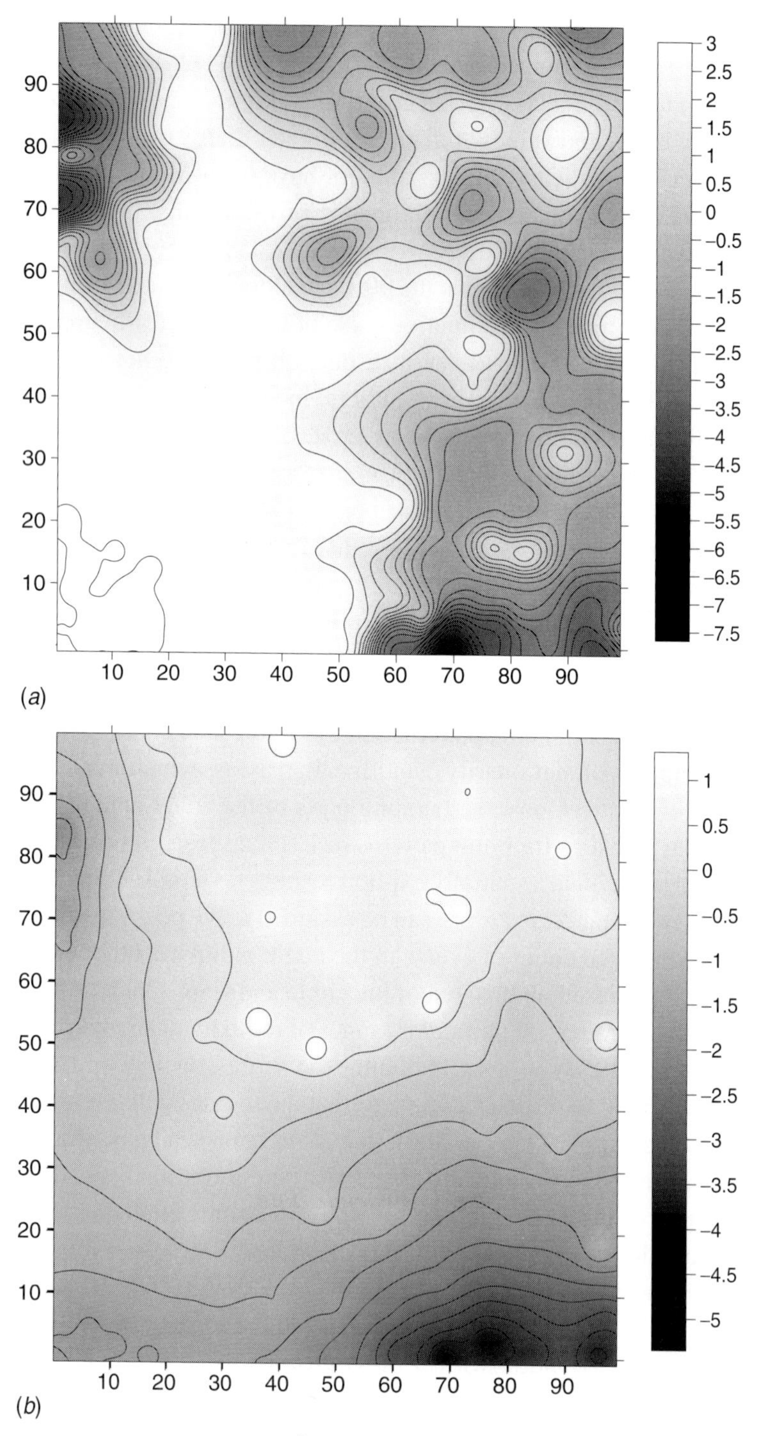


Figure 2.6 (*a*) A 'Getis' map of $\hat{L}$-function scores for artificial data (100 × 100 m) with hard-core repulsion in the lower left corner: $t = 3$ m. (*b*) As (*a*) for $t = 10$ m. (*c*) As (*a*) for $t = 20$ m. Note that the ranges of values are different in the three parts of the figure.

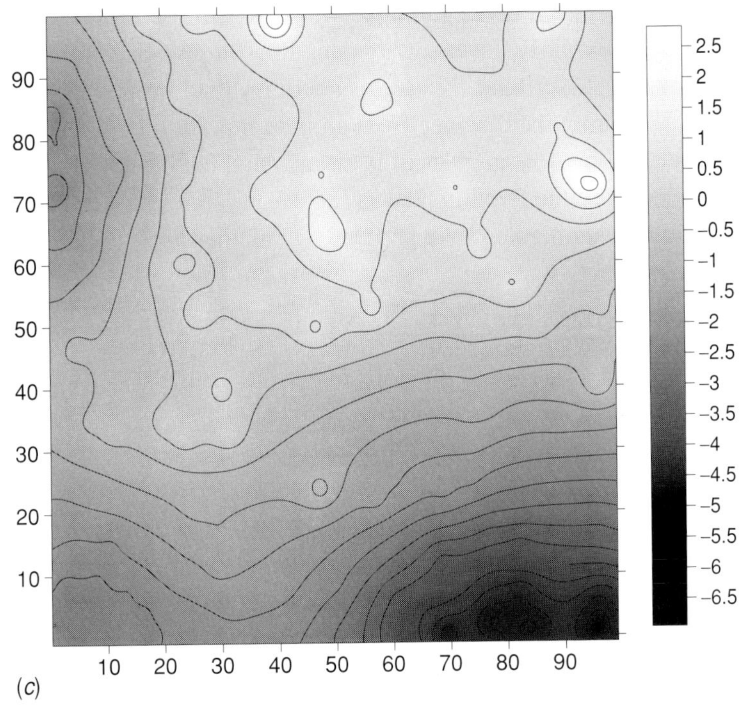


Figure 2.6 (*cont.*)

In a study of the demography of a palm tree species of the humid savanna of West Africa (Ivory Coast), Barot *et al.* (1999) demonstrated the advantages of using a complementary group of methods, rather than a single one, for spatial pattern analysis. They applied Diggle's $F$- and $G$-functions (nearest neighbour and empty space functions) together with Ripley's $K$-function. They commented that the simultaneous use of all three functions found significant departures from CSR that would not have been evident otherwise. Their findings confirm the suggestion that it is often a good idea to use some kind of combined analysis, using more than one method to characterize the spatial structure being investigated.

### 2.1.4 Bivariate data

The approaches described above can be easily modified for bivariate data, in which we want to analyse the spatial relationship of two different kinds of events, such as males or females, flowering or vegetative plants, diseased or healthy individuals and so on. For example, Diggle's nearest neighbour function, $G$, can be adapted for bivariate data by examining the distance from events of type 1 to nearest neighbour of type 2, which gives $G_{12}$, separately from examining the distance from events of type 2 to nearest neighbours of type 1, which gives $G_{21}$. This permits the detection of asymmetric associations, which can be very useful in some circumstances, such as studying the association between mature female palm trees and seedlings (see Barot *et al.* 1999). Similarly, for bivariate data, the 'empty space function' can be divided into two: $F_1$ describing the distance from a random point to an event of type 1 and $F_2$ describing the distance from a random point to an event of type 2.

For a Ripley's $K$-function analysis of bivariate data, the basic question becomes this: At what scales are the two kinds of events segregated from each other, and at what scales are they aggregated? We proceed by calculating:

$$\hat{K}_{12}(t) = A \sum_{\substack{i=1 \\ i \neq j}}^{n_1} \sum_{\substack{j=1 \\ j \neq i}}^{n_2} w_{ij} I_t(i, j)/n_1 n_2$$

and (2.9)

$$\hat{K}_{21}(t) = A \sum_{\substack{i=1 \\ i \neq j}}^{n_1} \sum_{\substack{j=1 \\ j \neq i}}^{n_2} w_{ji} I_t(j, i)/n_1 n_2.$$

The two are both estimates of the same function and are combined to compare the observed and the expected:

$$\hat{L}_{12}(t) = t - \sqrt{[n_2 \hat{K}_{12}(t) + n_1 \hat{K}_{21}(t)]/\pi(n_1 + n_2)} \tag{2.10}$$

(cf. Upton & Fingleton 1985; Andersen 1992). Values greater than 0 indicate segregation and values below 0 indicate aggregation of the different kinds of points (Figures 2.7*a*, *b*). This example shows the relationship between canopy trees (mainly pine) and understorey seedlings, mainly spruce, near Grande Cache, Alberta (Dale & Powell 2001). The trees and seedlings are segregated from each other at small scales but aggregated at a range of larger scales.

A different, but closely related, approach was suggested by Diggle & Chetwynd (1991), based on Ripley's $K$-function, calculated separately for the two types of events:

$$\hat{K}_{11}(t) = A \sum_{\substack{i=1 \\ i \neq j}}^{n_1} \sum_{\substack{j=1 \\ j \neq i}}^{n_1} w_{ij} I_t(i, j)/n_1^2$$

and (2.11)

$$\hat{K}_{22}(t) = A \sum_{\substack{i=1 \\ i \neq j}}^{n_2} \sum_{\substack{j=1 \\ j \neq i}}^{n_2} w_{ij} I_t(i, j)/n_2^2.$$

The statistic of interest is then the difference between these two:

$$\hat{D}_{12}(t) = \hat{K}_{11}(t) - \hat{K}_{22}(t). \tag{2.12}$$

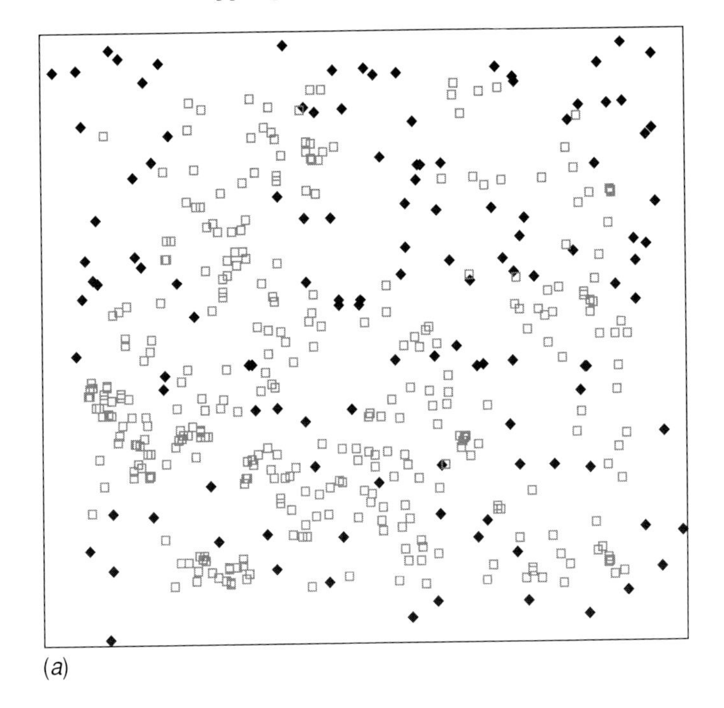


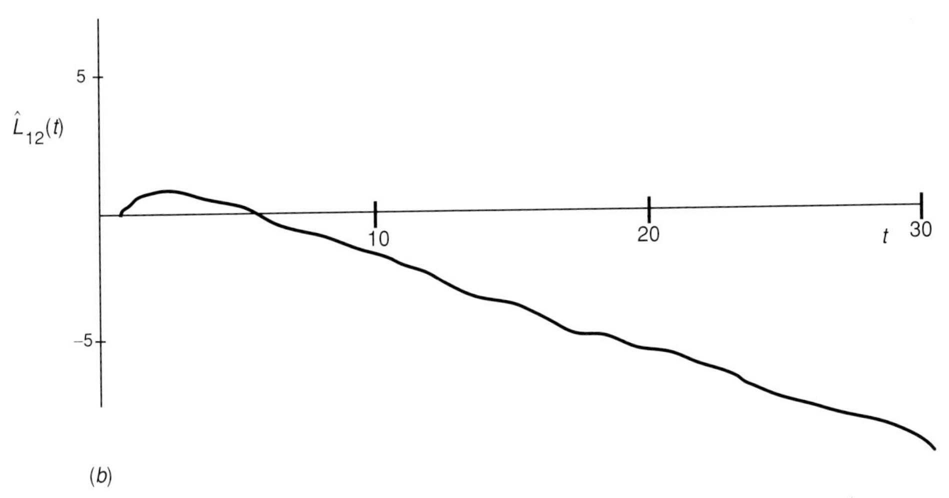


Figure 2.7 (*a*) Map of canopy trees and seedlings in a 50 × 50 m plot near Grande Cache, Alberta, where closed diamonds are canopy trees and open squares are seedlings. (*b*) Bivariate Ripley's $K$ analysis of canopy trees and seedlings shown in (*a*), where $\hat{L}$ is the standardized Ripley's $K$ and $t$ is distance. (*c*) Bivariate data re-labelling for significance testing by randomization: the events' positions are fixed but the labels are redistributed (left: original data; right: labels randomized).

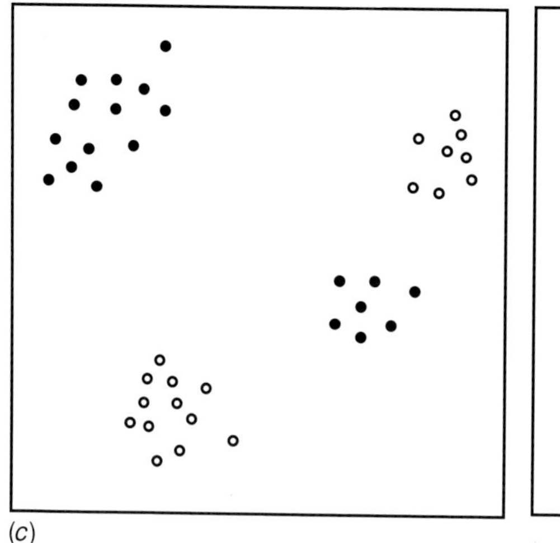
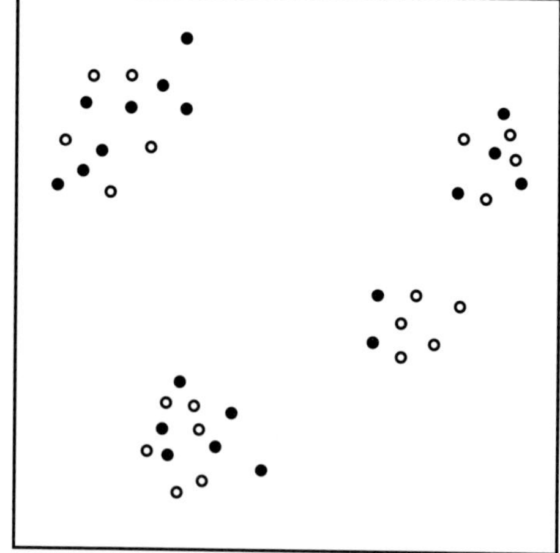


Figure 2.7 (*cont.*)

Assessment of significance is again by a Monte Carlo approach. The distances at which the statistic exceeds its envelope are the scales at which the events of the first type are clustered more than all the events combined. The properties and usefulness of this statistic require further investigation; our own informal trials suggest that the statistic might be more stable if the square-roots of the $K$-functions were used. In addition, this statistic is probably most useful when used in combination with other methods. For example, if the statistic takes values close to zero at all scales, all that you learn is that the two types of event have similar spatial patterns, but what those patterns are like is not detected.

Bivariate patterns are amenable to significance testing by randomization. The positions of events are retained but their 'labels' (the type to which they belong) are randomized (Figure 2.7*c*). This makes it possible to determine whether the events of one kind, for example, diseased organisms, are more clustered than can be explained by the overall non-randomness of the pattern as a whole. Dale & Powell (1994) provided the example of the positions of plants of *Solidago canadensis* L., growing at the edge of a hay field, classified into two categories depending on whether the plant had obvious signs of insect attack, in particular stem galls. In the *Solidago* data, based on a comparison of $K_{12}$ with the results expected from CSR, in quadrat 5, the two kinds of plants appear aggregated over a range of scales, but a randomization of the labels shows that this is a result of overall clumping of plants of either kind (Figure 2.8*a*). In contrast, quadrat 10 from the same data set gives results for $K_{12}$ that seem compatible with independence (values close to 0), but random re-labelling shows that the two kinds of plants are actually segregated (Figure 2.8*b*).

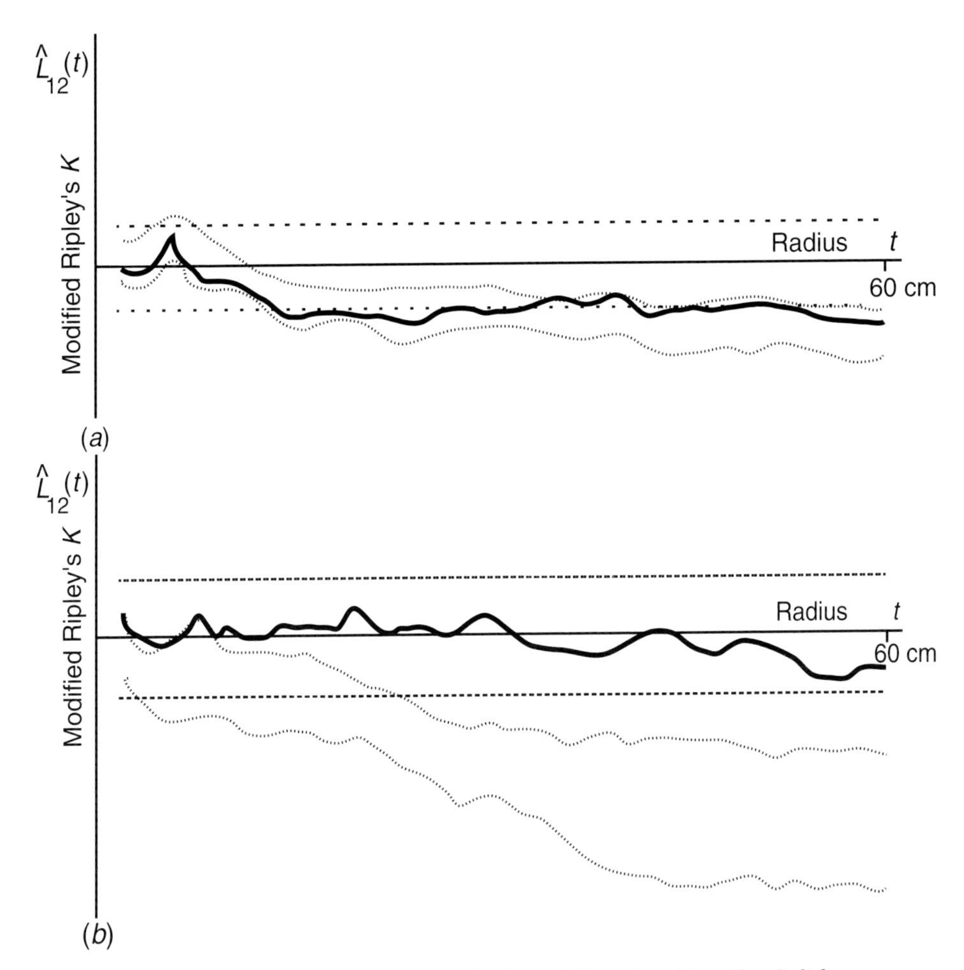


Figure 2.8 (*a*) Bivariate Ripley's *K* analysis on infected and healthy *Solidago* plants, where $L_{12}$ is standardized Ripley's *K* and *t* is distance (Quadrat 5). (*b*) Bivariate Ripley's *K* analysis on infected and healthy *Solidago* plants with randomization envelope (Quadrat 10).

### 2.1.5 Multivariate point pattern analysis

An obvious extension of the analysis of bivariate data is to consider several types of events using some sort of multivariate analysis. In their assessment of the analysis of multivariate point pattern, Lotwick & Silverman (1982) suggested that there were two basic approaches:

(1) methods based on nearest neighbour and empty space functions (described above under 'refined nearest neighbour' methods, Section 2.1.2), and
(2) methods based on second-order analysis (such as Ripley's *K*-function, Section 2.1.4).

There is a close conceptual relationship between the two approaches, which we can describe informally in the following way:

(1) Nearest neighbour methods are based on the following question. How big can a circle centred on an event (or on a random point) get before it encounters another event?
(2) Second-order methods are based on the following question. Given a circle of a given size, centred on an event, how many other events does it contain?

We will describe methods based on nearest neighbours first before proceeding to discuss the second-order approach.

Summary statistics for quantifying several forms of dependence between events of different types in a multivariate point pattern were introduced by van Lieshout & Baddeley (1999). In the univariate context, we introduced the 'event-to-nearest-event' function, $G(t)$, and the 'point-to-nearest-event' function, $F(t)$, also called the 'empty space' function. These can be examined separately, or they can be combined to produce an univariate index of spatial interaction:

$$H(t) = \frac{1 - G(t)}{1 - F(t)}. \tag{2.13}$$

The index takes the value 1 under CSR, values less than 1 for clustered patterns and values greater than 1 for overdispersed patterns.

To adapt this index for multivariate pattern, we consider having $S$ types (e.g. species) and revise the notation for the two kinds of functions (using $I$ and $J$ to denote types, reserving $i$ and $j$ to denote individual events):

$G_{IJ}(t)$ is the distance function for events of type $I$ to events of type $J$.
$G_{..}(t)$ is the distance function for events of any type to events of any type.
$F_J(t)$ is the empty space function from random points to events of type $J$.
$F_{.}(t)$ is the empty space function from random points to events of any type.

These can then be used to define two different $H$-functions:

$$H_{IJ}(t) = \frac{1 - G_{IJ}(t)}{1 - F_J(t)}, \quad \text{for a particular pair of types } I \text{ and } J$$

and

$$H_{..}(t) = \frac{1 - G_{..}(t)}{1 - F_{.}(t)}, \quad \text{for pairs of any type.} \tag{2.14}$$

Last, with $\lambda_I$ being the intensity of the $I$th type, and overall intensity $\lambda_{.}$, an overall index can be defined:

$$I(t) = \sum_{I=1}^{S} \frac{\lambda_I}{\lambda_{.}} H_{II}(t) - H_{..}(t). \tag{2.15}$$

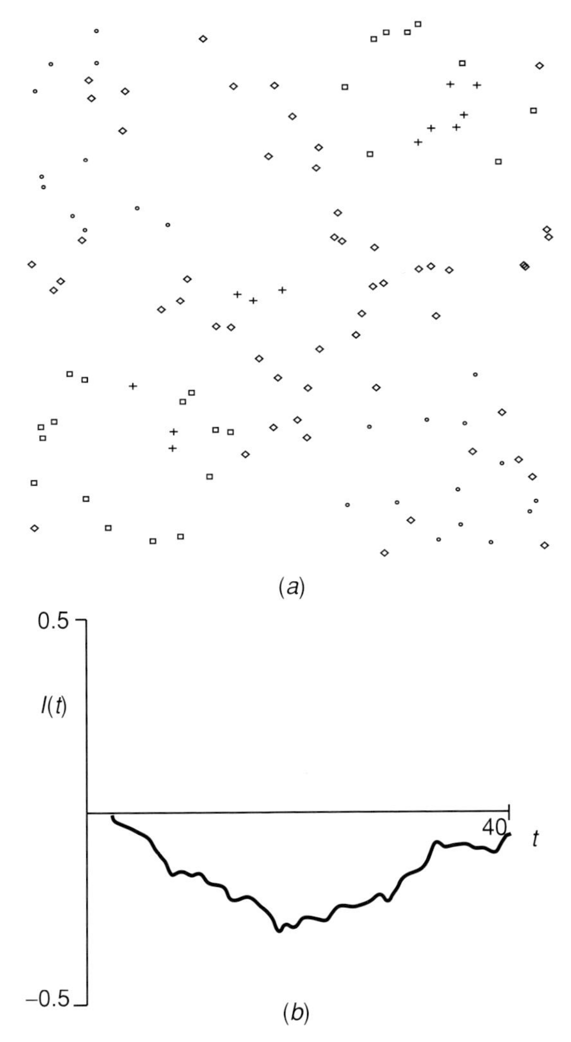


Figure 2.9 (*a*) Artificial multispecies point pattern data for analysis illustration. (*b*) Multivariate point pattern analysis on artificial multispecies: $I(t)$ is the multivariate pattern index and $t$ is distance.

Figure 2.9 illustrates the use of this index. The artificial data are shown in part (*a*) of the figure; Figure 2.9*b*, the index shows that the four species are, in general, segregated from each other, most strongly at a distance that is 20% of the length of the size of the sample plot.

A second approach to multivariate point pattern analysis using nearest neighbours has recently been elaborated by Dixon (2002). With $S$ types of events, a $S \times S$ contingency table is created, with the entry in the $I$th row and $J$th column, $m_{IJ}$, recording the number of times that the nearest neighbour of an event of type $I$

was an event of type $J$. This approach is very much like the analysis of neighbour contact data to detect interspecific association, as discussed by a number of authors including Yarranton (1966), de Jong *et al.* (1980) and Dale *et al.* (1991). In addition to being able to perform a test of the departure from expectation for the table as a whole, individual entries can be tested, using the normal approximation:

$$Z_{IJ} = \frac{m_{IJ} - E(m_{IJ})}{\sqrt{\mathrm{Var}(m_{IJ})}}. \tag{2.16}$$

Formulae for the calculation of the expected values and variances are given in Dixon (2002).

Table 2.1 gives an example of this technique, re-analysing the data provided in Reich *et al.* (1997) from a $3 \times 3$ m study plot in short grass prairie near Fort Collins, Colorado. The dominant species were *Bouteloua gracilis* ('Bogr'), *Agropyron smithii* ('Agsm'), *Oryzopsis hymenoides* ('Orhy') and *Stipa comata* ('Stco'), with the dicots being lumped into the category 'forbs'. The major feature of the table is that individual species had a strong tendency to be their own neighbour, as indicated by the large positive $z_{II}$ values. This tendency was not observed for the category 'forbs'.

Dixon (2002) also provided an index of segregation of the $I$th type, based on the excess of within-type nearest neighbours:

$$S_I = \log\left[\frac{m_{II}/m_{I,\sim I}}{E(m_{II})/E(m_{I,\sim I})}\right] = \log\left[\frac{m_{II}/(n_I - m_{II})}{(n_I - 1)/(N - n_I)}\right]. \tag{2.17}$$

This measure looks straightforward and easy to apply and Dixon (2002) provided an example.

There are two features of Dixon's nearest neighbour method that are noteworthy in comparison with other techniques described in this section. The first comment is that while the method uses the nearest neighbours of events, it does not use the distances to those neighbours in any of the calculations. The second comment is that because the procedure involves the development of a contingency table of the frequencies of neighbouring pair types, the method is not necessarily limited to the use of the first nearest neighbour. It could be extended to include the first and second nearest neighbours together, for example, or separate contingency tables could be set up for each of the first, second and third nearest neighbour frequencies. In Section 2.3, we describe a hierarchy of neighbour networks, and the contingency table approach could be applied to a full range of definitions of which pairs of events are neighbours.

Table 2.1 *Dixon's nearest neighbour method: short grass prairie data from Reich et al. (1997)*

| $I^a$ | $J^a$ | Obs[b] | Exp | | |
|---|---|---|---|---|---|
| From: | To: | $m_{IJ}$ | $m_{IJ}$ | $z_{IJ}$ | Significance |
| Bogr | Bogr | **89** | 49.2 | 5.49 | * |
| | Agsm | 49 | 64.9 | −2.39 | * |
| | Orhy | 28 | 37.9 | −1.81 | |
| | Stco | 37 | 43.4 | −1.12 | |
| | Forbs | 12 | 18.5 | −1.63 | |
| Agsm | Bogr | 53 | 84.9 | −3.62 | * |
| | Agsm | **169** | 64.9 | 15.58 | * |
| | Orhy | 48 | 49.8 | −0.28 | |
| | Stco | 41 | 57.0 | −2.44 | * |
| | Forbs | 28 | 24.3 | 0.82 | |
| Orhy | Bogr | 28 | 28.9 | −0.15 | |
| | Agsm | 44 | 37.9 | 1.12 | |
| | Orhy | **89** | 49.8 | 6.31 | * |
| | Stco | 18 | 33.3 | −3.01 | * |
| | Forbs | 16 | 14.2 | 0.52 | |
| Stco | Bogr | 40 | 38.0 | 0.31 | |
| | Agsm | 44 | 43.4 | 0.10 | |
| | Orhy | 15 | 57.0 | −6.40 | * |
| | Stco | **101** | 33.3 | 13.34 | * |
| | Forbs | 19 | 16.2 | 0.78 | |
| Forbs | Bogr | 14 | 6.8 | 2.33 | * |
| | Agsm | 22 | 18.5 | 0.88 | |
| | Orhy | 21 | 24.3 | −0.72 | |
| | Stco | 22 | 14.2 | 2.23 | * |
| | Forbs | **16** | 16.2 | −0.06 | |

[a] The species are Bogr = *Bouteloua gracilis*, Agsm = *Agropyron smithii*, Orhy = *Oryzopsis hymenoides* and Stco = *Stipa comata*.
[b] Bold font highlights intraspecific frequencies and results.

The data used to illustrate Dixon's method were originally used in the description of another method for multivariate point pattern analysis, developed by Reich *et al.* (1997), based on all the distances between events of the same type. Let the total number of events be $N$ (but not all of them can be assigned to known taxa, as is often the case in plant ecology). We have $S$ identified groups, $G_1$ to $G_S$, and one group for the unidentified, the 'other' category. If there are $N'$ identified events and $n_I$ in the $I$th group then:

$$C_I = n_I/N'. \tag{2.18}$$

The analysis is based on the average distance between members of a group:

$$\xi_I = \sum_{i=1}^{N'-1} \sum_{k=j+1}^{N'} (d_{jk} | j \in G_I \ \& \ k \in G_I) \Big/ \binom{n_I}{2}. \tag{2.19}$$

The test statistic is then the weighted within-group average:

$$\delta = \sum^{S} C_I \xi_I. \tag{2.20}$$

An analytic evaluation of this statistic is possible, based on the number of possible assignments of $N$ events to $S + 1$ groups, but a Monte Carlo test or a randomization test based on re-labelling the events is straightforward and easy to implement. One advantage of this approach to analysis is that different species can be examined separately using the observed values of $\xi$. Another advantage is that CSR is not the only null model that can be used, and, as noted above, these authors made use of the Neyman–Scott model for comparison. A disadvantage is that if there is more than one scale of pattern in the data, the average within-group distance may not be informative because it includes distances related to different scales in the pattern. This disadvantage is related to the difference between local and global measures of spatial pattern; a local version of this technique exists in which the average distances between events of the same type is calculated and then mapped. Reich *et al.* (1997) consider using an upper limit to the distances that is somehow related to 'cluster size'. It is not clear how effective this modification would be if there were several scales of pattern or if different species had markedly different cluster sizes. Prior analysis using some version of Ripley's $K$ or other exploratory technique might be a useful preliminary step. We will now proceed to describe the multivariate version of Ripley's $K$.

It is clear that, given multivariate point patterns, each type or species can be analysed separately using the univariate version of Ripley's $K$-function. Pairs of species can be analysed using the bivariate version too, but it is not clear what a truly multivariate analysis would involve. As always, the technique used will depend on the hypothesis that is of interest: for example, there is a difference between 'Do all species tend to be segregated from any other species considered individually?' and 'Do all species tend to be segregated from all other species considered together?'. It is a matter of 'partitioning' the overall pattern of events into those attributable to individual types and those attributable to the relationships among types.

In their review of approaches to multivariate point pattern analysis, Lotwick & Silverman (1982) made the interesting comment that 'Not surprisingly, description and estimation of the second order structure of a multitype process requires consideration of only two of the types at a time.' This statement is not quite true, in our opinion, because while there are insights to be gained from looking at all possible intraspecific statistics of type $K_{II}(t)$, there is also much to be learned from examining interspecific statistics of type $K_{I,\sim I}(t)$ or $K_{I\cdot}(t)$. In effect, we suggest partitioning $K_{\cdot\cdot}(t)$, which includes all events pairs of any species combination, into $K_{XX}(t)$, which considers all conspecific pairs of events, and $K_{X,\sim X}(t)$, which considers all interspecific pairs of events. $K_{XX}(t)$ can be partitioned into $S$ possible $K_{II}(t)$; and $K_{X,\sim X}(t)$ can be partitioned into all $\binom{S}{2}$ possible $K_{IJ}(t)$. As in the original version of Ripley's approach to point pattern analysis, these estimated partitioned $K$ statistics would be transformed into the equivalent $L$ statistics for easier interpretation. Figure 2.9*a* shows the artificial data for four species used to illustrate the multivariate pattern index $I(t)$ described above. Figure 2.10*a* analyses the same data and plots the estimates of $L_{X,\sim X}(t)$ (upper), $L_{\cdot\cdot}(t)$ (middle) and $L_{XX}(t)$ (lower). These results show that while the arrangement of events is random, events of the same type are clustered, on average, and segregated from events of a different type. The next part of the figure illustrates the different scales and intensities of clumping of the individual species, with type 2 having the smallest scale of strong clumping and type1 differing little from random (Figure 2.10*b*). Figure 2.10*c* summarizes the interspecific analyses for individual species, showing that species 3 and 4 have a high degree of negative association with events of other types over all scales. These could be further partitioned into particular species pairs, 1 and 2, 1 and 3, and so on. Although this analysis is more complicated than other approaches, it is also able to provide the most detailed information on the characteristics of the multispecies pattern.

Condit *et al.* (2000) suggested analysing multispecies point pattern using a modification of Ripley's $K$, based on counts in circular bands or annuli of width $\Delta t$ centred on individual events, rather than in circles. The statistic they suggested is:

$$\Omega_I(t) = [K_{II}(t + \Delta t) - K_{II}(t)] \div \{\lambda_I[\pi(t + \Delta t)^2 - \pi t^2]\}. \qquad (2.21)$$

While the divisor looks complicated, it is just the area of the circular band used to make the counts multiplied by the unit density of species $I$.

The use of rings rather than circles for event counts allows the isolation of specific distance classes, rather than including the short distances with the larger, as occurs with large-diameter circles. One disadvantage of this method is that the distance classes need to be broad, particularly for rare species, in order to avoid erratic-appearing curves due to zero counts. In addition, the choice of the width of the circular band used may be somewhat subjective. The difference between

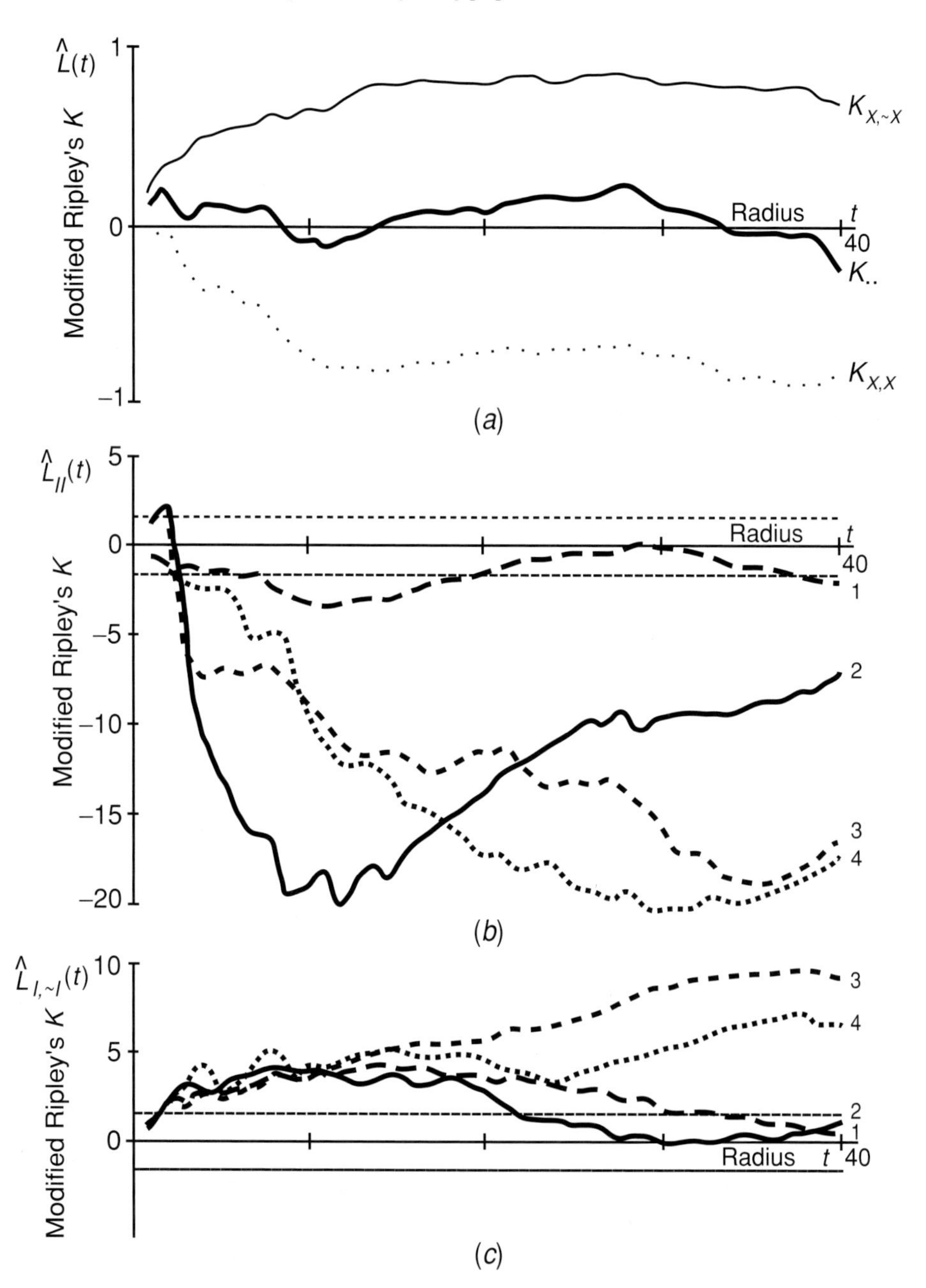


Figure 2.10 (*a*) Multivariate standardized Ripley's *K* analysis ($\hat{L}$): bottom line, conspecific pairs; middle line, all pairs; top line, interspecific pairs. (*b*) Standardized Ripley's *K* analysis for each species: conspecific pairs. (*c*) Multivariate standardized Ripley's *K* analysis ($\hat{L}_{I,\sim I}$): interspecific pairs.

using circular bands rather than complete circles is that what is being studied is distance, rather than scale of pattern, as in the relationship between PQV and TTLQV described later in Section 2.5.1.

Given the large range of methods available, a choice of method may seem difficult. Methods based on Ripley's $K$ are popular, and for good reason: they are easy to use and to interpret and there is a range of such techniques that cover most kinds of point data. In general, they look at scale of pattern and they can deal with situations in which several scales of pattern are present. The hierarchy of neighbours approach (described in Section 2.3 below) does not use distance explicitly in the way that Ripley methods do and would provide a good complement to that set of methods. No single method can tell us everything we may want to know and so the use of two or three complementary methods is recommended, as usual.

## 2.2 Mark correlation function

The following methods presented are designed to investigate the interactions of neighbouring trees in a forest and appear in the works of Penttinen *et al.* (1992), Gavrikov & Stoyan (1995) and Stoyan & Penttinen (2000). This approach follows on from the bivariate methods described above, to take account of a quantitative characteristic associated with events, $m_i$, for example for the diameter of a tree.

If $\mu$ is the mean value of $m_i$, then

$$\hat{K}_m(t) = \sum_{i=1}^{m} \sum_{j=1}^{m} w_{ij} I(i, j) m_i m_j, \tag{2.22}$$

and the observed value is compared with the expected value by calculating

$$\hat{L}_m(t) = t - \sqrt{\hat{K}_m(t)/\pi\mu^2}. \tag{2.23}$$

When $\hat{L}$ is plotted as a function of $t$, large positive values indicate overdispersion of the marks and large negative values indicate their aggregation. Parallels with the interpretation of Ripley's $K$-function are obvious. The authors also pointed out that replacing '$m_i m_j$' with '$(m_i - m_j)^2$' in (2.22) produces what is essentially the equivalent of the sample, experimental, variogram (see Chapter 3).

Figure 2.11 presents an artificial example (data in part (*a*)) in which the events are not clumped, but the values associated with them (diameter) are. Ripley's $K$-function analysis (Figure 2.11*b*) shows no evidence of clumping, but the mark correlation analysis does (Figure 2.11*c*). The authors suggested using a 'kernel' (smoothing) function as part of the calculation, but that may not be necessary since its only effect is to smooth the differences exhibited by different distances.

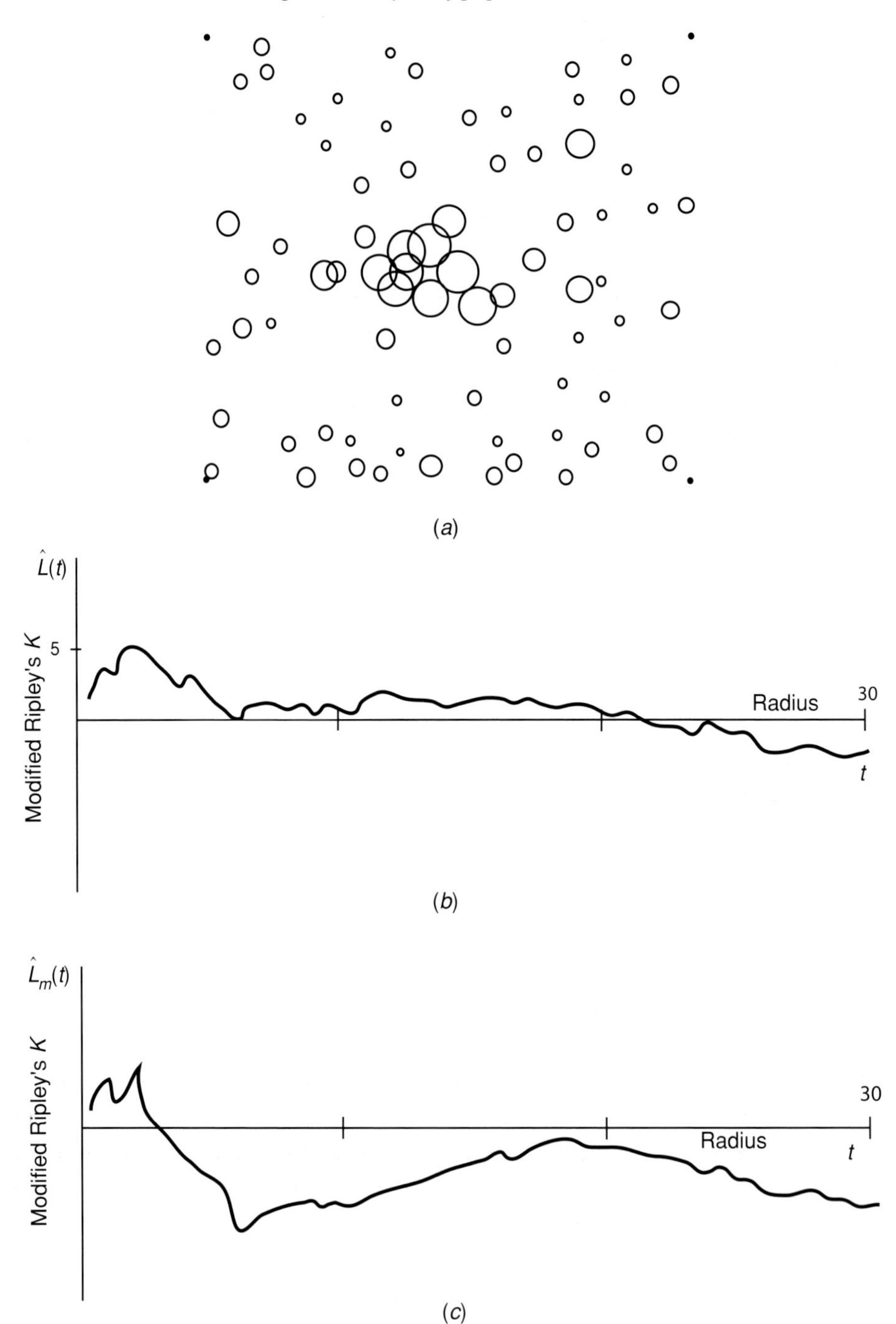


Figure 2.11 (*a*) Artificial univariate point pattern data with diameter characteristics (the dots indicate the size of the study area). (*b*) Standardized Ripley's *K* analysis on artificial univariate diameter data (positions only). (*c*) Mark correlation analysis on the same artificial univariate data, using diameter and position, where $\hat{L}_m(t)$ is the mark correlation statistic.

Goulard *et al.* (1995) provide an excellent example of the usefulness of this approach in a study of the clumps of sprouts of sweet chestnut, *Castanea sativa* Mill., in the Limousin region of France. In addition to the locations of coppiced clumps, four variables were measured: diameter, number of shoots before cutting, height at one year after cutting and height at three years after cutting. These were the 'marks' used in the analysis. The authors also measured soil depth to the granite beneath, at 120 locations, and produced an interpolated (kriged) estimated surface (see Chapter 3) of soil depth for the whole study area. They found that the clumps were regularly dispersed, with diameter and number of shoots displaying negative autocorrelation at smaller distances. Heights were not strongly correlated. The analysis showed that small clumps were aggregated in gaps between larger clumps and that heights could be related to their spatial correlation with local soil variables. This study provides a good example of the sophisticated and detailed analysis that the marked point process approach can offer.

## 2.3 Networks of events

In Section 2.1, we began the discussion of analysing point patterns with the concept of nearest neighbours. If the nearest neighbours are joined by lines on a map of the events' positions, a picture of the network of nearest neighbours emerges. Figure 2.12 shows the network of neighbours that results from the first nearest neighbour definition (artificial example). In technical terms, the network is a *graph*, consisting of the points or vertices, $v_i$, joined by lines or edges, $e_k = (v_i, v_j)$. A graph is *connected* if there is a sequence of vertices joined by edges (a *path*) between any two vertices in the graph. (Imagine tracing a route from one vertex to another following the edges.) In its purest form, a graph is a combinatorial entity, depicting only structure, so that the vertices do not have positions and the edges do not have lengths (or weights, or directions, or shape), except for the purposes of depiction. A graph as a geometric entity, drawn on a surface with vertex positions and edge lengths (that is to say *embedded* in the plane), is sometimes referred to as a *topological graph* (Harary 1967), but we will be informal and use the term graph for either kind of object. A graph of first nearest neighbours is almost never connected, as in Figure 2.12. A path that begins and ends at the same vertex without using any edge twice is a *cycle*; a graph that has no cycles is called a *tree*.

The nearest neighbour definition can be narrowed to include only those pairs of events (vertices in the graph) that are mutually nearest neighbours, producing a graph with fewer edges (the heavy lines in Figure 2.12). Some events will have no neighbour at all. This graph of mutually nearest neighbour pairs (MNNs) is a *subgraph* of the graph of all first nearest neighbours (NNs) because all edges in MNNs are also included in NNs. Under complete spatial randomness, about 62%

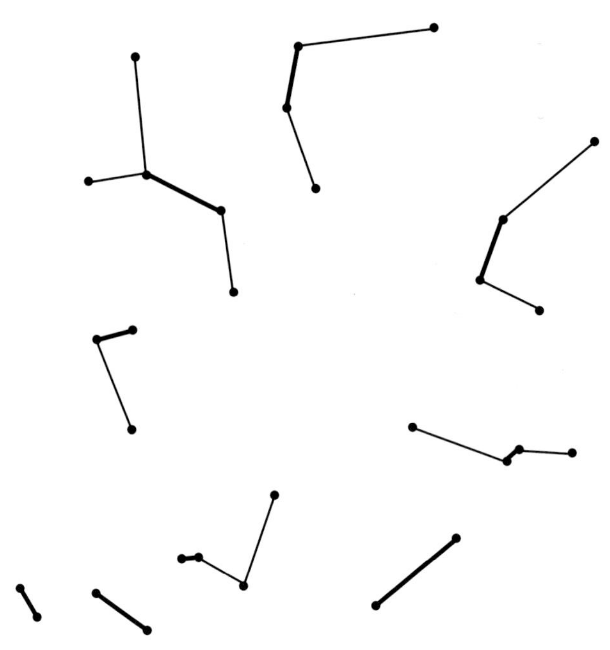

Figure 2.12 Network of first nearest neighbours for an artificial example of point pattern. Mutual nearest neighbour pairs have bold lines.

of the events are members of reciprocal nearest pairs (Pielou 1977) and so the expected number of neighbours per event (or edges per vertex) is about 0.62.

Table 2.2 *Hierarchy of neighbour networks*[a]

| |
|---|
| 1. Mutually nearest neighbour pairs (≈0.62) |
| 2. All nearest neighbours (≈1.4) |
| 3. Minimum spanning tree (≈2.0) |
| 4. Relative neighbourhood graph (≈2.4) |
| 5. Gabriel graph (≈4.0) |
| 6. Delaunay triangulation (≈6.0) |

[a] With approximate average number of neighbours.

The first nearest neighbour graph can be extended in several ways: for example, to include the first and second nearest neighbours. We can describe a neighbour hierarchy that begins with a graph of mutually nearest neighbours and ends with the Delaunay triangulation (Table 2.2). The advantage of using several networks for analysis is that a range of numbers of neighbours and a range of average distance to neighbours are used. This approach should provide more insight, for example, into the characteristics of segregation or aggregation of a bivariate pattern.

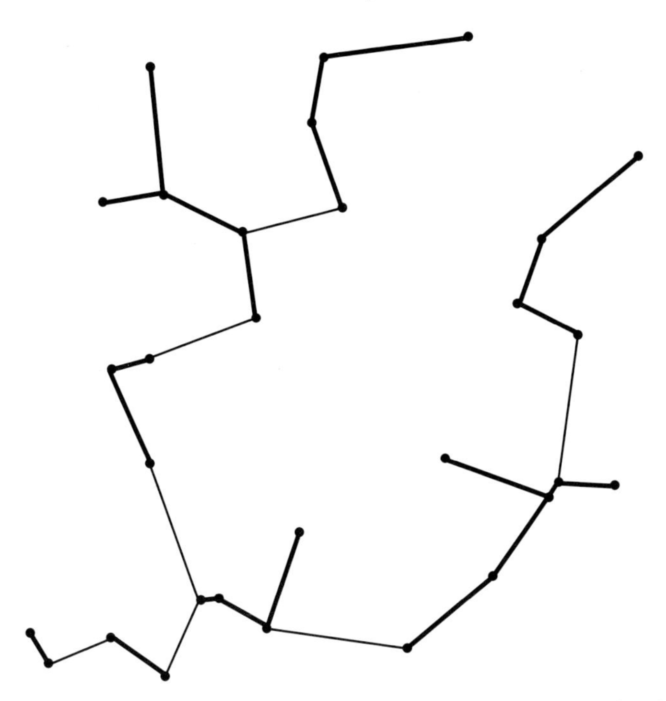

Figure 2.13 Minimum spanning tree for the same pattern as shown in Figure 2.12. The bold lines are those in the nearest neighbour graph.

The third network in the list is the minimum spanning tree (Figure 2.13). A spanning tree is a graph with no cycles that includes all vertices; the minimum spanning tree is the one with the smallest total length of edges. The minimum spanning tree (MST) contains all nearest neighbour edges and so the nearest neighbour network is a *subgraph* of it. One way to visualize the relationship is to consider that the minimum spanning tree is formed by connecting up the disconnected components of the nearest neighbour graph, using the shortest edges available. A spanning tree on $n$ vertices must have $n - 1$ edges, producing an average number of neighbouring vertices of $2 - 2/n$, or approximately 2 neighbours for each event.

The fourth network is the relative neighbourhood graph (RNG; Toussaint 1980), formed by joining all pairs of vertices, A and B, for which the lens formed by the radii of the two circles AB, centred on A and B, contains no other vertex (Figure 2.14). Figure 2.15 shows this fourth network and the minimum spanning tree, which is a subgraph of it. The average number of neighbours in this network seems to be about 2.4 under CSR.

Next is the Gabriel graph (GG; Gabriel & Sokal 1969). It is formed by joining all pairs of vertices, A and B, for which the circle with diameter AB is empty (Figure 2.16). The relative neighbourhood graph is a subgraph of the Gabriel graph

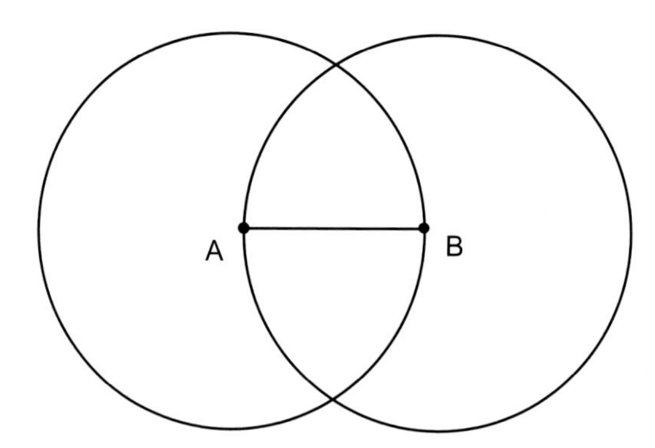


Figure 2.14 The definition of edges for the relative neighbourhood network. A and B are joined if the central lens contains no other points.

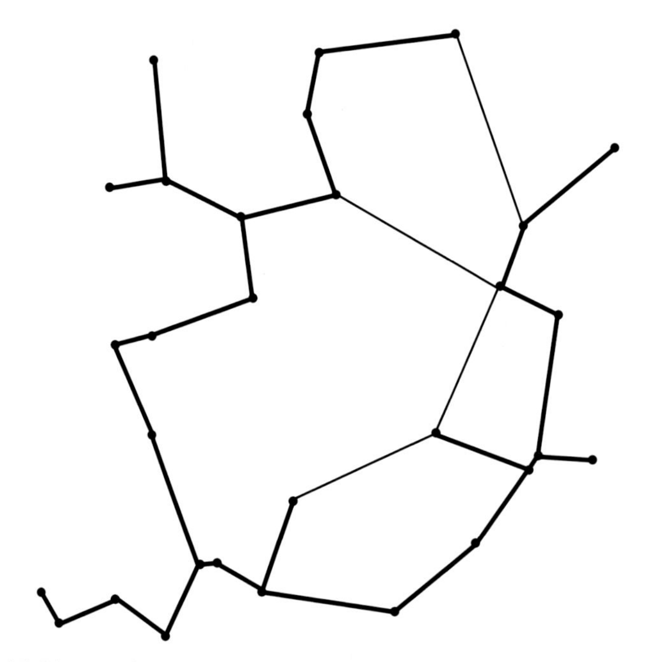

Figure 2.15 The relative neighbourhood graph for the same pattern.

(see Figure 2.17). Matula & Sokal (1980) showed that the average number of neighbours in the Gabriel network approaches four, under CSR.

The most complex neighbour network in the hierarchy is the Delaunay triangulation (DT; Okabe *et al.* 1992). It is formed by joining all triplets of vertices A, B and C, for which the circumcircle of the triangle ABC contains no other vertex (Figure 2.18). The Gabriel graph is a subgraph of this triangulation (see Figure 2.19). This triangulation is closely related to a familiar spatial structure, the tessellation of polygons variously known as Dirichlet domains, Thiessen polygons or Voronoi

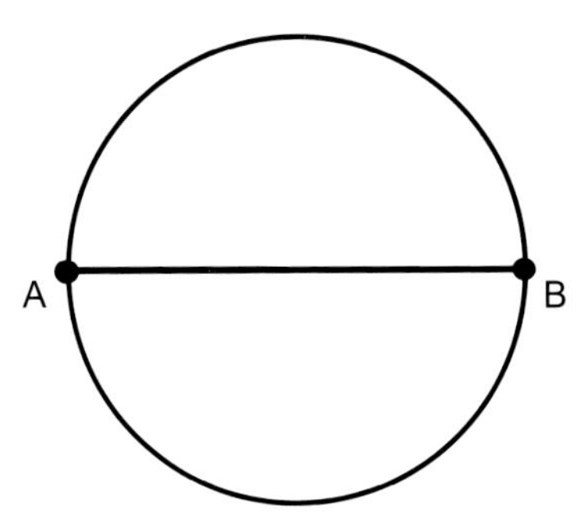


Figure 2.16 The definition of edges for the Gabriel graph network. A and B are joined if the circle on AB contains no other points.

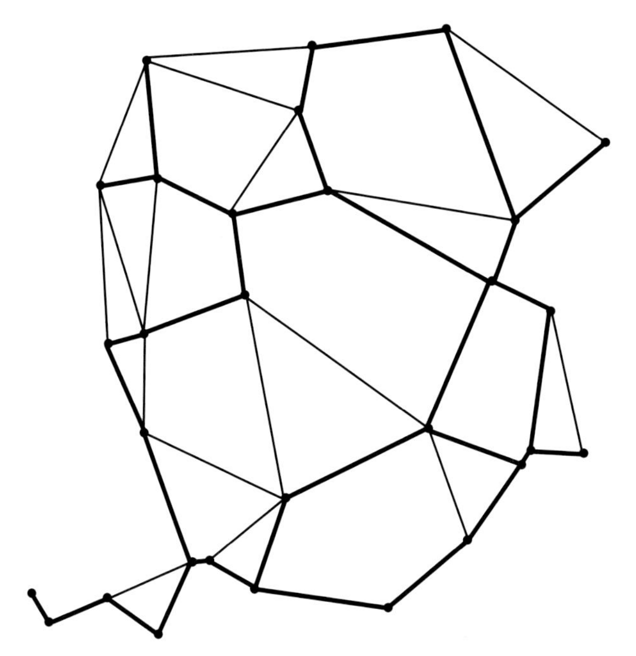

Figure 2.17 Network of a Gabriel graph for the same pattern as shown in Figure 2.12.

polygons (Figure 2.20). Any pair of events, the polygons of which have a common boundary in the tessellation, are joined by a line in the triangulation, indicating that they are first-order neighbours in that structure. The ecological application of the tessellation comes from the fact that each polygon contains all parts of the plane closest to its own event than to any other. Where the events are plants, the polygon associated with each plant determines the resources it can pre-empt and its success may depend on the size of its polygon (Mithen *et al.* 1984). It can be shown that the

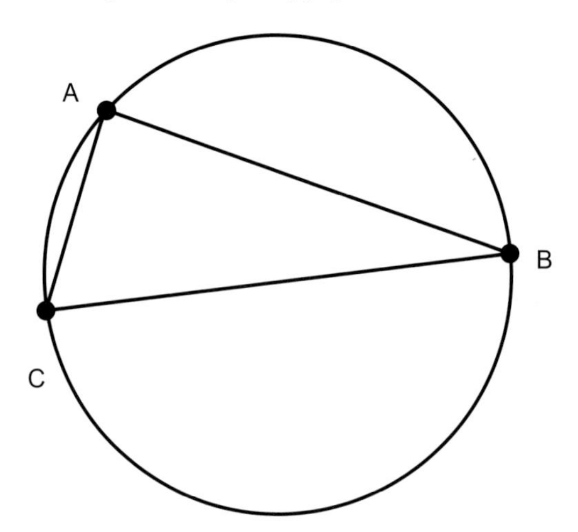


Figure 2.18 The definition of edges for the Delaunay triangulation network. A and B, B and C, and A and C are joined if the circumcircle on ABC contains no other points.

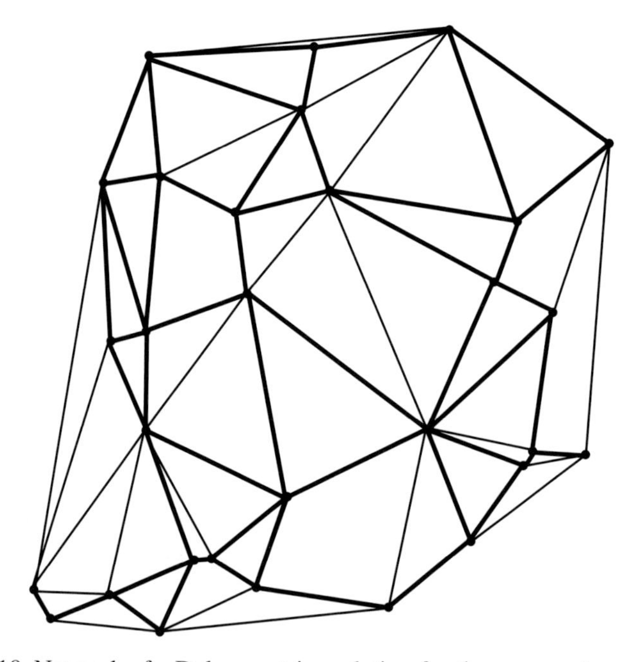

Figure 2.19 Network of a Delaunay triangulation for the same pattern.

average number of neighbours in the Delaunay network approaches six, no matter what the spatial arrangement of the events (Upton & Fingleton 1985).

The hierarchy of networks given in Table 2.2 is unified by the fact that each network is a subgraph of the next most complicated, so that in going up through the hierarchy, lines are only added, never removed or replaced. In ecological terms, this

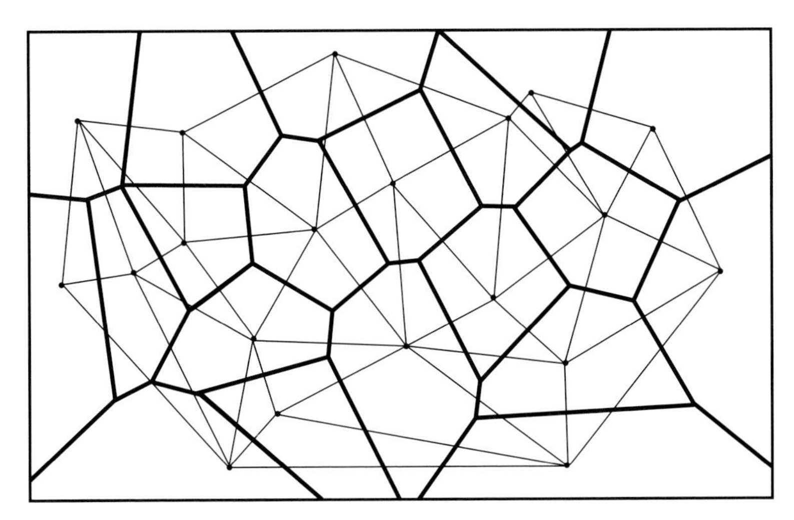

Figure 2.20 The Delaunay network (fine lines) as Dirichlet domains (bold lines), Thiessen polygons or Voronoi polygons.

means that more and more events are considered to be neighbours, but no event that was counted as a neighbour is removed from consideration when there are more.

For example, Figure 2.21 shows the positions of plants of *Solidago canadensis* L. growing at the edge of a hay field near Edmonton, Alberta (quadrat 2; Dale & Powell 1994). They are of two types: those that have obvious evidence of attack by insects (stem galls) and those that do not. We used the hierarchy of networks to define which pairs of plants are neighbours, and used a randomization procedure to determine whether there was a significantly large number of neighbour pairs that were of the same type. The results were not significant for the first member of the hierarchy (MNN), but they were for the next two (NN and MST); the results were not significant for RNG and GG, but they were for the Delaunay triangulation (DT).

This hierarchy of neighbour networks can be used in a number of other ways. It can be used for multivariate point pattern analysis and for the analysis of marked point patterns, for example by looking at the number of like–like joins (multivariate) or correlation coefficients (marked) in the range of networks available. In most applications, 're-labelling' randomization will be an obvious technique to evaluate the significance of any result. This description of the hierarchy is not intended to provide advice on which is the 'best' network for ecological analysis. The suggestion is to use the entire hierarchy of six networks because the differences among them can provide valuable insights. Even more useful to ecologists would be comparison of the results of this hierarchical analysis with the results from artificial data that are designed as realizations of the hypothesized underlying ecological processes, such as dispersal and mortality. Other applications of this hierarchy, as

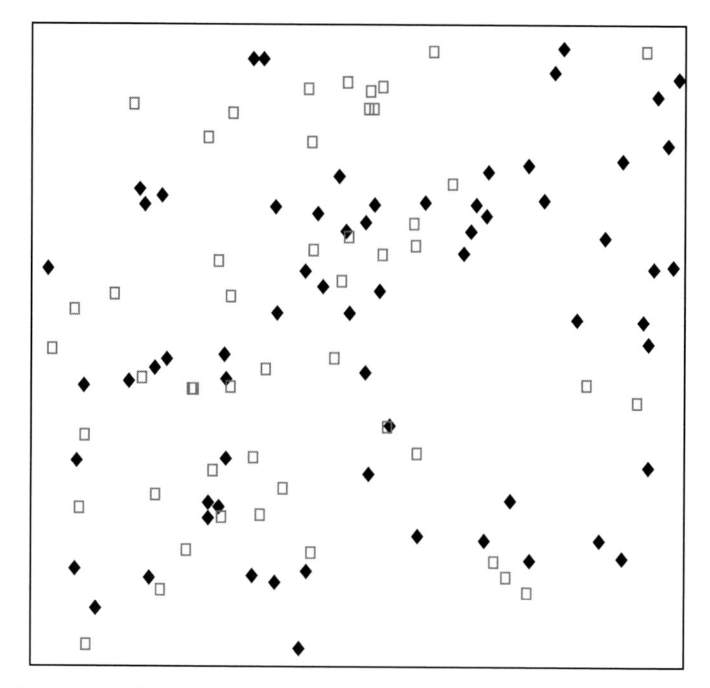

Figure 2.21 Map of *Solidago canadensis* in a 2 × 2 m study plot (quadrat 2) near Edmonton, Alberta. Closed diamonds are 'clean' plants, and open squares are those with galls or other forms of insect attack.

well as of extensions or modifications, will be found useful by ecologists for the spatial analysis of various kinds of point patterns.

## 2.4 Network analysis of areal units

There are two obvious extensions to the preceding discussion of the analysis of patterns of events in a two-dimensional plane using neighbour networks. The first is to look at areal units, such as habitat units ('patches') in a landscape, rather than events that can be treated as dimensionless points, and the second is to look at patterns of events in fewer, or in more, than two dimensions. We will follow that order.

Our treatment of analysing patches in the plane is, in part, prompted by the graph theoretic approach to studying landscape connectivity (Cantwell & Forman 1993; Urban & Keitt 2001, among others). That series of papers brought renewed attention to graph theory as a useful context for the evaluation of ecological structure, which is a favourite topic of the authors (cf. Dale 1977; Fortin 1994)!

A frequent underlying assumption in landscape ecology is that an area can be represented by a set of identifiable landscape units of habitat, referred to as 'patches'. Studies of fragmentation consider patches of habitat suitable for the focus organisms (e.g. an endangered species) situated within a matrix of unsuitable landscape elements (e.g. woodlots in an agricultural region). In the graph theoretic approach to evaluating this kind of structure, habitat patches are represented by the vertices of a graph and the edges that join them represent connections between the patches related to ecological processes. That is, the landscape is represented as a functional network, with colonization or dispersal often being the process of primary concern in conservation-oriented studies. In the preceding section we provided a non-technical introduction to some of the terminology used in graph theory. We need to provide a few more terms here. Recall that a graph is *connected* if at least one *path* exists (a sequence of vertices joined by edges) between any two vertices in the graph. The *degree* of a vertex is the number of edges that it has. The most extreme form of a connected graph is a *complete* graph, in which each vertex has an edge to every other vertex, so that all vertices have degree $n - 1$. A *planar* graph is one that can be drawn in the plane without the edges intersecting; the complete graph of order 4 is planar, but the complete graph of order 5 is not (try it!).

In a connected graph, a *cut-point* is any vertex, the removal of which causes the graph to be no longer connected. Similarly, any edge, the removal of which disconnects the graph, is called a *cut-edge* or *bridge*. Obviously, a complete graph has no cut-points and no cut-edges. A graph that is not itself connected will be made up of a number of connected subgraphs, called *components*. The *order* of a component is just the number of connected vertices it contains. For a connected graph, there are (at least) two measures of how strongly it is connected: *vertex connectivity* (sometimes called *node connectivity*) is the minimum number of vertices that must be removed to disconnect the graph. *Edge connectivity* is the minimum number of edges that must be removed to disconnect the graph. Figure 2.22 illustrates these various terms. There is an obvious analogy between the connectivity of a graph and the number of dispersal routes, made up of corridors and patches, available in the landscape it represents.

Graph theory is about the structure of connections and, in general, the edges of a graph do not have properties like length or weight, but in some applications of graph theory, it is useful for the vertices and edges to have properties of their own. For example, each edge in the graph may have a length associated with it, $d(e_k)$, which could be the physical distance between vertices which have locations, as drawn in a diagram or on a map, or some other property of the network, such as cost of transport or resistance to movement. The graph theoretical *distance* between two vertices in the graph, $\delta_{ij}$, is the minimum path length (smallest sum of edge lengths) of any path between the two vertices. This means that a minimum spanning tree (MST), which we constructed in the previous section using the criterion of minimum physical distance, could use the criterion of minimizing any other measure of edge length. For many applications in ecology, for example in the analysis of landscapes, we can think of a graph of landscape objects as being embedded in the plane, with the

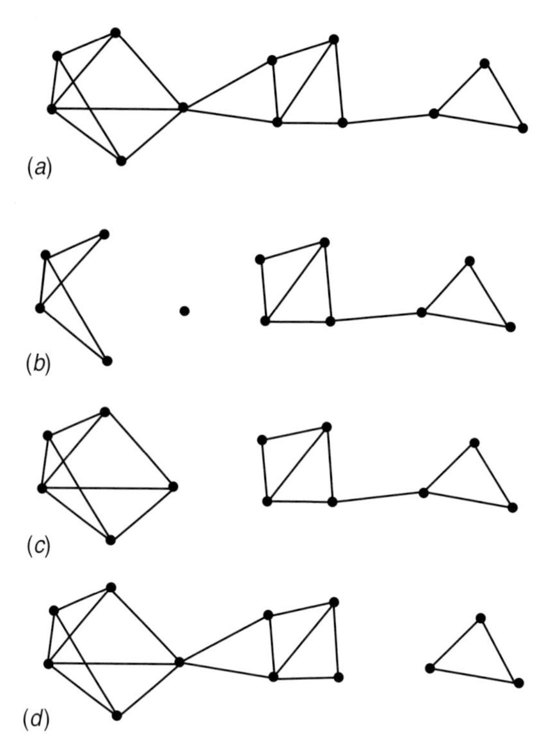


Figure 2.22 Illustration of graph theoretical definitions: (*a*) original graph, (*b*) the point now isolated is a cut-point because its removal disconnects the graph, (*c*) two lines removed disconnect the graph but (*d*) the line removed is a cut-edge because its removal disconnects the graph. The edge connectivity is 1 and the vertex connectivity is also 1.

vertices having geographic locations or positions, and the edges having the length of the (Euclidean) distances between the vertices they join. Whatever the measure of length, the *eccentricity* of any vertex, $\varepsilon(v_j)$, is the maximum graph theoretical distance to any other vertex in the graph. Last, the *diameter* of a graph is the maximum eccentricity of any node in the graph.

Urban & Keitt (2001) examine the properties of landscape graphs by considering the effects of removing edges from the graph and of removing vertices, analogous to the loss of dispersal corridors in the first case and of habitat patches in the second. The characteristics they suggested for evaluating edge removal are purely graph theoretic: the number of components that result, the diameter of the largest component and the order of the largest component. The procedure they suggested is to start with a complete graph (one in which all possible pairs of vertices are joined by edges) and then to use a series of threshold distances and to remove edges, leaving only those shorter than that threshold, thus creating a series of threshold distance networks. The response of the graph theoretical properties of these networks to the threshold distance then provides an evaluation of the landscape patch structure.

As an example, Figure 2.23*a* shows the approximate sizes and locations of 21 lakes in an extensive peatland near the Alberta–Saskatchewan border (55° 45′ N, 110° 45′ W) and their distance-based minimum spanning tree. There are no direct permanent surface-water connections between these lakes, and it is reasonable to take as a working hypothesis that dispersal between them is inhibited by distance. Figure 2.23*b* shows their Delaunay triangulation (DT) graph. Using the Urban & Keitt (2001) approach on this graph, Figures 2.23*c–f* show the series of threshold distance networks for 7, 6, 5 and 4 km. At 7 km, the graph is a single component of all 21 vertices, with no cut-points or bridges. The diameter of the single component is slightly larger than that of the original complete graph (30.3 rather than 26.5), because (for example) there is no longer an edge between Lakes 5 and 21, and the path between them goes through Lake 4. At 6 km, the graph is no longer connected, but the largest component includes 20 of the 21 lakes, and its diameter has decreased (27.6) due to the loss of the most isolated lake. The main component contains several cut-points (e.g. Lake 8) and cut-edges. The 5 km threshold network is similar to the 6 km graph, with only the loss of two more lakes (6 and 7) from the largest component. The change from 5 to 4 km, as the network threshold, is dramatic. There are now nine components, of which the largest has only five lakes and a diameter of only 6.1. This behaviour of a fairly abrupt transition from a few components, some of which are large, to many small ones is similar to Urban & Keitt's (2001) observations on a hypothetical but realistic landscape of patches.

If we return to the Delaunay triangulation of the example landscape depicted in Figure 2.23*b*, we can suggest a simple measure of the importance of any edge, $e_{ij}$, in the graph. That measure is the minimum 'cost' of its removal, either in absolute distance of edges required to replace the connection between $v_i$ and $v_j$, or as a proportion of its length:

$$c_d(e_{ij}) = l(e_{ik}) + l_{(e_{kj})} - l(e_{ij}) \tag{2.24}$$

or

$$c_r(e_{ij}) = \frac{l(e_{ik}) + l(e_{kj}) - l(e_{ij})}{l(e_{ij})}. \tag{2.25}$$

For example, the shortest replacement path for $e_{35}$ is through $v_4$, so that

$$c_d(e_{35}) = l(e_{34}) + l(e_{45}) - l(e_{35}) = 3.8 + 6.1 - 3.3 = 6.6;$$
$$c_r(e_{35}) = 6.6/3.3 = 2.0.$$

For $e_{9,11}$, the cost of its loss is much less:

$$\begin{aligned} c_d(e_{9,11}) &= l(e_{9,10}) + l(e_{10,11}) - l(e_{9,11}) \\ &= 1.7 + 2.1 - 2.2 = 1.6; \\ c_r(e_{9,11}) &= 1.6/2.2 = 0.73. \end{aligned}$$

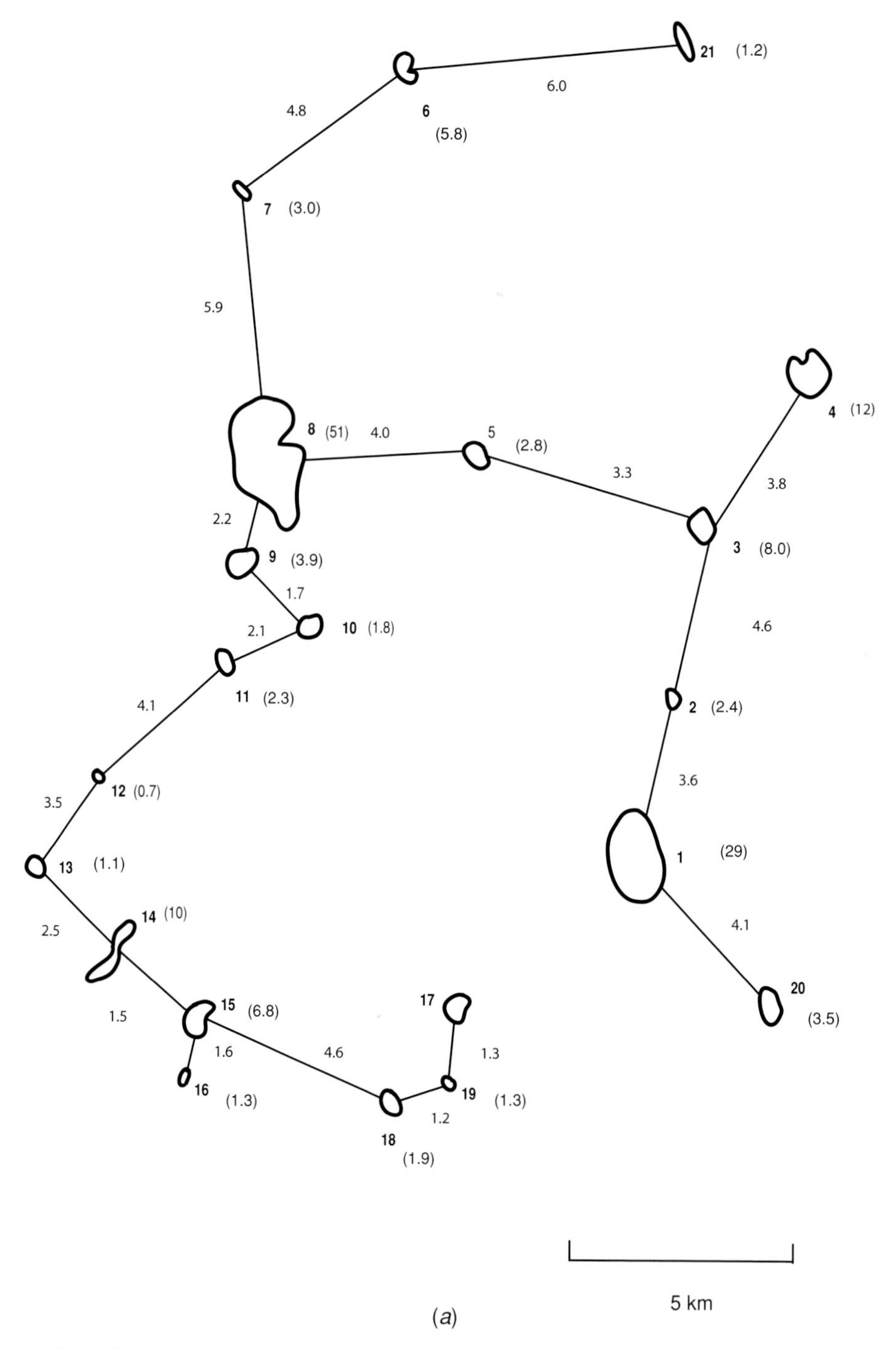


Figure 2.23 (*a*) The minimum spanning tree for 21 lakes in a peatland, with approximate distances between them (in kilometres) and their relative sizes (arbitrary scale). (*b*) Delaunay triangulation graph of the same set of lakes. (*c*)–(*f*) Network of lake neighbours using a 7, 6, 5 and 4 km threshold, respectively.

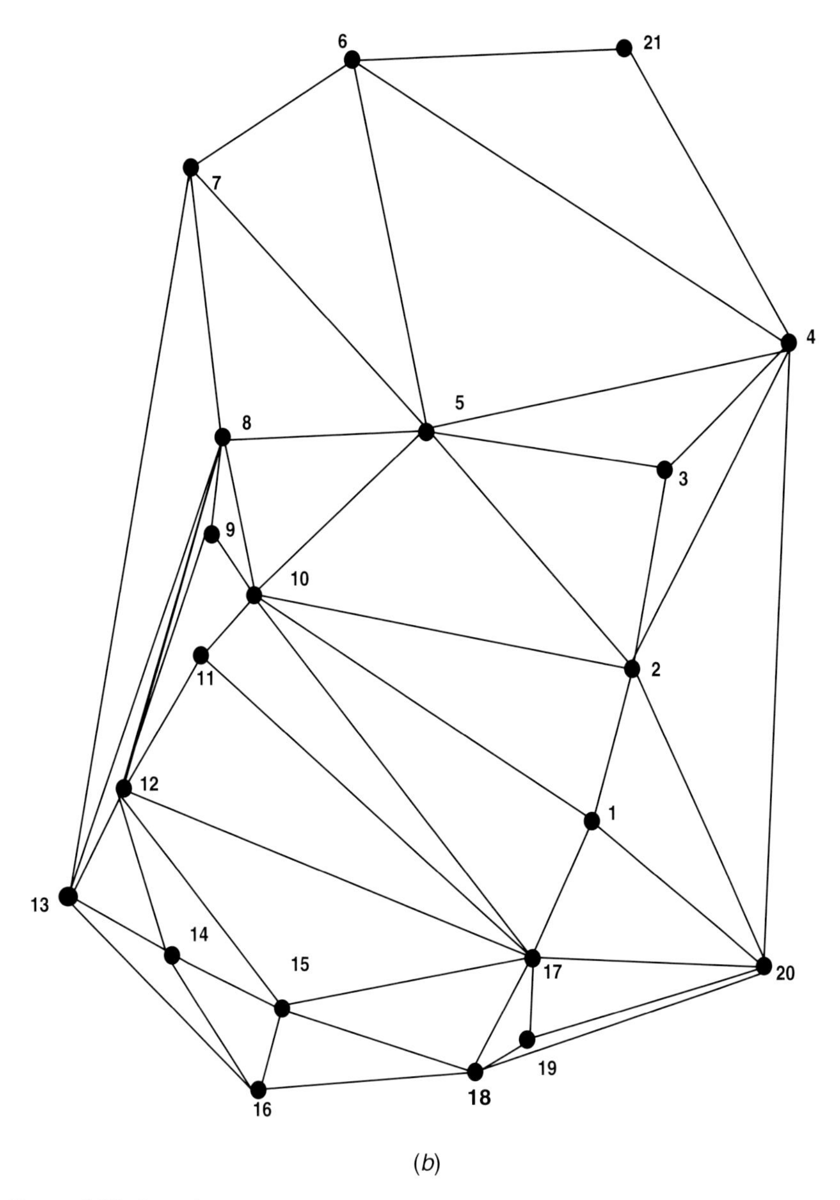


(*b*)

Figure 2.23 (*cont.*)

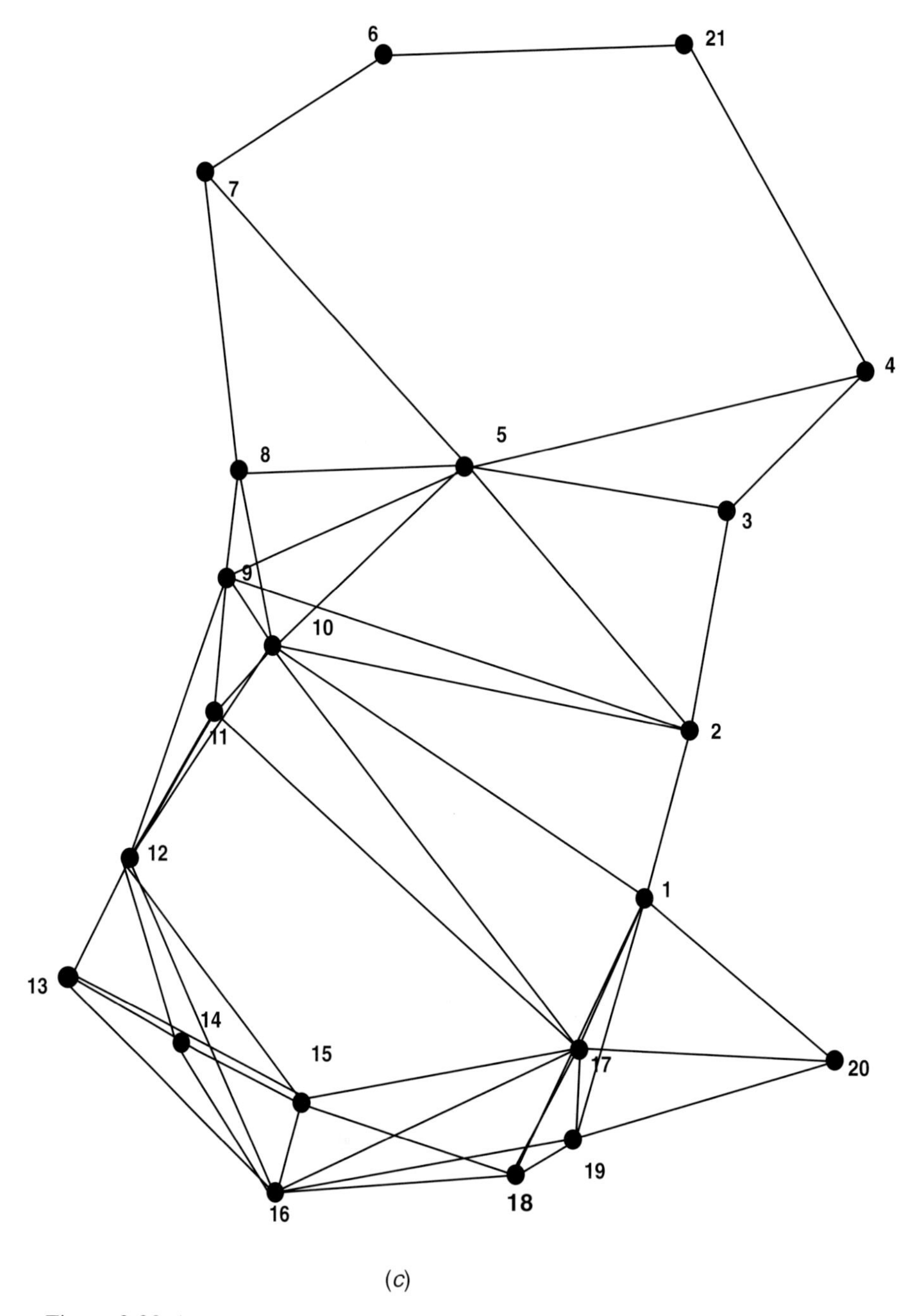


(c)

Figure 2.23 (*cont.*)

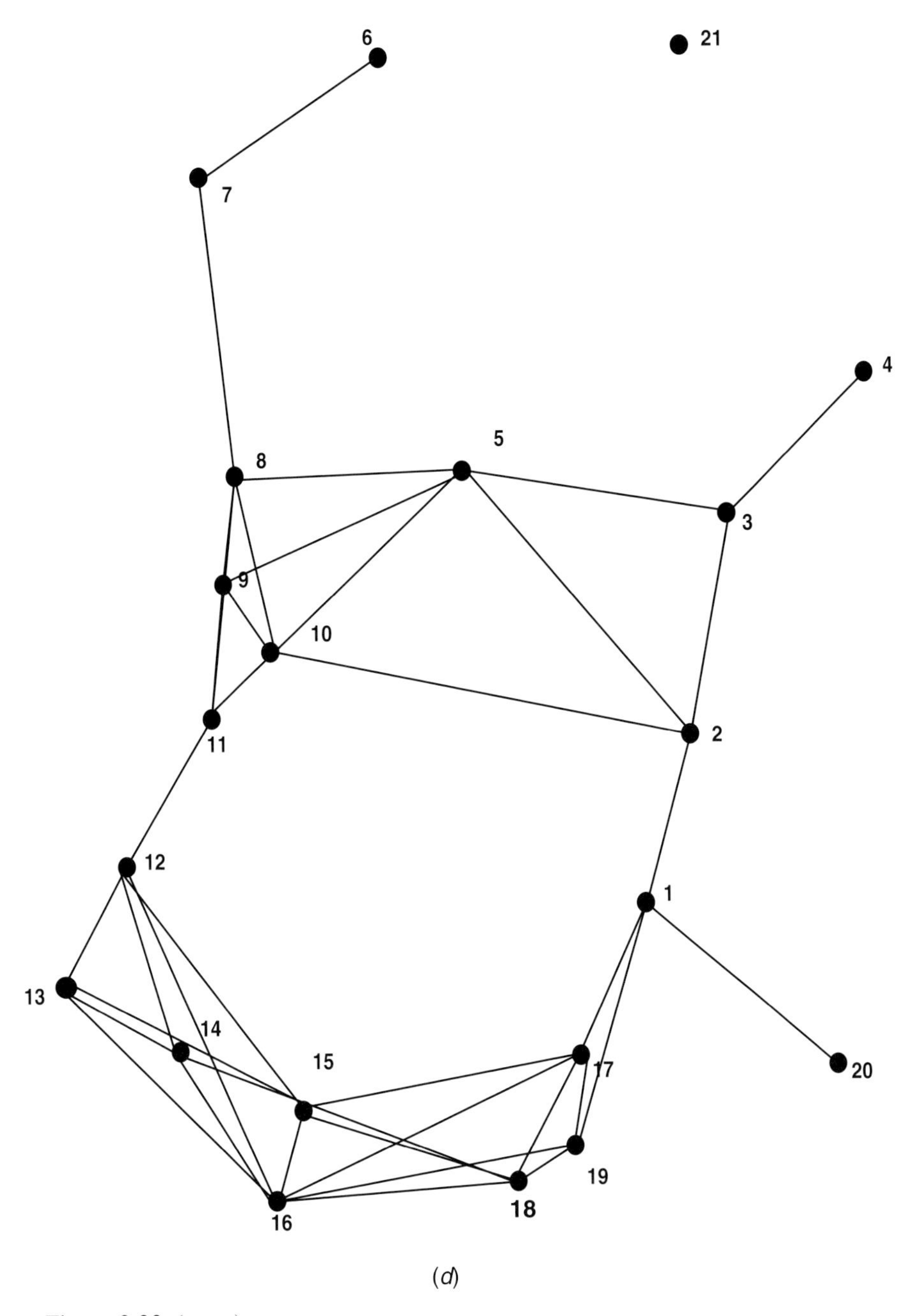


(*d*)

Figure 2.23 (*cont.*)

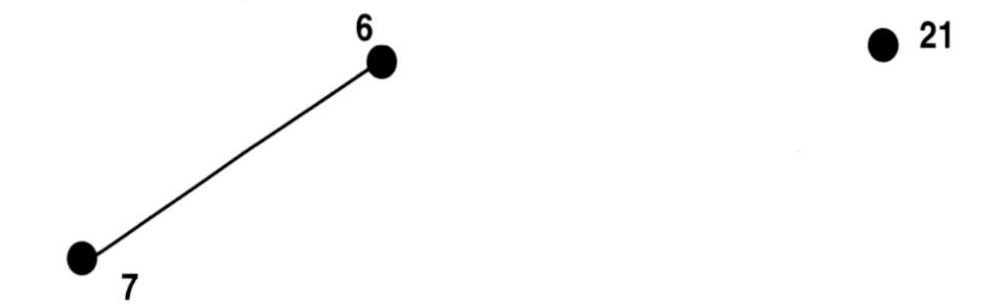


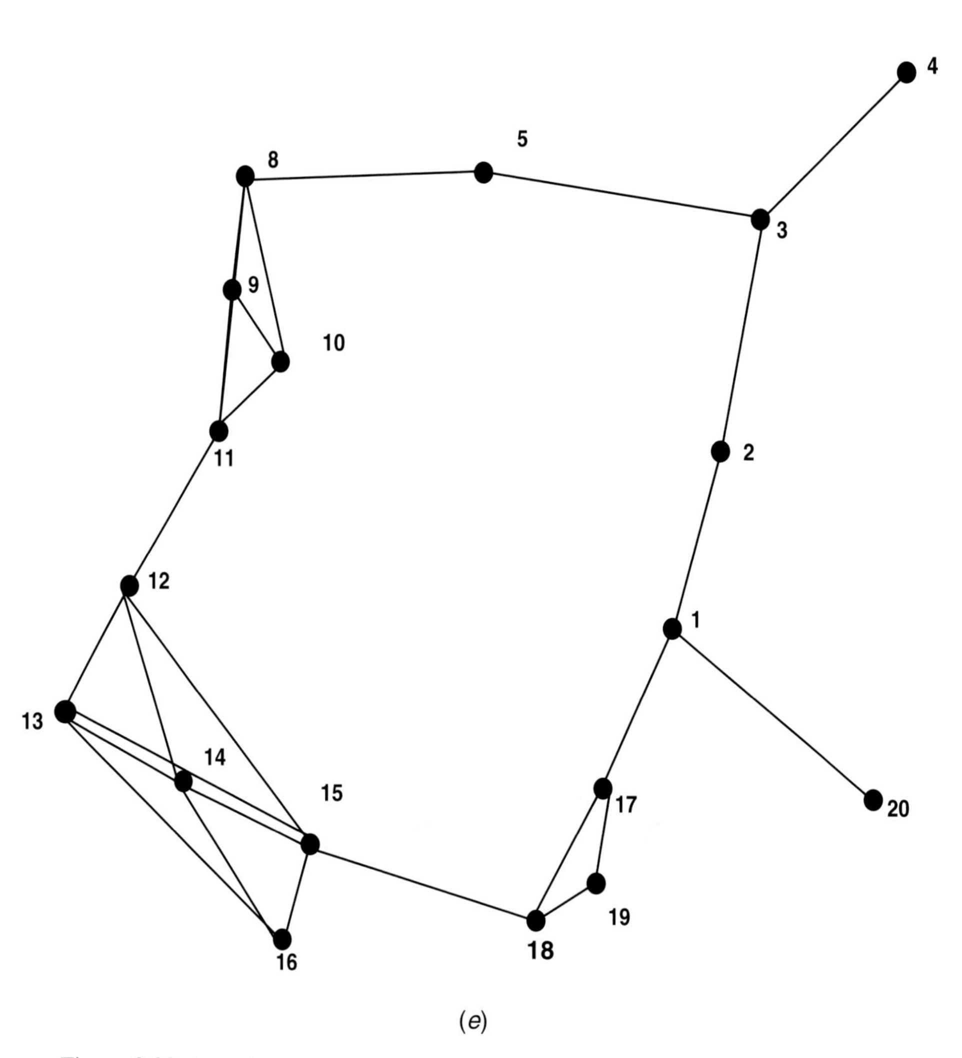


(e)

Figure 2.23 (*cont.*)

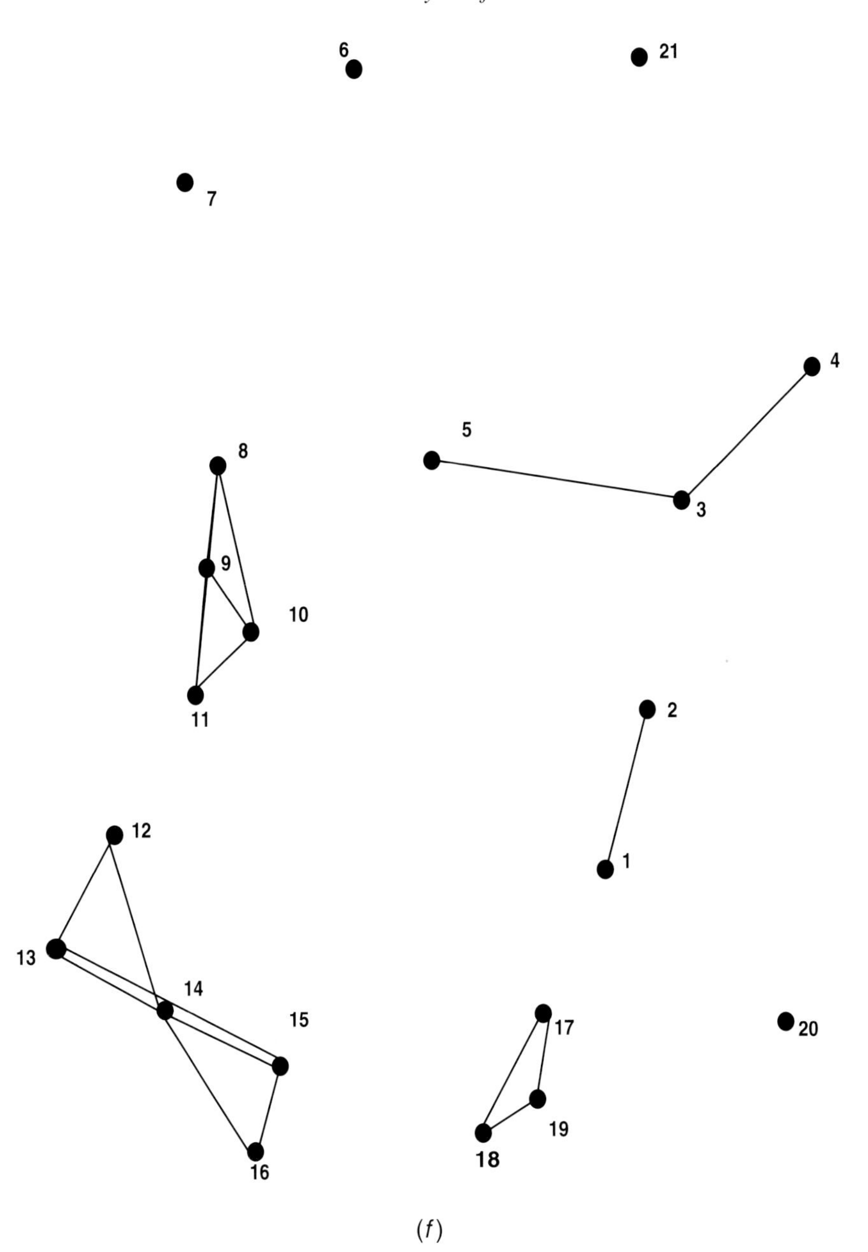


Figure 2.23 (*cont.*)

To evaluate vertex removal, the criteria that Urban & Keitt (2001) suggested are more explicitly ecological than those for edge removal:

(1) a recruitment index, *R* (a weighted sum of patch areas, with the weighting being a measure of patch quality);
(2) an index of dispersal flux, *F* (a weighted sum of patch areas, as for recruitment, but including the probability of dispersal from the focal patch to another); and
(3) a measure of traversability, *T* (the diameter of the largest component that remains after the vertex is removed).

They examined the iterative removal of nodes from the complete graph by three different procedures: random choice for removal; removal of the node with smallest patch area; and removal of the smallest area node i.e. a 'leaf' in the current MST (i.e. attached only to one other node, as Lake 4 in Figure 2.23*a*). Not surprisingly, they found that small-leaf removal degraded the characteristics of the network less rapidly than did random removal. The same three measures, *R*, *F* and *T*, could be used to evaluate the importance of individual nodes according to each criterion. This was done by examining the difference in the measure before and after the node's removal from the entire graph. Because this evaluation is based on the complete graph, which includes all possible edges, evaluation of importance may be very different from one based on the DT or on the MST itself. For example, in a complete graph of our landscape of lakes, removing Lake 8 would have little effect on the graph's diameter, whereas the removal of Lake 21 would make a big difference. In the MST of the same landscape, the removal of Lake 21 reduces the graph's diameter somewhat, but the removal of Lake 8 has a profound effect on the graph's characteristics.

Again, as with edges, other measures of the importance of individual nodes in the landscape can be considered. For example, we could compare the average length of edges attached to a node in the DT graph, with the average length of new edges in a DT graph when the node is removed. Thus, the average length of edges attached to Lake 5 in Figure 2.23*b* is 5.5 km; when Lake 5 is removed, and the DT graph is reformed, the average length of the all edges is 8.4, a ratio of 1.51. On the other hand, if Lake 14 is removed, no new edges are required in the DT graph, and the ratio is 0.

If we use a complete graph of the landscape and treat it as a purely combinatorial structure, there is no 'topological' distinction among the points or among the edges. Once the graph is embedded in the plane, as a map, the points are differentiated, with some more central in location and some more peripheral. More specifically, we can define the *perimeter* vertices of the graph as those that are in the convex hull (Lakes 21, 4, 20, 18, 16, 13, 7 and 6 in our example), with the rest being *interior* nodes. In a complete graph, there are few ways to evaluate the importance of a particular node to the overall connectedness of the graph, because there are no cut-points, cut-edges and so on. To discuss node importance, we should turn our attention to the minimum spanning tree, threshold distance network graph or Delaunay triangulation graph. The importance of a node will be related to its position in the graph (perimeter or interior), its degree (number of edges) and the distances to the neighbours to which it is joined. The removal of a perimeter node reduces the network extent (its 'footprint'); the removal of particular perimeter nodes will reduce the graph's diameter, and the removal of interior nodes will reduce the number of alternate paths between other pairs of nodes. In the DT graph, node removal does not result in disconnection, but it often does in the MST. The removal of an interior node with a degree of three or more from the MST will result in the creation of more components of considerably lower order and smaller diameter than the original graph. For example, the removal of Lake 8 in Figure 2.23*a* would have a large potential impact on dispersal or colonization in this group of lakes.

This is an interesting area of research and, clearly, there is more work needed, particularly on evaluating which features of habitat patch networks are most important to the dispersal of particular kinds of organisms.

## 2.5 Point patterns in other dimensions

### *2.5.1 One dimension*

The arrangement of events in one dimension can arise in many different ways in ecology: the positions of species boundaries along an environmental gradient, the heights of epiphytes up a tree trunk, encounters with plants or nests along a line transect and so on. Based on studies of boundaries on gradients, Dale (1999) describes several methods to study the arrangements of events in one dimension.

We can standardize the total length under study to the value 1, and then the $n$ events, the $x$s, divide the interval 0 to 1 into $n + 1$ pieces of length $u_i$ (see Figure 2.24*a*). Two statistics can then be used to help distinguish among different arrangements of the events:

$$W_n = \sum_{i=1}^{n+1} u_i^2 \tag{2.26}$$

and

$$h_n = \sum_{i=1}^{n} u_i u_{i+1}. \tag{2.27}$$

Tables of approximate critical values and guidance for the use of these two statistics are presented in Dale (1999). The statistic $W_n$ detects the pairwise clumping of

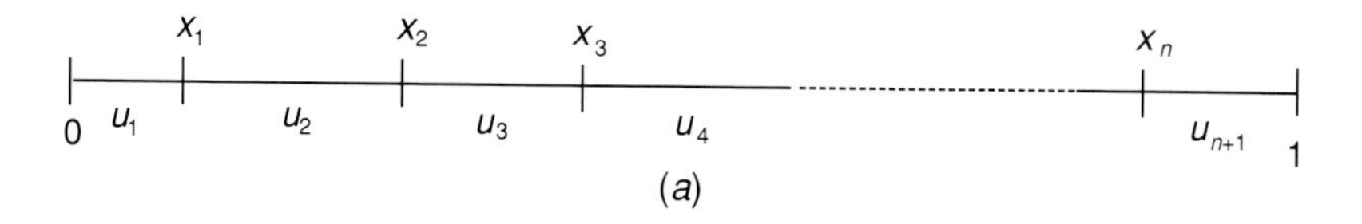


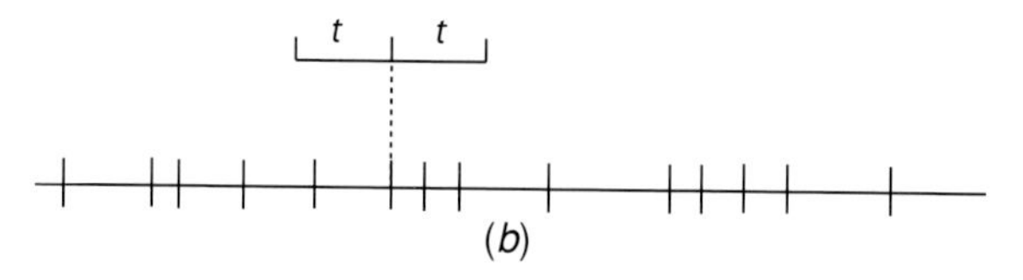


Figure 2.24 (*a*) Illustration of standardizing one-dimensional transect data. (*b*) Illustration of a line segment with $n$ events, and the '$t$-bar' used for calculating the one-dimensional version of Ripley's $K$-function.

boundaries that produce unexpectedly short segments, because large values indicate a greater inequality of segment lengths. The statistic $h_n$ measures the serial autocorrelation because large values indicate that short segments tend to be adjacent to other short segments and large segments tend to be adjacent to other large segments. A more detailed discussion of these statistics is provided in Dale (1999). For the purposes of this chapter, however, which has so far emphasized methods related to Ripley's $K$-function analysis, it is of interest to provide a one-dimensional version of that familiar technique.

Figure 2.24*b* shows a total sample line segment of length $B$ in which $n$ events occur. For a range of values, $t$, the number of events within distance $t$ of each event is counted:

$$\hat{K}(t) = B \sum_{\substack{i=1 \\ i \neq j}}^{n} \sum_{\substack{j=1 \\ j \neq i}}^{n} w_i(t) I_t(i, j)/n^2, \tag{2.28}$$

where $w_i(t)$ is a weight that corrects for edge effects. If a bar of length $2t$ centred on event $i$ is completely within the segment, $w_i(t) = 1$, otherwise it is the reciprocal of the proportion of that bar that is within the line segment (Figure 2.24*b*). Because the number of events expected to fall inside a bar of length $2t$ if the events occur randomly is $2nt/B$, we plot

$$\hat{L}(t) = t - \hat{K}(t)/2 \tag{2.29}$$

as a function of $t$. Values above 0 indicate overdispersion of the events and values below zero indicate clumping. An artificial example is given in Figure 2.25*a*, *b*, and a field example of the sequence of rapids on the Winisk River in Northern Ontario (53° 30′ N, 87° 20′ W) is given in Figure 2.25*c*, *d*. In this

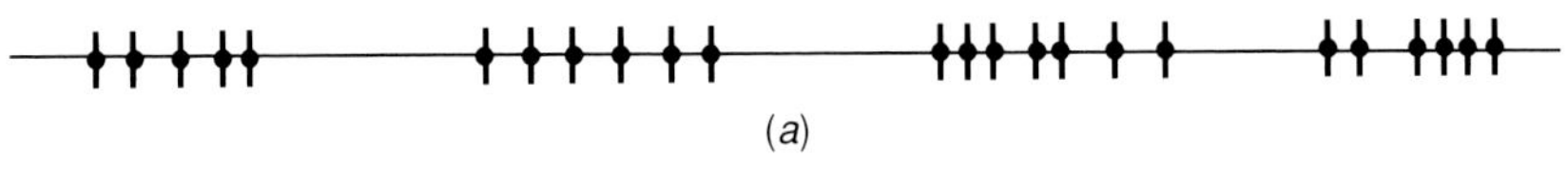


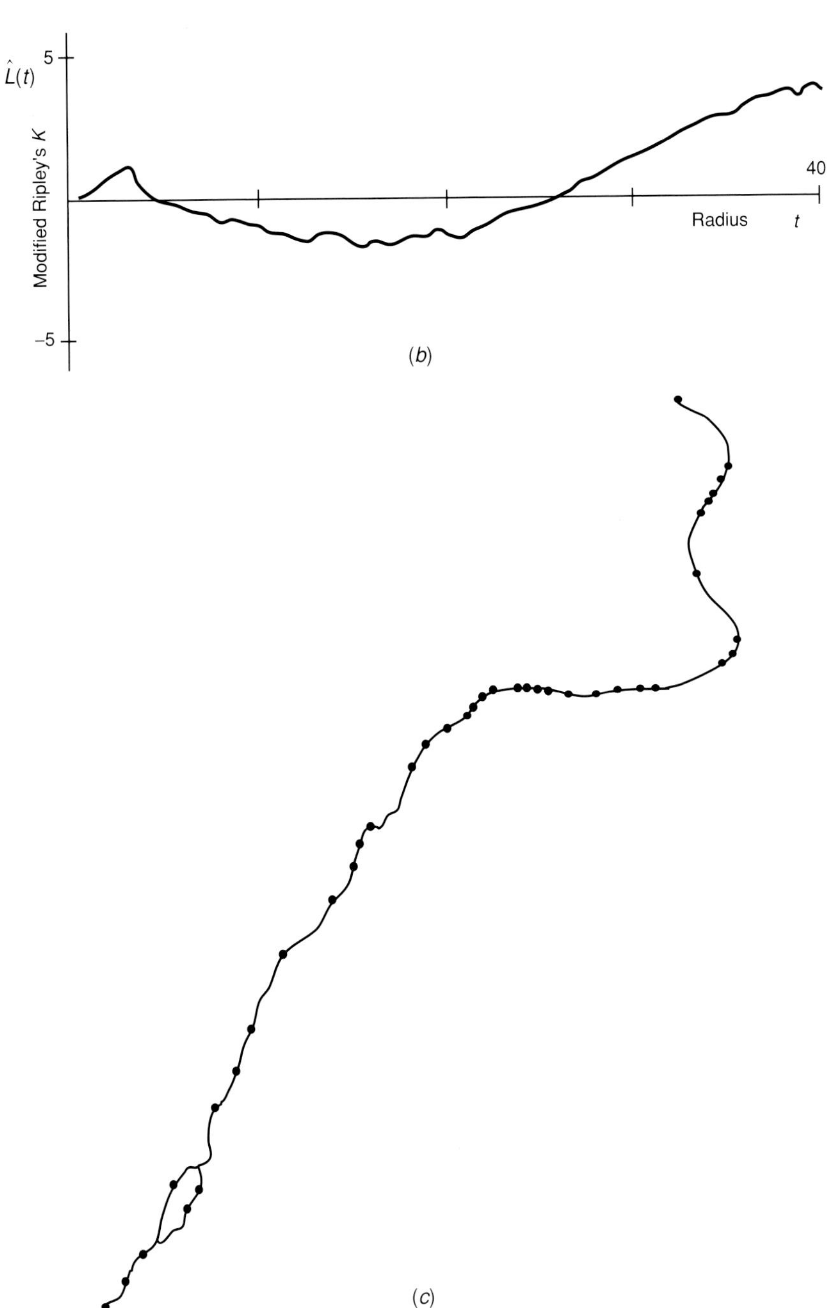


Figure 2.25 (*a*) Artificial one-dimensional transect data (clumps). (*b*) Standardized Ripley's *K* analysis on one-dimensional transect data. There is overdispersion at both very small and very large scales. (*c*) Rapids on the Winisk River. (*d*) Standardized Ripley's *K* analysis of the Winisk River rapids data. (*e*) Rapids on the Morris/Pipestone River. (*f*) Standardized Ripley's *K* analysis of the Morris/Pipestone River rapids data.

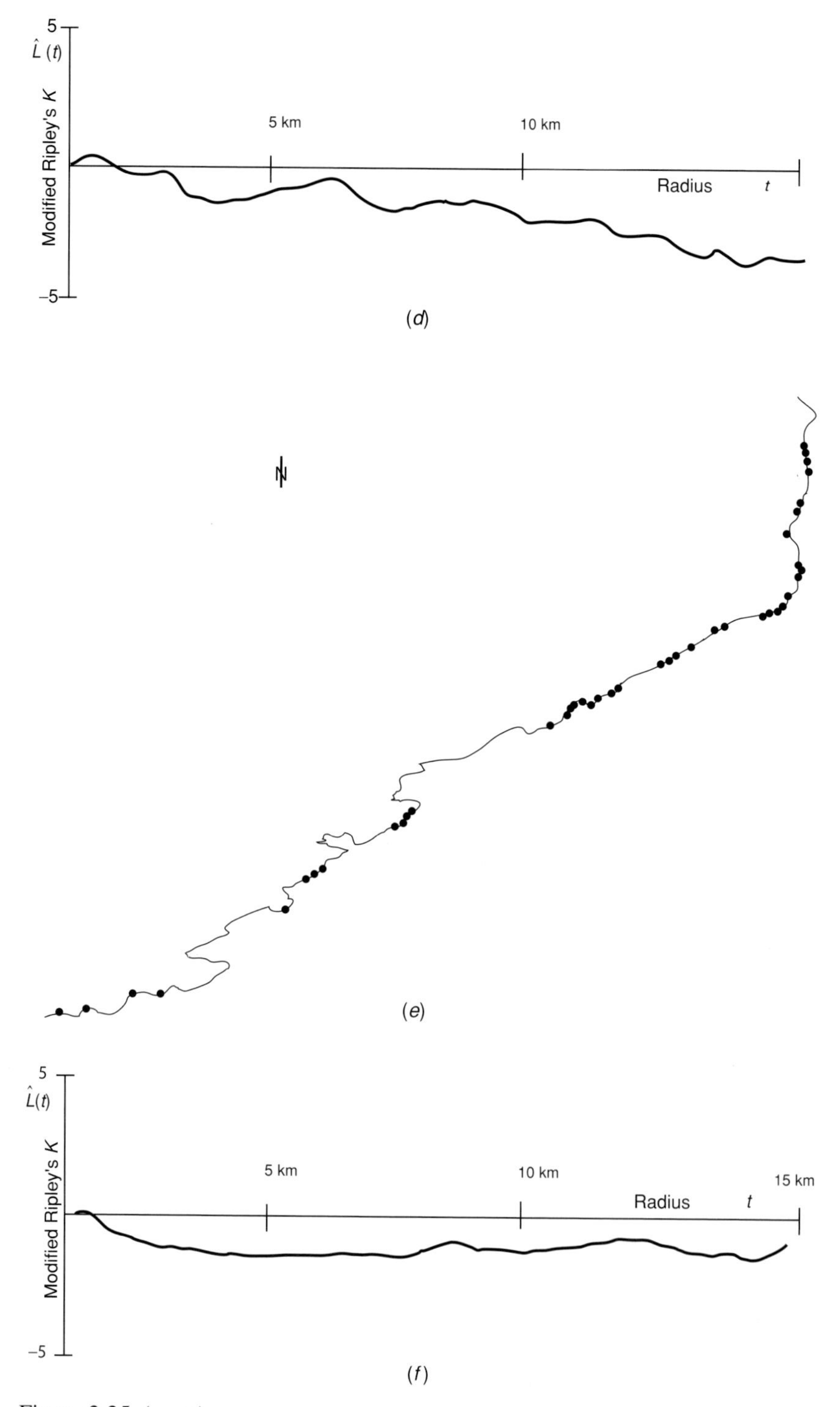


Figure 2.25 (*cont.*)

example, $W_m = 0.043$ and $h_m = 0.024$, neither of which are significant. Figure 2.25*e*, *f* give a second example from the Morris/Pipestone River, also in Northern Ontario (52° 15′ N, 90° 45′ W); here $W_m = 0.089$ and $h_m = 0.027$. The first is significant, indicating that the rapids are clumped, but the second is not, indicating that there is no tendency to have clusters of short inter-rapid sections of the river.

A simple and obvious modification of (2.28) produces a bivariate version of Ripley's $K$ in one dimension. As an extension to calculations in one dimension, Okabe & Yamada (2001) explained the univariate and bivariate forms of Ripley's $K$-function analysis when the events occur in a two-dimensional space, but are constrained to a network of one-dimensional structures. The example they gave is of fast-food stands on a network of streets, or fast-food stands and subway stations for the bivariate case. Applications to ecological studies such as the occurrence of species of aquatic macrophytes or aquatic invertebrates in the water channels of a delta are obvious. We are not aware of any so far in the ecological literature.

### *Lacunarity*

The term 'lacunarity' is derived from the Latin '*lacuna*', meaning a literal or metaphorical hole, and so the concept of lacunarity refers to the characteristics of the holes or gaps in a spatial structure. Several different methods for calculating a measure of lacunarity have been proposed, but Plotnick *et al.* (1996) recommend the 'gliding box' (or 'moving window') method of Allain & Cloitre (1991). Picture a string of events in one dimension; a box of length $r$ is placed at the beginning of the set and the number of events that lie within it are counted. The box is moved one unit along the series and the number counted again, and so on, as illustrated in Figure 2.26. The first and second moments for the frequency distribution of the number of events per box are then determined for boxes of size $r$, call them $m_1(r)$ and $m_2(r)$. The measure of lacunarity for box size $r$ is

$$\Lambda(r) = m_2(r)/[m_1(r)]^2. \qquad (2.30)$$

The graph of lacunarity as a function of box size is usually presented in a double-log form: $\log[\Lambda(r)]$ as a function of $\log(r)$, as in Figure 2.27.

Plotnick *et al.* (1996) provided guidelines for the interpretation of the lacunarity index as a function of box size. Random data produce a curve that is concave upward; clumped data of the same density produce curves that have greater lacunarity and that are initially concave downward; and regularly spaced data of the same density produce curves that are initially straight and have lower lacunarity (Figure 2.27). The shape of the double-log curve depends only on aggregation and is independent of the overall density; density determines the curve's maximum (Plotnick *et al.* 1996).

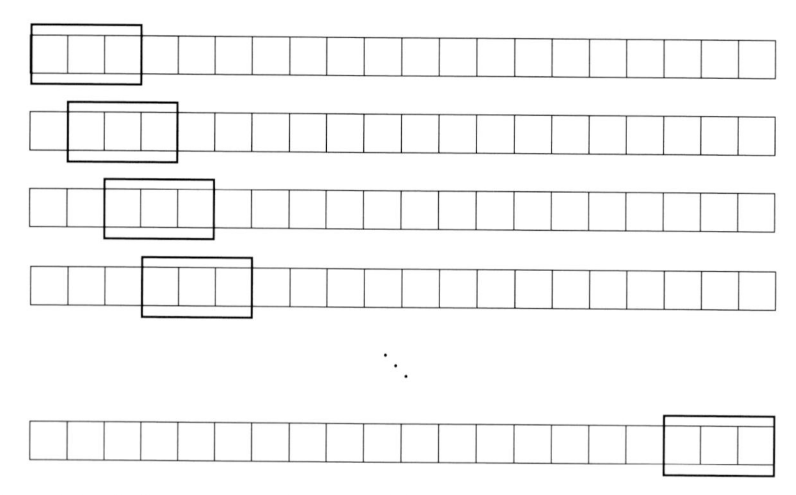

Figure 2.26 In lacunarity analysis, a 'gliding box' moves along a string of events in one dimension (redrawn from Dale *et al.* 2002).

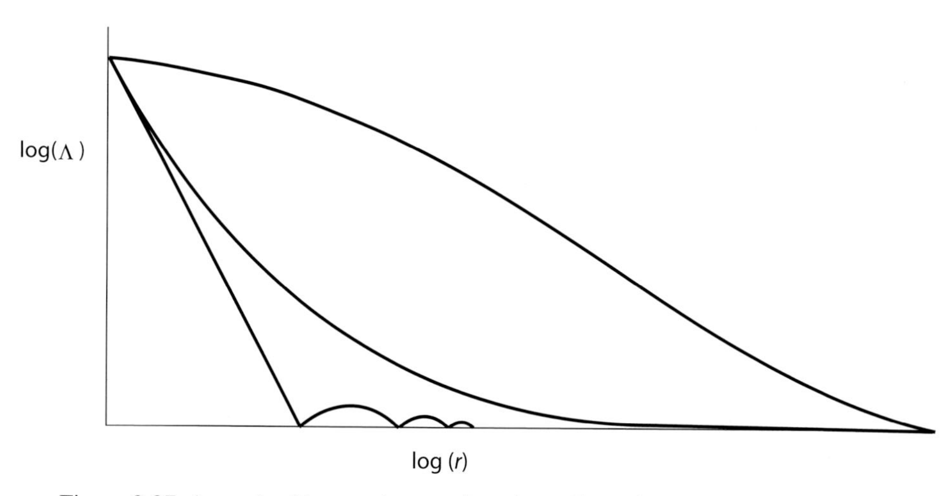


Figure 2.27 A graph of lacunarity as a function of box size. Top: clumped events; middle: random data; bottom: regularly spaced events.

Plotnick *et al.* (1996) suggested that for data consisting of randomly placed clumps, the log–log lacunarity curve declines gradually with increasing box size to a break point corresponding to the size of the clumps. Beyond this break point, the plot declines more rapidly and is concave upward. They suggested that lacunarity curves of one-dimensional sets have distinct breaks in slope corresponding to distinct scales within the sets. Our own investigations (Dale 2000) suggested that the method is not that precise in determining the scale or patch size in these kinds of patterns, but the method is still popular, particularly in its two-dimensional form, as we will illustrate below, and lacunarity remains part of the conceptual frameworks of spatial analysis.

### *2.5.2 Three or more dimensions*

The analysis of patterns in three dimensions has received less attention than two-dimensional analysis, in part because such data may be encountered less often. Many of the methods described for two dimensions can be adapted for use in three, for example, the comparison of event–event nearest neighbour distance distribution with the distribution of random point–event distances can probably be transferred directly. König *et al.* (1991) describe the adaptation of many of the methods described above for two dimensions for use with three-dimensional data. Mugglestone (1996) describes an approach that uses the Dirichlet tessellation for analysing such data. The tessellation is used as the basis for a randomization test of the arrangement of labels for bivariate data. In both these studies the subject of discussion was cellular: the position of cells in tissue or of the centromeres of chromosomes within cells.

We can also modify Ripley's $K$-function analysis for three dimensions, as described by Baddeley *et al.* (1987) and König *et al.* (1991) among others. With the usual notation, and $V$ being volume, calculate:

$$\hat{K}(t) = V \sum_{\substack{i=1 \\ i \neq j}}^{n} \sum_{\substack{j=1 \\ j \neq i}}^{n} w_i(t) I_t(i, j)/n^2. \tag{2.31}$$

That is, count the number of events centred on an event within a sphere of radius $t$ of each event. The edge correction factor $w_i(t)$ is 1 if the sphere, centred on $i$ and radius $t$, is completely within the study volume; otherwise, it is the reciprocal of the proportion of the sphere that lies within the study volume. Because the volume of the sphere is $4\pi t^3/3$, calculate:

$$\hat{L}(t) = t - \sqrt[3]{3\hat{K}(t)/4\pi}. \tag{2.32}$$

Under CSR, the expected value is 0 and significant departures from 0 are interpreted in the usual way. Extensions of this approach to bivariate and multivariate analysis will proceed with simple modifications as described for two-dimensional data.

In addition to the adapted Ripley's $K$, König *et al.* (1991) described three-dimensional versions of the 'event-to-nearest-event' and 'random-point-to-nearest event' statistics, the $G$-function and $F$-function described in Section 2.1.2. They also introduced a three-dimensional marked point process analysis such as we discussed in Section 2.2 above. These three-dimensional approaches have, so far as we know, been applied in the context of the positions of cells in tissue, and not to ecological examples. Once it becomes well known that methods for three-dimensional pattern analysis are available and easy to apply, we expect to see applications of these methods in ecological studies. As only one example, characterizing the three-dimensional positions of leaves in a forest canopy would be an important first step toward evaluating and modelling the infiltration of light to lower strata of the forest.

## 2.6 Contiguous units analysis

We will begin this section with a discussion of analysis in one dimension; the basic form of data is to have a series of values, $x_1$ to $x_n$, representing the density of a particular species in a transect of $n$ contiguous sampling units, here quadrats. The density will often be measured in per cent or may be expressed as a proportion, running from 0 to 1. In presence : absence form, the data take only the values 0 for absence and 1 for presence. The methods we describe in the following sections can apply to density data, presence : absence, or to 'count' data such as the number of stems found in each quadrat. Spatial pattern of a single species is often studied using one of several methods that examine the effects of distance or block size on a calculated variance, with low variance indicating similarity and high variance indicating dissimilarity.

### *2.6.1 Quadrat variance methods*

Two 'blocked quadrat' methods of pattern analysis were published by Hill (1973) in which the quadrats are combined in groups or blocks of a range of sizes. Both use a continuous range of block sizes and average over all possible starting positions for the blocking. The first of these is 'two term local quadrat variance', TTLQV (Hill 1973), which can be thought of as a method that uses a two-part window, each containing $b$ units for its calculation. The block size affects both the size of the window and the distance between the two parts of the template.

The variance in TTLQV is:

$$V_2(b) = \sum_{i=1}^{n+1-2b} \left( \sum_{j=i}^{i+b-1} x_j - \sum_{j=i+b}^{i+2b-1} x_j \right)^2 \Big/ 2b(n+1-2b). \qquad (2.33)$$

This variance is calculated for a range of block sizes and, when plotted, peaks in the variance are interpreted as being indicative of scales of pattern in the data (Hill 1973; Dale 1999), as illustrated in Figure 2.28. This method is most often used with density data or estimated cover, but it can be used for presence : absence data, or for counts.

An alternative is to have a two-part template for which only the spacing changes with each half containing only a single original sample unit; this method is known

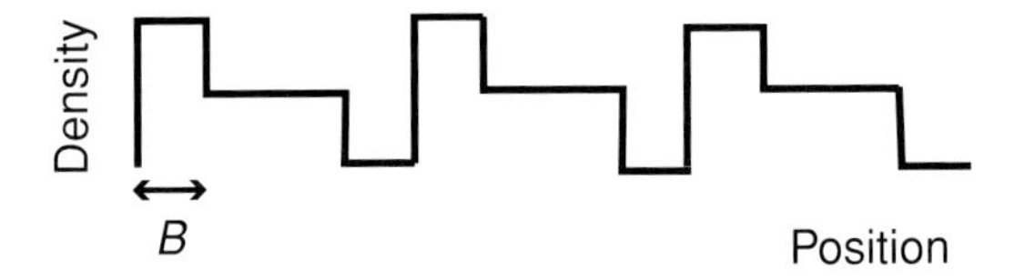


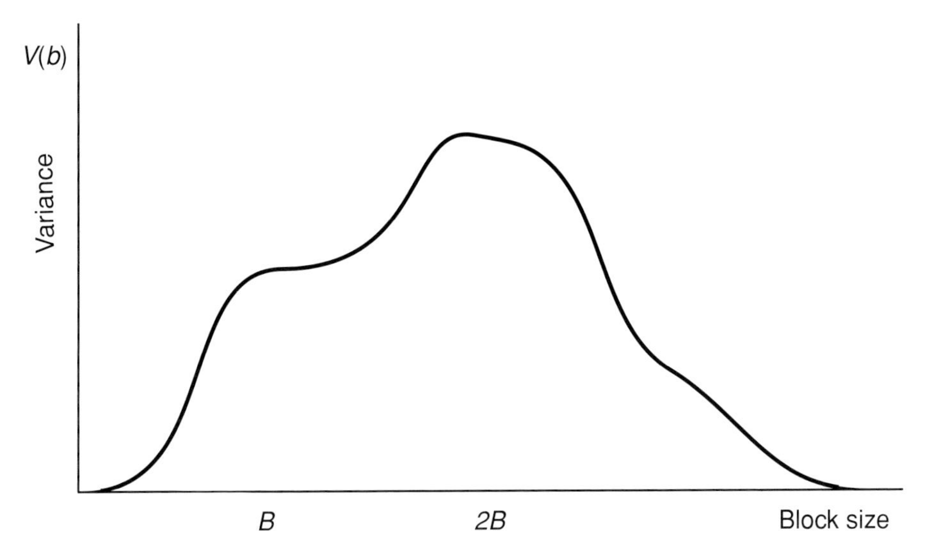


Figure 2.28 The data are derived by the addition of two square wave patterns, with scales of $B$ and $2B$ (as in Figure 2.24). Variance (TTLQV) as a function of block size; peaks (at scales $B$ and $2B$) are interpreted as corresponding to scales of pattern in the data.

as paired quadrat variance, PQV (Ludwig and Goodall 1978):

$$V_p(d) = \sum_{i=1}^{n-d} (x_i - x_{i+d})^2 / 2(n-d). \tag{2.34}$$

As with TTLQV, peaks in the plot of $V_p$ as a function of $d$ are interpreted as being scales of pattern in the data (cf. Ludwig and Reynolds 1988). TTLQV and PQV are similar in that they both use a two-part window, but in PQV, the size of the window does not change, only the spacing changes; whereas in TTLQV, both the size and distance between the centres of the two parts change. Ver Hoef *et al.* (1993) showed that there is a close, but not simple, relationship between TTLQV and PQV:

$$V_2(b) = V_p(b) + \frac{1}{b} \sum_{i=1}^{b-1} \{2(i-b)V_p(i) + (b-i)[V_p(b+i) + V_p(b-i)]\}. \tag{2.35}$$

Both TTLQV and PQV can be extended to three-part forms, 'three term local quadrat variance', 3TLQV (Hill 1973), and 'triplet quadrat variance', tQV (Dale 1999).

The equation for 3TLQV is:

$$V_3(b) = \sum_{i=1}^{n+1-3b} \left( \sum_{j=i}^{i+b-1} x_j - 2 \sum_{j=i+b}^{i+2b-1} x_j + \sum_{j=i+2b}^{i+3b-1} x_j \right)^2 \Bigg/ 8b(n+1-3b). \tag{2.36}$$

For tQV, it is:

$$V_t(d) = \sum_{i=1}^{n-2d} (x_i - 2x_{i+d} + x_{i+2d})^2 / 4(n-2d). \tag{2.37}$$

In both these methods, peaks in the variance indicate scales of pattern in the data, as in the previous two methods (cf. Figure 2.28*b*). The two-part window methods can filter out the addition of a constant and the three-part window methods can filter out a linear trend. Therefore, 3TLQV and tQV are less sensitive to trends in the data (Dale 1999).

The concept of lacunarity was introduced above in the discussion of one-dimensional point pattern analysis. The gliding box algorithm for calculating a measure of lacunarity can be applied to the kind of data we are describing here, whether they are densities counts or presence : absence in the quadrats (see Dale 2000). In fact, that approach can be seen as something like a one-part window equivalent of the methods we have just described (Figure 2.29). We commented above, however, that the lacunarity analysis does not give results as precise as those of the quadrat variance approach for patterns of known characteristics (cf. Dale 2000).

Another approach to the analysis of spatial pattern that is based on Hill's (1973) quadrat variance methods is Galiano's (1982) new quadrat variance, which was proposed in both two- and three-term form. In the two-term form it is:

$$V_{N,2} = \sum_{i=1}^{n-2b} |T_2(b,i) - T_2(b,i+1)|/(2b(n-2b),$$

where

$$T_2(b,i) = \left( \sum_{j=i}^{i+b-1} x_j - \sum_{j=i+b}^{i+2b-1} x_j \right)^2. \tag{2.38}$$

In the three-term form it is:

$$V_{N,3} = \sum_{i=1}^{n-2b} |T_3(b,i) - T_3(b,i+1)|/(8b(n-3b),$$

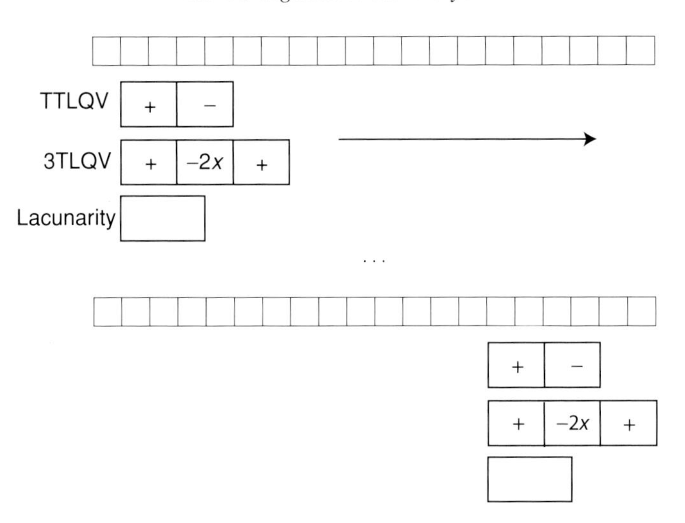


Figure 2.29 Gliding box techniques for one-dimensional data (TTLQV, 3TLQV and lacunarity).

where

$$T_3(b, i) = \left( \sum_{j=i}^{i+b-1} x_j - 2 \sum_{j=i+b}^{i+2b-1} x_j + \sum_{j=i+2b}^{i+3b-1} x_j \right)^2 . \tag{2.39}$$

Dale (1999) provided a discussion of the properties of these two statistics, but recommended against the use of the three-term version until its properties are better understood. In two-phase pattern of patches and gaps, the two-term version gives a variance peak at the average size of the locally smaller phase. As with many other methods, careful interpretation is necessary if there is non-stationarity in the data.

### *2.6.2 Significance tests for quadrat variance methods*

Before proceeding to consider adaptations of these quadrat variance methods to more than one species or in more than one dimension, we should discuss the interpretation of the results. In almost all other kinds of data analysis, the principal approach to interpretation is a test of statistical significance, whether in evaluating a single data set or in comparing data sets. In spatial pattern analysis, evaluating statistical significance is made difficult by several forms of lack of independence in the data and in the ways the data are used. First, the values found in adjacent quadrats will tend to be more similar than those at some distance from each other. That fact is part of the underlying logic of spatial pattern analysis, but it also represents spatial autocorrelation in the data which can make the evaluation of statistical tests difficult. Positive spatial autocorrelation, in general, tends to make tests too liberal: they give more apparently significant results than the data actually justify (see Chapter 5).

The second form of lack of independence is the fact that each piece of data may be used more than once in the analysis, both in calculating the variance at a single block size and in calculating the variances at different block sizes. The overall result of the lack of independence is that it is difficult to provide statistical tests to evaluate the results of a single analysis or to compare results.

Randomization procedures can be used to help evaluate the 'significance' of detected pattern in data, but because that kind of assessment requires both the data and a considerable number of re-analyses, it is not always feasible. Second, complete randomization destroys the spatial structure of the data, so that we can test only the null hypothesis that there is no pattern at all (often not very interesting!) In some cases, restricted randomizations in which the spatial structure is preserved are possible, as we will describe in more detail in Section 2.4.3. In this section, however, we will evaluate the results of quadrat variance analysis, based on the characteristics of the analysis of random data in which there is no pattern. This provides an alternative to permutation methods for which the data themselves and re-analysis are necessary for an evaluation of the results.

Because of its mathematical simplicity, we will begin by considering how we might develop statistical tests for PQV. We start with a string of data, $x_i$, $i = 1, 2, 3, \ldots, n$, which we assume are independent and random following some particular probability distribution. There is a number of possible distributions to be considered, including the normal distribution, but for the purposes of illustration we will use the uniform probability distribution. PQV averages a number of terms of the form $(x_i - x_{i+d})^2$ for a range of distances, $d$. For independent and random data, $(x_i - x_{i+d})^2$ will have the same mean and variance for any value of $d$ and, therefore, in PQV, $V_p(d)$ will have a constant mean and variance.

For the uniform probability distribution, we can show by integration that:

$$E[(x_i - x_j)^2] = 1/6,$$

and

$$E[(x_i - x_j)^4] = 1/15,$$

so that

$$\mathrm{Var}[(x_i - x_j)^2] = 1/15 - (1/6)^2 = 7/180. \tag{2.40}$$

Because $V_p(d)$ is half the average of $n - d$ similar terms, we might be tempted to suggest that: (1) $E(V_p(d)) = 1/12 = \mu$ and $\mathrm{Var}(V_p(d)) = 7/720(n - d) = \sigma^2$; and (2) being an average of independent terms, $V_p(d)$ should approach the normal distribution, with the given mean and variance, so that a significance test of the variance for any distance could be based simply on the interval $\mu \pm 1.645\sigma$. Unfortunately, both parts, i.e. points (1) and (2) of the preceding sentence, are not quite true.

Point (1) is wrong because the terms contributing to $V_p(d)$ are not completely independent. For example, for $d = 3$, the calculation includes both $(x_1 - x_4)^2$ and $(x_4 - x_7)^2$, which share the variate $x_4$, and therefore have a non-zero covariance. For uniform distribution, those non-zero covariances increase the variance from $7/720(n - d)$ to approximately $9/720(n - d)$.

Point (2), also, is not quite right, because of the re-use of the same data. Because calculations of $V_p(d-1)$ and $V_p(d)$ use the same variates, they are not independent and with some effort, we can show that:

$$\mathrm{Cov}[V_p(d-1), V_p(d)] = \frac{(2n - 3d + 1)(1/90)}{4(n - d)(n - d + 1)} \approx 1/180n. \qquad (2.41)$$

This value may seem small, and it is; but not only will the variances of successive distances have non-zero covariance, such as $d = 4$ and $d = 5$, but so will $d = 4$ and $d = 6$, $d = 4$ and $d = 7$, and so on. This lack of independence is an important obstacle to the development of statistical tests.

That being said, however, we can use our understanding to provide some guidance to the interpretation of PQV plots, specifically to do with the position of the first variance peak. Because $E(V_p(d))$ is constant and the distribution of $V_p(d)$ is independent of $d$, $V_p(d_j) > V_p(d_k)$ with probability of about 0.5 for all values of $d_j$ and $d_k$. For $d = 1$ to be the first peak, all that is required is $V_p(1) > V_p(2)$ and, with random data, that occurs with a probability of approximately 0.5. For $d = 2$ to be the first peak, we need $V_p(2) > V_p(1)$ and $V_p(2) > V_p(3)$, which occurs with a probability of $0.5 \times 0.5 = 0.25$. In general, then, the first peak for random data occurs at distance $d$, with probability $(0.5)^d$. The probabilities are not exact because of non-zero covariances, but this provides a direct explanation of the observation by Campbell *et al.* (1998) that, for random data, variance peaks occur at block sizes 1, 2 and 3 with frequencies of approximately 50, 25 and 12% (their Figure 1). A second comment is that there is an important distinction to be made between testing whether the data as a set are non-random and have significant spatial pattern in them and providing significance tests for the variance at a particular block size or spacing.

If we turn our attention from PQV to TTLQV, it is obvious that the problems of derivation will be greater. The blocking of data introduces more dependence and more covariance to the calculations for any single block size and more dependence and more covariance to the calculations at successive block sizes. It could be done, but is it worth the effort? NO!

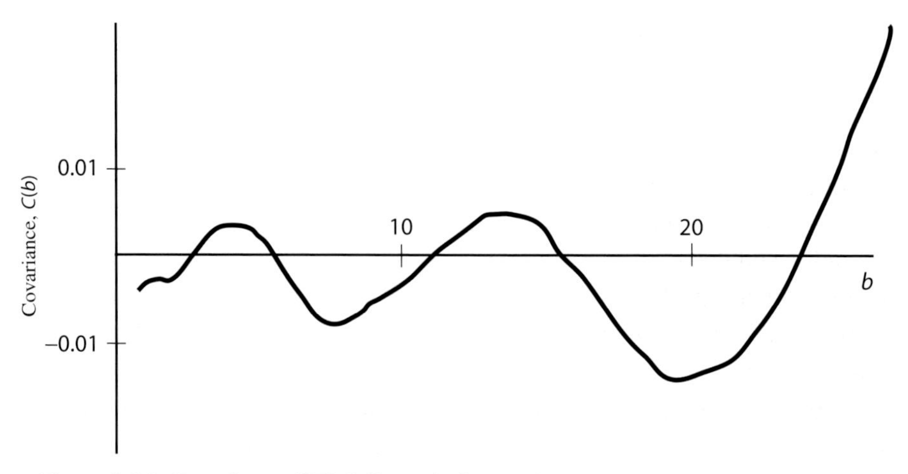


Figure 2.30 Covariance (3TLQC) analysis on Ellesmere Island sedge meadow data. The association of *Carex aquatilis* and *Eriophorum triste* cycles between positive and negative. *b* is the block size.

It is not worth the effort for TTLQV; and it is not worth the effort and ingenuity we put in for PQV either. The reason is that the values derived for the mean and variance depend strongly on the underlying distribution of the data. The real problem for deriving statistical tests of the significance of results from pattern analysis is that we never know for certain the true underlying distribution or whether the distribution is stationary for the length of the transect. For that reason alone, notwithstanding analytical complexities, these methods will remain techniques for data exploration, not for statistical testing unless some meaningful restricted randomization approach is available.

### 2.6.3 Adaptations for two or more species

An obvious extension of single species pattern analysis is to examine the scales of the association of pairs of species by looking at the effect of scale on species covariance (Greig-Smith 1961, 1983; Dale & Blundon 1991). Adoption of these quadrat variance methods for examining covariance proceeds on the basis that the covariance of two variables can be derived from their individual variances and the variance of their sum (cf. Dale 1999):

$$\text{Cov}(A, B) = [\text{Var}(A + B) - \text{Var}(A) - \text{Var}(B)]/2. \tag{2.42}$$

This formula can be applied to any of the methods just described to produce TTLQC, 3TLQC, PQC and tQC. For example, Figure 2.30 shows an example of the application of 3TLQC to presence : absence data from a study of sedge meadows on Ellesmere Island (Young *et al.* 1999). The relationship between *Eriophorum triste* and *Carex aquatilis* changes with block size from negative to positive and back again. The negative association at small block sizes can be attributed to the ecological 'preferences' of the two species, one more mesic and one semi-aquatic, coexisting in a wet hummock–hollow habitat.

In many cases, it may be of interest to study the spatial pattern of the whole community, and the analysis of all possible pairs of species may be confusing and difficult to interpret. Therefore, to evaluate the spatial pattern of all the species at once, researchers have proposed a variety of methods with which to examine multispecies pattern in vegetation.

The concept of multispecies pattern in vegetation may have arisen because of a perception that a plant community is like a mosaic of distinguishable phases or vegetation types, defined by the combinations of species densities or occurrence. Each phase need not be homogeneous, but there must be greater similarity within a phase than among phases, even if the boundaries between phases are not sharp. Local similarity is not sufficient to give true multiple species pattern with a detectable scale; it requires the repetition of similar species combinations. It is possible for different phases of a mosaic to have different scales and so the scale of multiple species pattern for the whole community can be defined as half the distance that maximizes the probability of finding the most similar combination of species' densities (Dale & Zbigniewicz 1995).

The recommended method for the evaluation of multispecies pattern is that introduced by Noy-Meir & Anderson (1971) and modified by Ver Hoef & Glenn-Lewin (1989) and by Dale & Zbigniewicz (1995). The method is based on a combination of quadrat (co)variance calculations and ordination methods of principal components analysis (PCA), and is usually called multiscale ordination (MSO).

For analysing data for $k$ species, we need first to calculate the $k \times k$ variance–covariance matrix for each block size, $b$, from 1 to a maximum, $B$: $\mathbf{C}(1), \mathbf{C}(2), \ldots, \mathbf{C}(B)$. The variances and covariances can be based on any of the methods described above, and Ver Hoef & Glenn-Lewin (1989) used TTLQV and its covariance TTLQC. We recommend using 3TLQV and 3TLQC because they are less affected by trends in the data. The matrices, $\mathbf{C}(b)$, for block sizes 1 to $B$ are all added together and the sum matrix is eigenanalysed as in principal components analysis. Eigenanalysis can be described as creating linear combinations of the original variables, the species densities, $x_1$ to $x_s$, that are all mutually orthogonal and with the condition that the new linear combinations, $y_1$ to $y_s$ in order, explain as much of the total variance in the data as they can. For each new $y_i$, its eigenvalue measures the proportion of the total variance that it explains, usually designated as $\lambda_i$ and its 'eigenvector' is the vector of the weights of the linear combination of the $x$s used to produce $y_i$.

Having completed the eigenanalysis, the largest eigenvalues are each partitioned into the amounts of variance contributed by each block size using the weights of the eigenvectors (for details see Ver Hoef & Glenn-Lewin 1989 or Dale & Zbigniewicz 1995). Peaks or plateaux in each plot of variance as a function of block size are interpreted as being to scales of pattern in the vegetation (cf. Ver Hoef & Glenn-Lewin 1989).

Because larger block sizes tend to dominate the analysis, the covariance matrices should be weighted prior to summing and Dale & Zbigniewicz (1995) suggested weighting of the variance–covariance matrices calculated for each block size, $b$, by the factor $6b^2/(b+2)$. Dale (1999) provided a detailed example of this multispecies analysis, using artificial data, which illustrates the ability of the method to recover the important properties of the data.

One feature of multispecies pattern may be the relative strengths of the contributions of the various species to the pattern. If one species overly dominates an eigenvector, the pattern detected is not truly multispecies. A measure of species' contributions can be based on the fact that each new variable produced by the eigenanalysis is a linear combination of the original species densities with weights given by the eigenvector:

$$y_i = \sum_{j=1}^{k} u_{ij} x_j, \quad \text{with} \quad \sum_{j=1}^{k} u_{ij}^2 = 1. \tag{2.43}$$

We can propose a measure of how evenly the species contribute to pattern using the variance of the absolute values of the weights, $u_{ij}$. If all species have equal weights, the variance will be 0, and if one weight is 1.0 and the rest are 0, then the variance is $(k-1)/k^2$. Let $C$ be the coefficient of variation, the square-root of the variance divided by the mean. Its maximum value is the square-root of $k$, and a measure of the evenness of the weights in the $i$th eigenvector is therefore:

$$E_i = 1 - C_i/\sqrt{k}. \tag{2.44}$$

We established six 50 m transects at a site near Fort Assiniboine, Alberta, dominated by Jack Pine and Aspen, and sampled the understorey in each with 200 contiguous 25 × 25 cm quadrats. The species list is typical of the boreal forest including vascular plants, such as *Linnaea borealis*, *Maianthemum canadense* and *Aralia nudicaulis*, species of *Vaccinium*, and feather mosses, such as *Ptilium crista-castrensis*, *Pleurozium schreberi* and *Hylocomium splendens*. We analysed the data with a version of multiscale ordination based on 3TLQV/3TLQC, as described above. We have chosen one of the six transects, the southern east–west transect, as an example to discuss. Based on the 12 most common species, the first three axes explained 22.6, 16.8 and 15.2% of the variance (55.6% for the first three axes, which in our experience is a high proportion). The evenness of the eigenvector weights was also high, 0.868, 0.801 and 0.819, indicating true multispecies pattern, not dominated by any one species. Figure 2.31*a* shows the data for this transect and Figure 2.31*b* shows the partitioned variances as a function of block size for the first three eigenvalues. The three axes show clear evidence of pattern in the range of 17–27 units (8.5–13.5 m). This is in agreement with an informal evaluation of the data, which seem to have a pattern of about 10 m.

### 2.6.4 Two or more dimensions

The basic concepts of the quadrat variance methods can also be extended to two-dimensional data collected on a grid of contiguous units. Again, any of the methods can be adapted for this purpose.

For example, tQV had the formula:

$$V_t(d) = \sum_{i=1}^{n-2d} (x_i - 2x_{i+d} + x_{i+2d})^2 / 4(n - 2d). \tag{2.45}$$

Expanded to two dimensions, it could become:

$$V_5(d) = \sum_{i=1}^{n-2d} \sum_{j=1}^{m-2d} (x_{i,j} + x_{i+d,j} - 4x_{i+d,j+d} + x_{i+2d,j} + x_{i,j+2d})^2 / 20(n - 2d)(m - 2d). \tag{2.46}$$

The other possible extensions to two dimensions are similar and equally simple in concept, but the equivalents of TTLQV and 3TLQV require somewhat long and complicated equations (see Dale 1999). They may be more intuitively understood from diagrams showing the templates used for their calculation. Figure 2.32*a* shows the template used for the calculation of $V_5(d)$ and compares it with the template for $V_9(b)$, which is the two-dimensional equivalent of 3TLQV, Figure 2.32*b*. Expansion of the concept of $V_5(d)$ to compare any sample unit with all units that are very approximately a distance $d$ away from it, as in Figure 2.32*c*, comes very close to the estimation of the omni-directional variogram, as described in Chapter 3.

Dale (1995, 1999) described a 'random paired quadrat frequency' method for the analysis of two-dimensional mosaics such as communities of crustose saxicolous lichens. As the name suggests, it compares the frequency of particular species combinations in randomly chosen pairs of quadrats, as a function of the $x$ and $y$ displacement between them, with the expected value based on occurrence in the entire data set. This approach provides an easy assessment of anisotropy; for details see Dale (1995) or (1999).

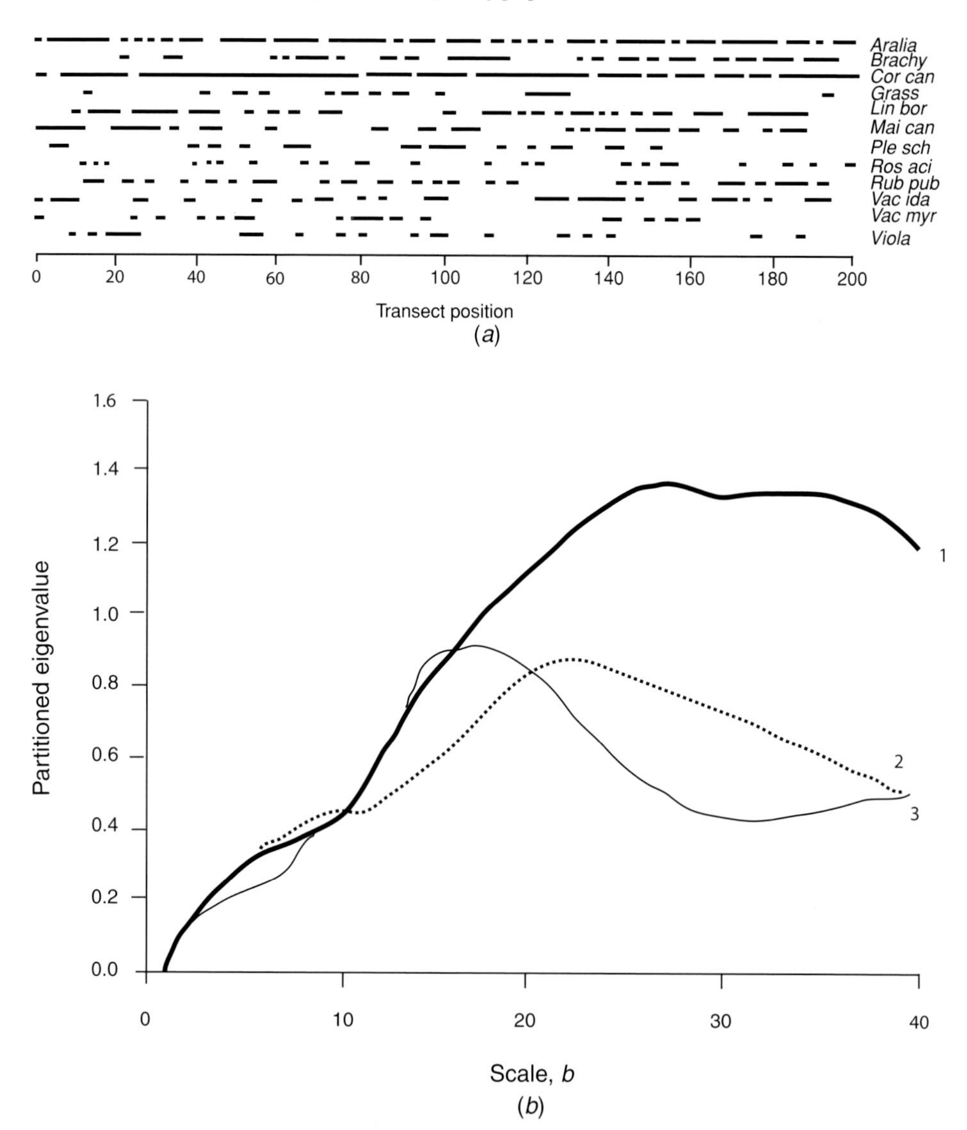


Figure 2.31 (*a*) Presence: absence of 12 understorey species along a 50 m transect at Fort Assiniboine, Alberta. (*b*) Partitioned variances as a function of block size for the first three eigenvalues in multiscale ordination of the data in (*a*). Key to species abbreviations: *Aralia, Aralia nudicaulis* L.; *Brachy, Brachythecium* sp.; *Cor Can, Cornus canadensis* L.; *Grass, Poaceae*; *Lin bor, Linnaea borealis* L.; *Mai can, Maianthemum canadense* Desf.; *Ple sch, Pleurozium schreberi* (Brid) Mitt.; *Ros aci, Rosa acicularis* Lindl.; *Rub pub, Rubus pubescens* Raf.; *Vac ida, Vaccinium vitis-idaea* L.; *Vac myr, Vaccinium myrtilloides* Michy; *Viola, Viola* sp.

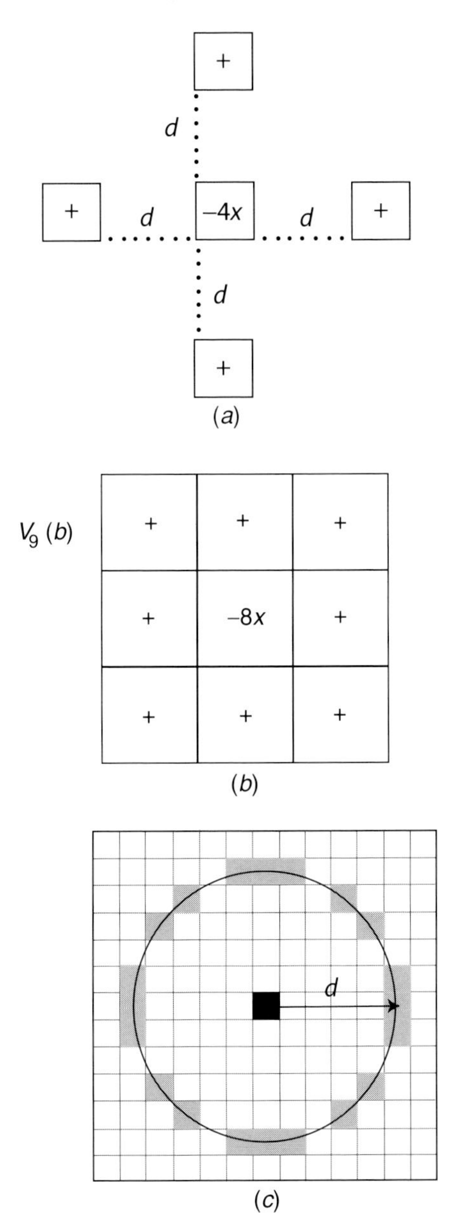


Figure 2.32 (*a*) Template used for the calculation of $V_5(d)$. (*b*) Template used for the calculation of $V_9(b)$. (*c*) An illustration of extending the $V_5(d)$ concept to compare a sample unit to all others at a distance of approximately *d*.

### SADIE

One set of methods we have yet to introduce is that referred to as the SADIE techniques, from '*s*patial *a*nalysis by *d*istance *i*ndic*e*s', developed by Perry and coworkers in a series of papers (Perry 1994, 1995, 1996, 1999). The basic concept begins with the counts of individual events (e.g. insects) in a grid of contiguous sampling units (e.g. a field divided into quadrats). The approach then uses either the total distance that individuals would have to move in order to for each to occupy a single quadrat (distance to crowding) or the total distance they would have to move in order to have equal numbers in all quadrats (distance to regularity). The two different versions can be seen as maximizing or minimizing the variance-to-mean ratio of the counts, but the spatial arrangement of the counts is used to evaluate the distance characteristics of those processes. To consider the 'distance to regularity' approach in greater detail, let $m$ be the mean of the counts of $N$ events in a grid of $n$ quadrats. Each unit that has a count of more than $m$ events, call it $c_i$, must lose $c_i - m$ events, and each unit that has a count of fewer than $m$ events, call it $c_j$, must gain $m - c_j$ events. If there are $p$ units with counts greater than $m$, and $q$ units with counts less than $m$, there are $pq$ pairs of units between which a movement of numbers might take place from a source unit, $i$, to a sink unit, $j$, with magnitude $v_{ij}$ (which is not necessarily an integer). Although there are many different ways in which the flow of numbers from sources to sinks could produce complete regularity, the SADIE algorithm finds the combination that has the smallest total flow distance. That is, where $d_{ij}$ is the distance between the two units, the algorithm minimizes the value of the sum, $D$:

$$D = \sum_{i=1}^{p} \sum_{j=1}^{q} v_{ij} d_{ij}. \tag{2.47}$$

The observed value of $D$ from data can be tested for statistical significance using a randomization procedure. There are several elaborations and modifications of the basic method available, as described in the literature. For example, a spatially explicit result can be obtained by either creating a diagram of the flow of numbers or by colouring the hot spots of high density one colour (red) and the cold spots of low density another (blue), producing 'red–blue' plots of the data (see Perry 1999). The method can also be modified to deal with point pattern data, rather than the quadrat count data described here.

Another method for the analysis of two-dimensional grid data is the technique of lacunarity analysis which was introduced above in Section 2.5.1 in the context of a one-dimensional point pattern, and was then alluded to in Section 2.6.1 in the discussion of one-dimensional contiguous unit arrays. The technique is most popular for use with two-dimensional arrays of contiguous units, the pixels of satellite images, in particular. For example, there have been several recent publications using the 'gliding box' algorithm on air photographs or Landsat images to calculate measures of lacunarity for the study of fragmentation in tropical landscapes (Peralta & Mather 2000; Wu *et al.* 2000; Weishampel *et al.* 2001).

The extension of these methods to three dimensions is certainly possible, although we have not found many examples in the ecological literature. The units in a three-dimensional sample array are usually referred to as 'voxels', to parallel the term 'pixel'. Fukushima *et al.* (1998) counted the leaves of trees in 972 $1.8 \times 1.8 \times 0.9$ m (vertical) cells in order to test methods of estimating foliage profiles. They did not apply the kinds of methods we have described here; they looked at foliage height diversity. As we noted above, however, in the discussion of three-dimensional point pattern analysis, as these approaches become better known, ecologists will quickly see their usefulness and take them up for application in ecological studies.

### *2.6.5 Spectral analysis and related techniques*

Spectral analysis is a technique that detects repeating pattern in spatial density data by fitting sine and cosine functions to the data, thus determining which frequencies or wavelengths best fit the data (Ripley 1978). The data to which this analysis is usually applied are quantitative measures in continuous or evenly spaced series. This approach is most suitable for 'rich' data sets, with large numbers of observations, and in situations in which the assumption of stationarity is justified. One technique for spectral analysis is the Fourier transform, which decomposes the 'signal' into sine waves of various frequencies and positions (Figure 2.33; see Legendre & Legendre 1998). Spectral analysis has been applied to two-dimensional ecological data by Renshaw & Ford (1984) and although originally developed for the analysis of continuous signals, it can also be applied to point pattern data (see Mugglestone & Renshaw 1996).

For a transect of $n$ values, $x_i$, the values are expressed as a weighted sum:

$$x_i = \bar{x} + \sum_{p=1}^{n/2-1} c_p \cos(2\pi ip/n) + s_p \sin(2\pi ip/n), \tag{2.48}$$

where

$$c_p = \frac{2}{n}\sum_{i=1}^{n} x_i \cos(2\pi ip/n) \quad \text{and} \quad s_p = \frac{2}{n}\sum_{i=1}^{n} x_i \sin(2\pi ip/n). \tag{2.49}$$

A closely related technique, is the Walsh transform which decomposes the signal into square waves instead of sine waves, of various frequencies and positions (Figure 2.34; see Ripley 1978).

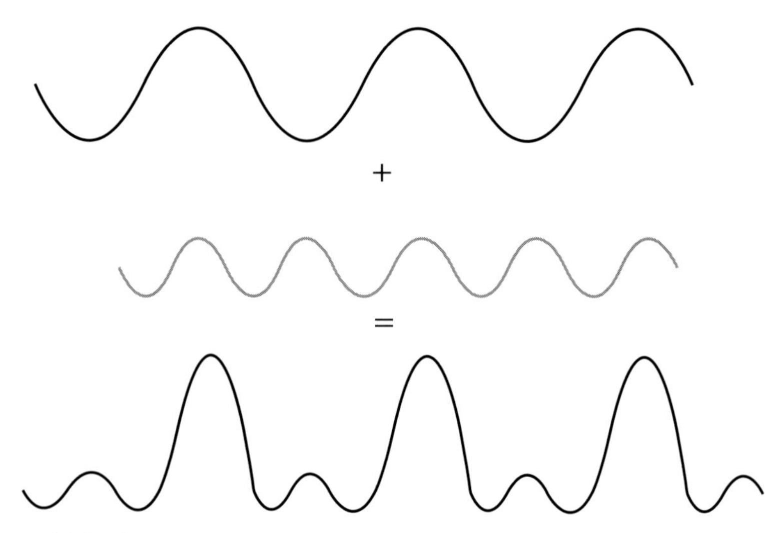


Figure 2.33 The concept of Fourier transform: the combined pattern of sine and cosine waves is resolved back into its original components.

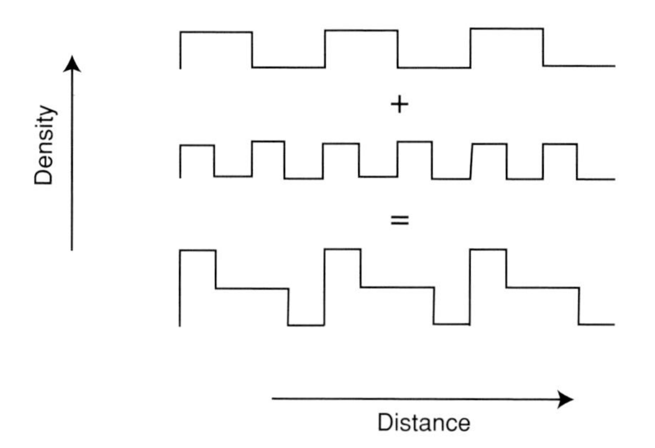


Figure 2.34 A Walsh transform resolves a combination of different frequencies of square waves into its components.

### 2.6.6 Wavelets

Wavelet analysis, as an approach to analysing spatial data, is closely related to spectral analysis, but it uses a finite wavelet as the template rather than sine or cosine functions applied over the entire data sequence. This approach provides both local evaluation of spatial structure and a global analysis using the wavelet variance, which gives variance peaks at scales of pattern in the data like those in the quadrat variance methods. The analysis evaluates how well the wavelet template, of different sizes, matches the data over a range of positions. The data used for this

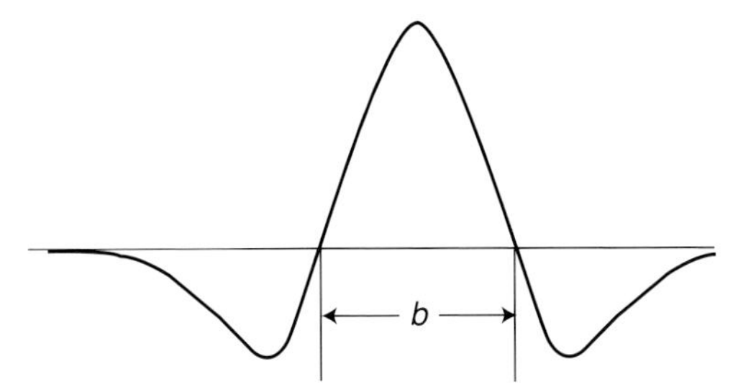


Figure 2.35 Mexican hat wavelet with relative width $b$.

are typically quantitative in a continuous or evenly spaced series. Where $g$ is the wavelet function, its transform, $T$, is a function of the wavelet size and position:

$$T(b, u_i) = \frac{1}{b} \sum_{j=1}^{n} y(u_j) g[(u_j - u_i)/b], \tag{2.50}$$

where $b$ is the wavelet's relative width (Figure 2.35), $y(u_j)$ is the data value at position $u_j$. The transform is essentially calculating the inner product of $y(u)$ with the wavelet function localized in size and position (Daubechies 1993). $T(b, u_i)$ takes large positive values when the match of the wavelet function and the data centred at $u_i$ is very good and large negative values when the match is very poor (see Dale 1999, Figure 9.8, or Dale *et al.* 2002, Figure 6). As a technical note, the wavelet transform given in (2.50) is the discrete form, using summation, whereas a continuous wavelet transform would use integration.

Wavelet analysis can use a range of different functions, but the 'Mexican hat' template (Figure 2.36) is one of the most common, e.g. Bradshaw & Spies (1992). For $b = 1$, its equation is:

$$g_M(u) = \frac{2}{3^{0.5}} \pi^{-0.25} (1 - 4u^2) \mathrm{e}^{-2u^2}. \tag{2.51}$$

We can define a wavelet variance, based on the transform, as:

$$V_W(b) = \sum_{i=1}^{n} T^2(b, u_i)/n. \tag{2.52}$$

A great variety of shapes are possible, but they must have an integral of zero; three more are shown in Figure 2.36: the Haar, the French top hat (FTH) and the Morlet. The wavelet variance based on the Haar wavelet is equivalent to TTLQV and that based on the French top hat wavelet is equivalent to 3TLQV (Dale & Mah 1998). Both are also related to spectral analysis using the Walsh transform, which

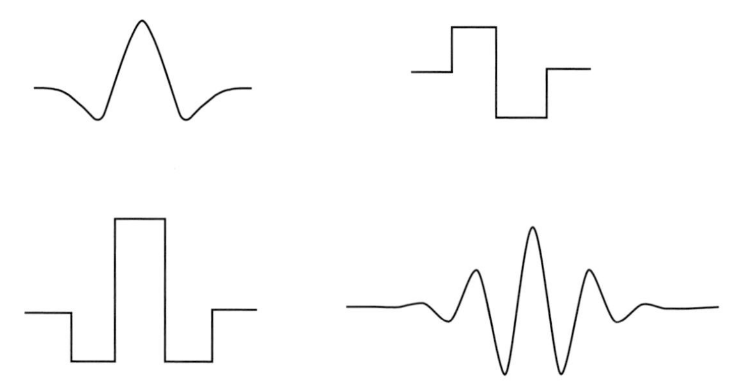

Figure 2.36 Wavelet shapes: Mexican hat, Haar, French top hat, Morlet.

decomposes a signal into combinations of square waves of different frequencies (cf. Ripley 1978).

We can also use the wavelet approach to perform a local equivalent of spectral analysis by using a sine wavelet:

$$g_S(u) = \begin{cases} \sin(\pi u), & \text{if } -1 \le u \le 1; \\ 0, & \text{otherwise.} \end{cases} \tag{2.53}$$

If the sine wavelet was expanded into more cycles, indefinitely, the resulting very long wavelet would produce something very much like Fourier analysis.

Whatever wavelet is used, wavelet variance analysis can be modified to give a wavelet covariance for bivariate data using (2.42), and thus to multivariate analysis, but we will not describe this feature in detail. Wavelet analysis can also be applied to two-dimensional data such as densities measured in a plane, e.g. vegetation cover in an area of grassland (Csillag & Kabos 1996). One wavelet for such an analysis is that created by rotating the Mexican hat wavelet in Figure 2.35 about its centre, which resembles a true three-dimensional sombrero. In this three-dimensional form the ‘brim’, the negative part of the template, must be narrower so that the integral of the whole remains zero.

## 2.7 Circumcircle methods

The next set of methods presented are conceptually closely related to several of the approaches already described, including Ripley’s $K$-function, neighbour networks and wavelets. Continuing with the idea of counting events in circles for completely mapped point data, we can consider other ways of locating the circles, as well as centring them on single events in the pattern. Each triplet of events defines a triangle and each triangle has a circle that goes through all three corners, called the circumcircle (Dale & Powell 2001). If we count the events in circumcircles, there is a close relationship with Ripley's approach because both count events in circles. There is also a relationship between the use of the circumcircle with the neighbour networks described in Section 2.3, because the Delaunay triangulation is defined by empty circumcircles (Section 2.3, Figure 2.18).

### 2.7.1 Univariate analysis

For a total area of mapped plot, $A$, and $n$ events in it, the average density is $\lambda = n/A$. There are approximately $n^3/6$ triplets of events and therefore the same number of circumcircles. For the $k$th circumcircle, let $n_k$ be the number of events inside it (exclusive of those that define it) and let $a_k$ be the area of the circle that is within the sample plot. The expected number of events in the circle is $e_k = (n-3)\,a_k/A$, based on the usual hypothesis of CSR. The Freeman–Tukey standardized residual can be used to compare the observed and expected numbers of events:

$$z_k = \sqrt{n_k} + \sqrt{n_k + 1} - \sqrt{4e_k + 1}. \tag{2.54}$$

Values of $z$ less than $-1.96$ can be considered to indicate gap circles and values greater then 1.96 indicate patch circles, but any region of low density may have many overlapping 'gap' circles and any region of high density may have many overlapping 'patch' circles. To distinguish among the overlapping patch or gap circles indicated by high or low values of $z$, we can seek the 'best' patch and gap circles defined as those that have the greatest contrast with their immediate surroundings. To find these, count the number of events in a ring around circle $k$ that has an area equal to the circle; if the radius of the circle is $r_k$, then the ring has width $(\sqrt{2} - 1)r_k$. If the number of events in the ring is $p_k$ with expected value $f_k$, then the standardized residual for the outer ring is:

$$\varsigma_k = \sqrt{p_k} + \sqrt{p_k + 1} - \sqrt{4f_k + 1}. \tag{2.55}$$

The circle's residual, $z_k$, can then be combined with the ring's residual, $\varsigma_k$, to produce a measure of the contrast between the densities of the circle and the ring:

$$Z_k = (z_k - \varsigma_k)/\sqrt{2}. \tag{2.56}$$

$Z$ is calculated using a double-circle template, which can be considered to be a wavelet, closely related to the French top hat, but in one more dimension, which we have termed the 'boater' wavelet (Figure 2.37, Dale & Powell 2001). (The boater is not just a rotation of the FTH; the 'brim' is narrower so that the integral of the entire template is zero.) As in other forms of wavelet analysis, the value of $Z$ for a given circle measures how well the data match the shape of the template. The average $Z^2$ can be plotted as a function of the circle radius, and peaks in this graph will reflect

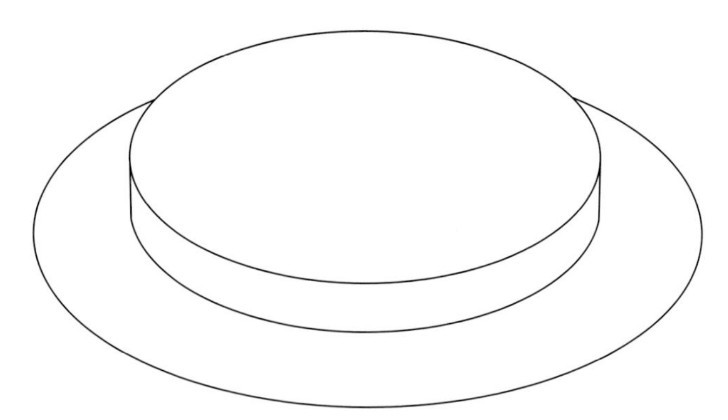

Figure 2.37 The 'boater' wavelet used in circumcircle analysis.

the sizes of patches and gaps in the pattern. To distinguish between the two, we can plot $Z^2$ as a function of the radius for positive residuals only, $Z_P^2$, which will show patch sizes, and then separately for negative residuals, $Z_N^2$, which will show gap sizes. Dale & Powell (2001) gave an example of this technique.

In some situations, we will want to achieve results that are spatially explicit. To do so, the $z$ or $Z$ score of each circle is associated with the centre of the circle, and a contour map of the scores can be produced for each of several size classes or scales. Figure 2.38 provides an illustration of the circumcircle technique where the shaded polygons indicate clusters of the centres of significant gap circles and empty polygons indicate clusters of the centres of significant patch circles.

### *2.7.2 Bivariate analysis*

The bivariate version of this approach is a straightforward adaptation of the univariate version: type 1 and type 2 events are counted in circles based on triplets of type 1 events, and type 1 and type 2 events are counted in circles based on triplets of type 2 events. The four different counts are kept distinct; this procedure makes the analysis truly asymmetric, unlike Ripley's $K$-function approach. For example, when used with the positions of canopy trees and seedlings, we can determine whether seedlings tend to be found in canopy gaps and whether their occurrence is affected by gap size. We can also use this approach to determine whether there are tree gaps that have significantly few seedlings in them. This method provides somewhat different information from the bivariate $K$-function, which detects the scales at which the two types of events are aggregated and the scales at which they are segregated.

Let $n_i$ be the number of type 1 events in a circumcircle based on type 1 events, with area $a_i$, and a standardized residual of $z_i$. Let $m_i$ be the number of type 2 events in the same circle and with $y_i$ as the standardized residual. The response of type 2

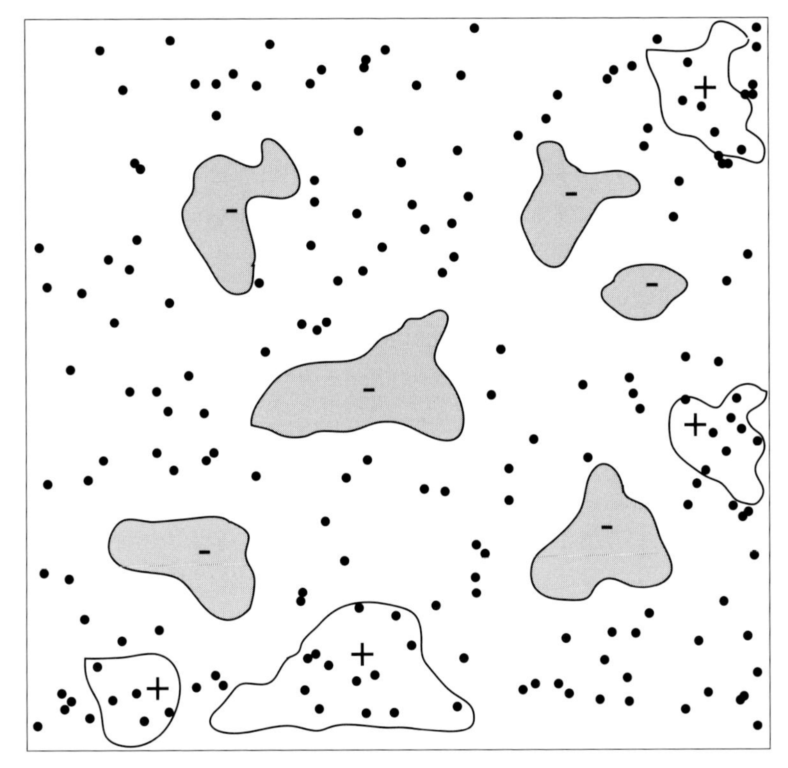


Figure 2.38 Contour map of significant patches ('+') and gaps ('−') based on $Z$ scores of artificial univariate data.

events to type 1 gaps can be examined by graphing the mean and variance of $y_i$ as a function of $a_i$ for circles with $n_i = 0$, or by graphing the same values for circles with $z_i < -1.96$. This asymmetric analysis reflects the asymmetry of the biological processes; the positions of the canopy trees, for example, are expected to influence the presence of seedlings, but not the reverse. For different ecological situations, for example in studying the spatial relationships of two different species of canopy tree or alpine meadow forb, the counts in both kinds of circle (based on types 1 and 2) could be used in the analysis.

A second evaluation of the relationship between the two kinds of events is to calculate covariance as a function of circle area, using either the basic circumcircle or the 'boater' wavelet. For the latter, if there were no edge effects, the covariance could be based on the squares of the differences between the observed counts of events in the inner circle and outer ring, such as $(n_i - p_i)^2$ for type 1 events, $(m_i - q_i)^2$ for type 2 events and $(n_i + m_i - p_i - q_i)^2$ for both types, using (2.42). Because the circles can intersect the edge, we need to adjust for the area within the study area, again using standardized residuals.

In the univariate case, the circle and ring counts, $n_i$ and $p_i$, were converted to standardized residuals, $z_i$ and $\zeta_i$. In the same way, counts of the second type of event can be converted to standardized residuals, $y_i$ and $\eta_i$, and the total counts of both kinds to $w_i$ and $\xi_i$. For an individual circumcircle, a covariance score can be calculated, using only the inner circle, as:

$$c_i = \left(2w_i^2 - y_i^2 - z_i^2\right)/2. \tag{2.57}$$

Using the boater wavelet, in the univariate case, the standardized residuals, $z_i$ and $\zeta_i$, are combined to form a residual of their difference, $Z_i$. For the bivariate situation, the other residuals are also combined to form difference residuals:

$$Y_i = (y_i - \eta_i)/\sqrt{2} \tag{2.58}$$

and

$$W_i = (w_i - \xi_i), \tag{2.59}$$

and the overall covariance score is:

$$C_i = \left(W_i^2 - Y_i^2 - Z_i^2\right)/2. \tag{2.60}$$

For a particular range of radius values, $R$, the variances are then:

$$\begin{aligned} V_Z(R) &= \text{average of } Z_i^2 \text{ for all } i \text{ with } r_i \text{ in } R, \\ V_Y(R) &= \text{average of } Y_i^2 \text{ for all } i \text{ with } r_i \text{ in } R, \\ V_W(R) &= \text{average of } W_i^2 \text{ for all } i \text{ with } r_i \text{ in } R. \end{aligned} \tag{2.61}$$

The covariance for the range $R$ is then:

$$C(R) = [V_W(R) - V_Z(R) - V_Y(R)]/2. \tag{2.62}$$

The covariance of the two kinds of event is graphed as a function of $R$, and positive and negative peaks in that graph are interpreted as indicating the scales of aggregation and segregation of the events of the two kinds (cf. Dale & Powell 1994).

Again, we may wish to generate spatially explicit results. As described above, Getis & Franklin (1987) made contour maps of the $K$-function values of the events and interpolated points, for each of several radii. Extending that approach, the $z$ or $y$ score of each circle is associated with the centre of the circle, and a map of the scores is produced for each radius range. More simply, we can plot the centres of circles of a particular radius class, using two different symbols or colours, one for values less than $-1.96$ and one for values greater than 1.96. The resulting maps, 'circle score maps', show the positions and extent of the centres of significant patches and gaps, and using $Z$, $Y$ or $C$, instead of $z$, $y$ or $c$, will reduce the number of scores to be plotted and will make the maps more informative.

### 2.7.3 Multivariate analysis

As with other univariate and bivariate analyses described in this chapter, the extension of circumcircle analysis to multivariate data is both straightforward in application and complex in interpretation. For the analysis using Ripley's $K$, we recommended partitioning the total counts into 'conspecific' and 'interspecific', and then into components '$I$, $I$', '$I$, $\sim I$', and '$I$, $J$'; the same recommendation applies to the circumcircle approach to multivariate spatial analysis.

Figure 2.39 shows two examples of multivariate circumcircle analysis. The data are from one quadrant of the Lansing Woods data set (cf. Gerrard 1969; Diggle 1983). For maples we give the range and average value of $z$ as a function of circumcircle area for $I$, $I$ and $I$, $\sim I$ combinations. The results consistently show larger numbers of congenerics in smaller circles and fewer stems of different genera in the smaller circles. This approach provides some insights that complement the results from different kinds of analysis.

## 2.8 Concluding remarks

Although, in this chapter, a wide range of data types and approaches to analysis have been discussed, the coverage is not exhaustive. It is very possible that the practising ecologist will encounter situations in which different kinds of data are met or different techniques of analysis are needed. These situations may be completely novel, in which case ingenuity and advice from analytical specialists may be required, or they may have been encountered before, but are not easy to find in the literature. We will describe a few such situations for the purposes of illustration.

Many of the methods used for the analysis of the positions of events in two dimensions are based on the assumption that the events can be adequately represented as dimensionless points in a plane. In the analysis of a particular set of data, it may become clear that that assumption is not sufficiently realistic and the events should be treated as circles of non-zero radius: what then? This is one situation that has been encountered before, and the researcher can follow the treatment by Simberloff (1979), who studied the nearest neighbour assessment of the patterns of events that are circles, not points. The approach is illustrated using data such as the positions of ant-lion pits and ant nests. There is even a discussion of solving the

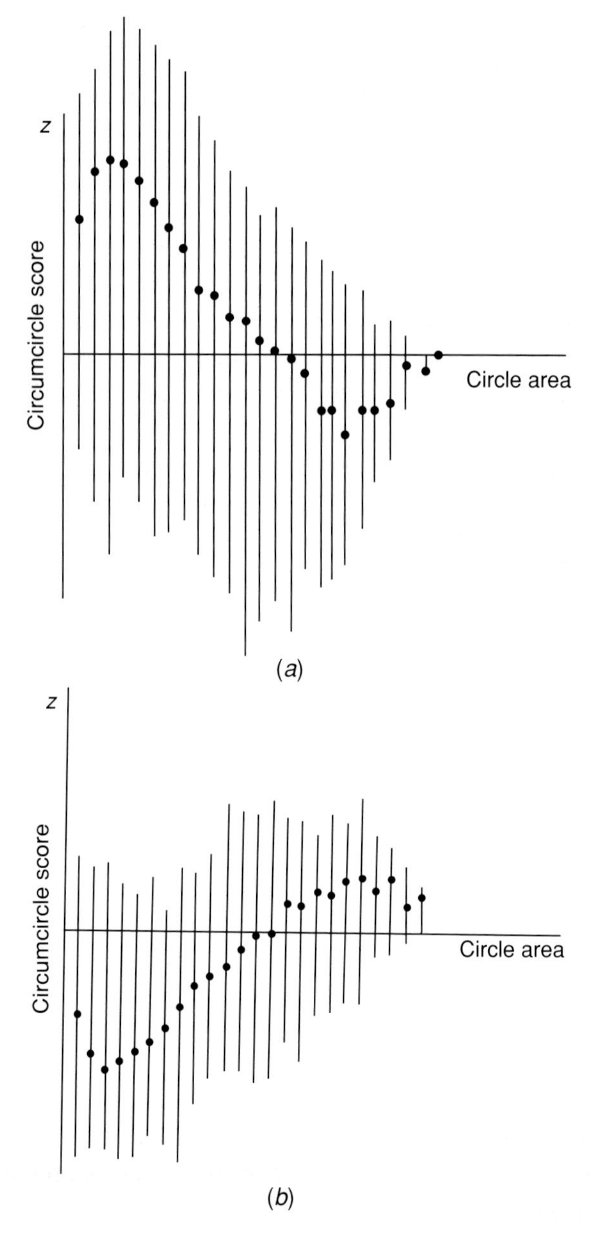


Figure 2.39 (*a*) The $z$ scores, mean and range, as a function of circle size, for maples: congeneric analysis. (*b*) The $z$ scores, mean and range, as a function of circle size, for maples: intergeneric analysis.

same problem in three dimensions, considering the dispersion of spheres, instead of circles.

Another unconventional type of data an ecologist might encounter would be linear objects such as downed logs on the forest floor or ungulate paths across a grassland. Clearly, the positions of such objects cannot be reduced to dimensionless points or isotropic circles. Such data can be considered as planar 'fibre processes' and a helpful discussion of these is provided by Stoyan *et al.* (1995). Chadoeuf *et al.* (2000) provided an interesting illustration, looking at the relationship between the pattern of plant roots and the density of the soil in which they occur. Where the linear features are streams or rivers, there is already an abundance of literature on how these can be characterized (cf. Morisawa 1985), and the applications in aquatic ecology are obvious.

In some situations, a modification of the way in which the data are collected may help solve an apparent problem. For example, if the objects under study have no well-defined centre or patch boundary, such as in a lush grassland, it may be possible to adapt grid sampling to deal with them (see Fehmi & Bartolome 2001) and then proceed to analysis with one of the methods already described here.

We have not included, in this chapter, a discussion of how data that are a mosaic of polygons or other shapes are to be analysed, although such data might be considered for inclusion here. Instead, we have treated them in Chapter 3 (simple analysis) and Chapter 6 (changes through time). We have also not discussed the spatial aspects of species diversity, for example in a plant community, although it would fit well in the discussion of multispecies point pattern analysis. We will save that topic for another time and place.

In considering the material that has been included in this chapter, we can identify three themes:

The first is the 'relatedness' of the various approaches used, both conceptually and mathematically. The basic concept of wavelets, the comparison of the data with some kind or shape of template, recurs and unifies many of the approaches. Not just wavelets, but Ripley's $K$-function, TTLQV and many of the other techniques (almost all!) can be considered to follow this approach. Table 2.3 illustrates this fact. (Mathematically this can be expressed as using a cross-product approach, see Getis 1991; Dale *et al.* 2002.)

The second is the concern for using null hypotheses other than CSR for the evaluation of spatial structure. We noted some examples where the Poisson–Poisson distribution or a Markov inhibition model seemed appropriate for comparison. More needs to be said about this theme and more work needs to be done on this topic, especially in areas other than univariate point pattern analysis.

The third is the usefulness of Monte Carlo and randomization techniques to solve or circumvent problems with analytical approaches. Again, more could be said

Table 2.3 *Summary of the spatial analysis methods presented in Chapter 2*

| Spatial analysis method | Template | |
|---|---|---|
| Points | | |
| Nearest neighbour | 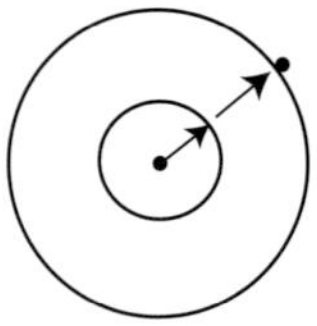 Expanding circle based on event |  Simple link |
| Refined nearest neighbour | 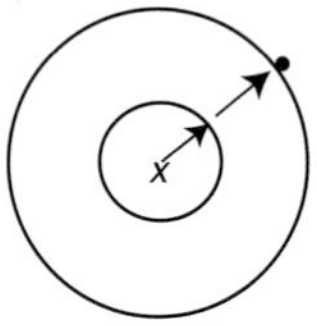 Expanding circle based on event or random point |  Simple link |
| Univariate $K$ | 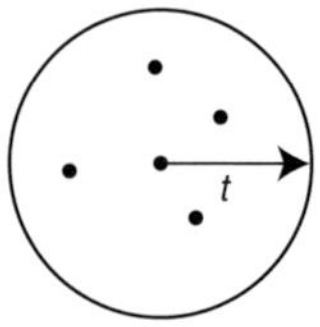 Expanding circle based on event | |
| Getis | 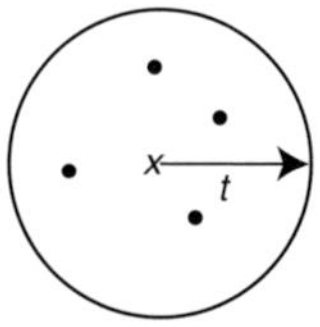 Expanding circle based on event or random point | |
| Condit | 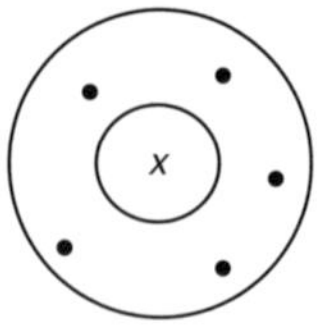 Rings centred on events | |

Table 2.3 (*cont.*)

| Spatial analysis method | Template |
|---|---|
| Bivariate $K$ | 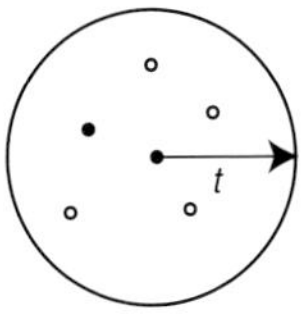 $t$<br>Expanding circle |
| Multivariate $K$ | 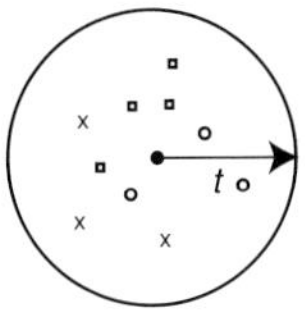 $t$<br>Expanding circle |
| Dixon |  Simple link; 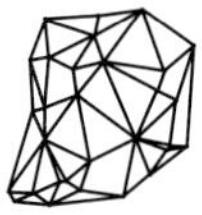 Network of neighbours |
| Reich *et al.* | 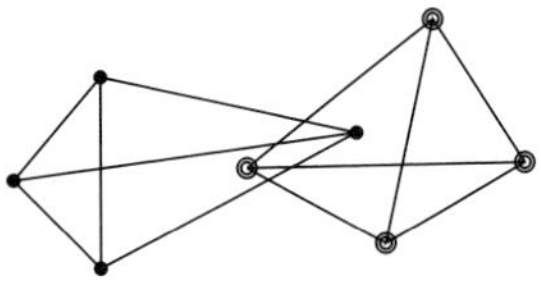 Links to own type |
| Mark correlation | 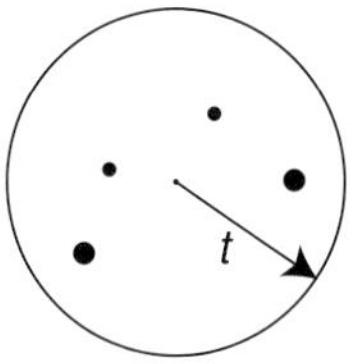 $t$<br>Expanding circle |
| Events networks |  Links |
| One-dimensional Ripley's $K$ | 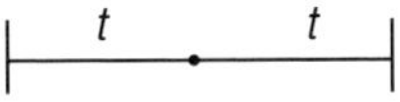 $t$ $t$ '$t$-bar' |
| One-dimensional lacunarity | 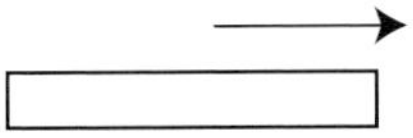 Moving window/gliding box |

Table 2.3 (*cont.*)

| Spatial analysis method | Template |
| --- | --- |
| Three-dimensional nearest neighbour | 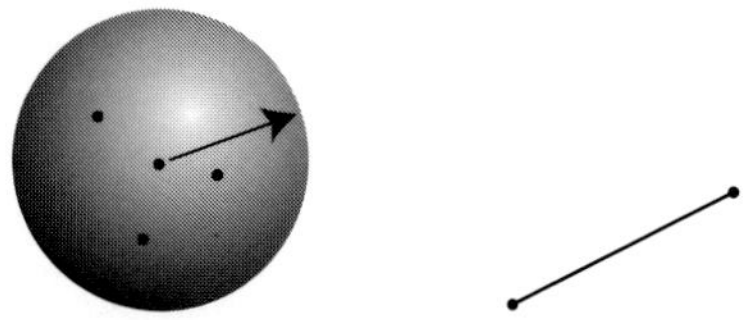 Expanding sphere; Simple link |
| Three-dimensional Ripley's *K* | 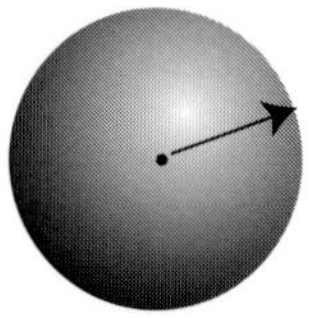 Expanding sphere |
| TTLQV and TTLQC | 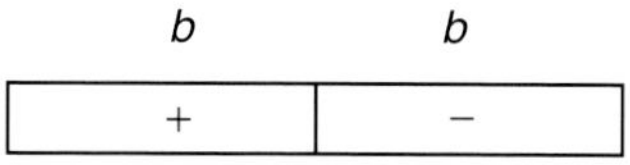 *b* *b*; + − |
| PQV and PQC | 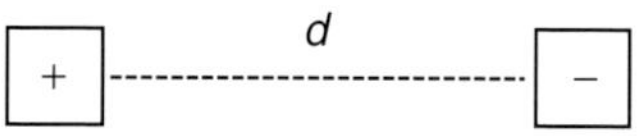 + *d* − |
| 3TLQV and 3TLQC | 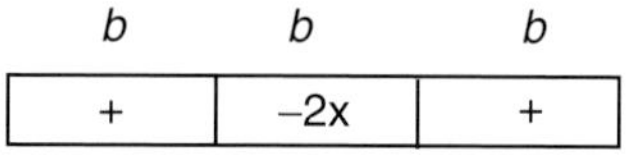 *b* *b* *b*; + −2x + |
| tQV and tQC | 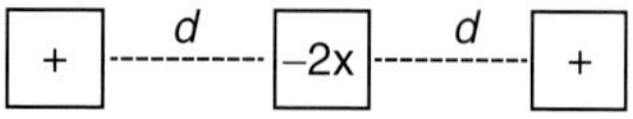 + *d* −2x *d* + |
| Lacunarity | 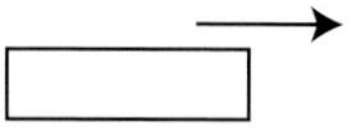 One-part window |
| 2DtQV = 5QV | 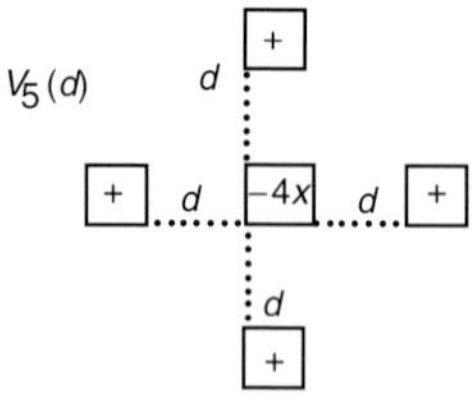 $V_5(d)$; + *d*; + *d* −4*x* *d* +; *d* + |
| 2D3TLQV = 9TLQV | 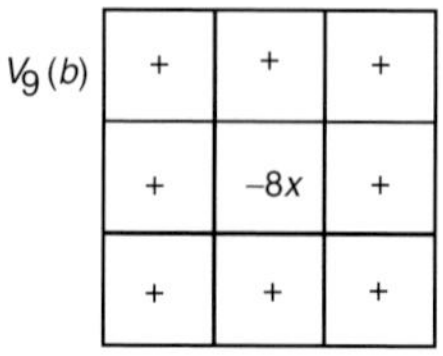 $V_9(b)$; + + +; + −8*x* +; + + + |

Table 2.3 (*cont.*)

| Spatial analysis method | Template |
|---|---|
| Two-dimensional lacunarity | 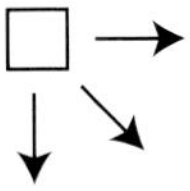 Gliding box |
| Spectral analysis: Fourier |  Sine wave |
| Walsh transform |  Square wave |
| Wavelets<br>One-dimensional data |  Mexican hat |
| 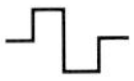 | Haar |
| 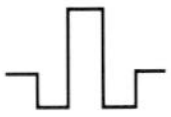 | French top hat |
|  | Morlet |
| | Sine |
| Two-dimensional data | 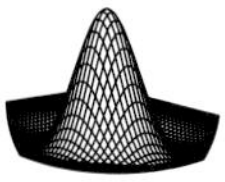 Sombrero |
| Circumcircle | 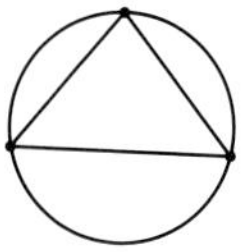 Simple |
| 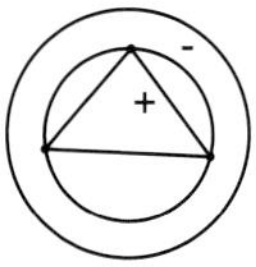 | 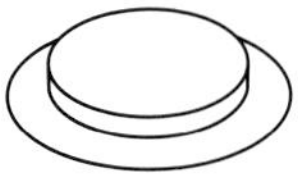 Double = boater wavelet |

on this and more work followed up, but it is clear that this approach has found its time and place in this age of fast and easy computation. This comment applies not only to the methods described in this chapter, of course, but is a common thread through the entire book.

In developing recommendations on which methods to use, it will have become clear to the reader of this chapter that the set of methods based on the concept of Ripley's $K$-function covers a lot of different kinds of data and a range of situations. That set of methods has much to recommend it. On the other hand, the set of methods based on wavelets has flexibility, based on the choice of the wavelet, and a conceptual sophistication that is also very appealing. The decision on the method to use will depend, of course, on the data and the question being asked, but our recommendation is to suggest the use of two or more complementary approaches, so as not to miss important features of the data.

# 3

# Spatial analysis of sample data

## Introduction

Chapter 2 presented the spatial methods used to analyse patterns generated by point pattern processes, i.e. point pattern methods. Point pattern processes generate a spatial distribution of events, which can then be analysed for their spatial pattern. The $x$–$y$ coordinates of all events (e.g. individuals, objects, entities) in a given study area are required for analysis. In ecological studies, a number of point pattern processes take place, one of the most common being seed or spore dispersal. There are, however, several other types of ecological and environmental data that are not discrete, like individual trees, but rather are continuous, like soil moisture. The processes that generate such continuous variables are called surface pattern processes and the spatial statistics that analyse them are called 'surface pattern' methods or 'area pattern' methods. There are, however, grey zones between point pattern and surface pattern analyses, as when point pattern data are transformed into surface pattern data by summarizing the number of events per sampling unit as density. In doing so, point data are converted into quantitative data such that surface pattern methods can be used to analyse their spatial structure. When the entire study area is surveyed using contiguous sampling units, these quantitative data represent the entire population of data in the study area and can be analysed using the spatial methods for contiguous sample unit data presented in Chapter 2. However, given the sampling effort required to census an entire area in the field, much of our understanding of natural complexity is based on sample data. The term 'sample data' refers to the fact that, within a study area, not all the area was studied explicitly, but only portions of it. Hence, the aim of this chapter is to present a wide range of spatial statistics that explore, characterize and quantify spatial patterns from sample data, and then to model their spatial structure. Although most surface pattern statistical methods were developed to analyse sample data, they can also be used with population data to characterize and quantify their spatial pattern.

The use of sample data implies that information about a variable is measured using sampling units distributed according to some sampling design. By using a sampling design, researchers are consciously or unconsciously making implicit and explicit assumptions about the process under investigation. The most important underlying assumption that is shared by many of the spatial statistical methods presented in this book is the notion that nearby values of a variable are more likely to be similar than distant ones (Tobler's first law of geography). In Chapter 2, the definition of 'near' was sharing an edge, because the sampling units were contiguous. In this chapter, sampling units are not contiguous and there is a spatial distance between them, the separation being determined by the sampling design used (e.g. random, systematic, stratified random, etc.). Hence, in order to perform spatial analysis with sample data, it is important to understand how various spatial statistics methods determine and treat the spatial relationships among sampling units. Given that this is at the heart of most of the spatial statistical analyses introduced in this chapter, we will begin by describing how spatial neighbourhoods can be determined for sampling units (see Section 3.1).

Another often forgotten assumption is that the smallest spatial scale to which a measured value of a variable (quantitative or qualitative) can apply is limited to the size of the sampling unit (spatial resolution). Therefore, a measured value is assumed to represent the entire sampling unit, as if there can be no spatial heterogeneity within sampling units. In fact, the sampling design can distort our ability to identify spatial pattern due to confounding effects of:

(1) the size of the sampling unit, i.e. setting the spatial resolution (also known as the grain) limits the precision with which we can identify spatial pattern (Jelinski & Wu 1996; Qi & Wu 1996; Bellehumeur *et al.* 1997; Fortin 1999a);
(2) the shape of the sampling unit, which can minimize the within-sample unit variance, reducing the effects of environmental gradients, and which can over- or underestimate the intensity of spatial structure (Fortin 1999a);
(3) the number of sampling units and their spatial configuration, which have a direct relationship with the spacing between sampling units (also known as the spatial lag), affecting our power to detect significant spatial pattern (Fortin 1999a; Webster & Oliver 2001); and
(4) the size of the study area (also known as the extent), which can incorporate several subregions, each having different underlying ecological processes and environmental conditions, making the response variable non-stationary over the entire area (Fortin 1999a).

Given that these issues are shared by several spatial statistics, we discuss the effects of sampling design decisions (i.e. the number of sampling units, their size and shape as well as their spatial layout) on the detection of spatial patterns in Section 3.3.

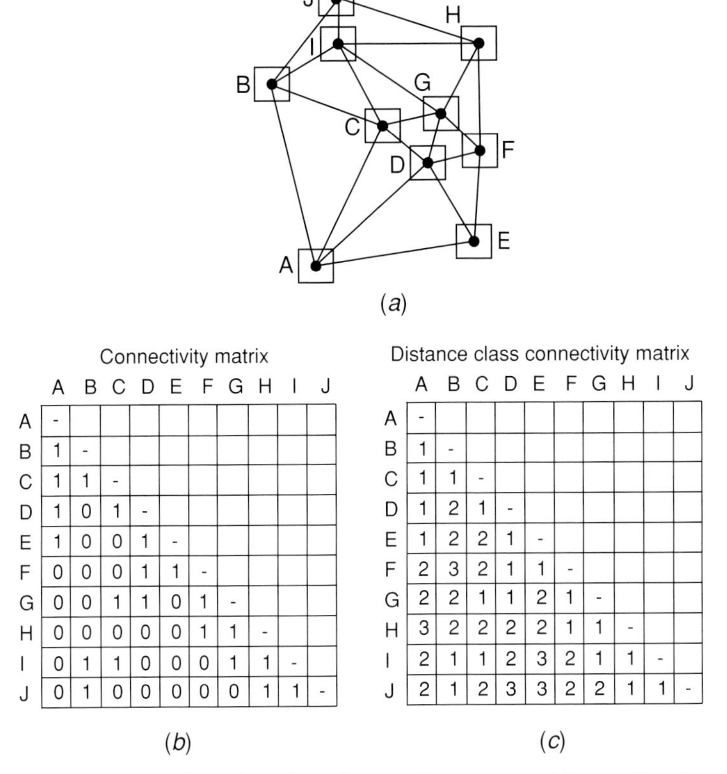


Connectivity matrix

| | A | B | C | D | E | F | G | H | I | J |
|---|---|---|---|---|---|---|---|---|---|---|
| A | - | | | | | | | | | |
| B | 1 | - | | | | | | | | |
| C | 1 | 1 | - | | | | | | | |
| D | 1 | 0 | 1 | - | | | | | | |
| E | 1 | 0 | 0 | 1 | - | | | | | |
| F | 0 | 0 | 0 | 1 | 1 | - | | | | |
| G | 0 | 0 | 1 | 1 | 0 | 1 | - | | | |
| H | 0 | 0 | 0 | 0 | 0 | 1 | 1 | - | | |
| I | 0 | 1 | 1 | 0 | 0 | 0 | 1 | 1 | - | |
| J | 0 | 1 | 0 | 0 | 0 | 0 | 0 | 1 | 1 | - |

(*b*)

Distance class connectivity matrix

| | A | B | C | D | E | F | G | H | I | J |
|---|---|---|---|---|---|---|---|---|---|---|
| A | - | | | | | | | | | |
| B | 1 | - | | | | | | | | |
| C | 1 | 1 | - | | | | | | | |
| D | 1 | 2 | 1 | - | | | | | | |
| E | 1 | 2 | 2 | 1 | - | | | | | |
| F | 2 | 3 | 2 | 1 | 1 | - | | | | |
| G | 2 | 2 | 1 | 1 | 2 | 1 | - | | | |
| H | 3 | 2 | 2 | 2 | 2 | 1 | 1 | - | | |
| I | 2 | 1 | 1 | 2 | 3 | 2 | 1 | 1 | - | |
| J | 2 | 1 | 2 | 3 | 3 | 2 | 2 | 1 | 1 | - |

(*c*)

Figure 3.1 (*a*) Connectivity based on the topology among 10 sampling locations (A–J). Here, the *x*–*y* coordinates of the centroid of each sampling unit are used to determine a tessellation network linking all the sampling locations. From this series of links, a binary connectivity matrix (*b*) of first nearest neighbours (1, connected; 0, not connected) can be defined as well as a distance class connectivity matrix (*c*) of *k* nearest neighbours (1, first nearest neighbour; 2, second nearest neighbour; and so on).

## 3.1 How to determine 'nearby' relationships among sampling units

To determine the spatial neighbourhoods of sampling units, we will use the concepts and terminology already outlined in Chapter 1 (see Figures 1.11 and 1.12) and Chapter 2 (see Figures 2.12–2.20). Specifically, two different representations of space can be used to describe the spatial relationship among sampling units: topological space or Euclidean space. In topological space, sampling units can be reduced to simple points in the plane, using the *x*–*y* coordinates of their centroids (Figure 3.1), and then any of the neighbour network algorithms presented in Chapter 2 can be used to establish the 'connectivity matrix', which lists the links between sampling units. For example, the first-order neighbours of sampling unit A are B, C, D and E, which is indicated by code 1 in the connectivity matrix (Figure 3.1). Such a

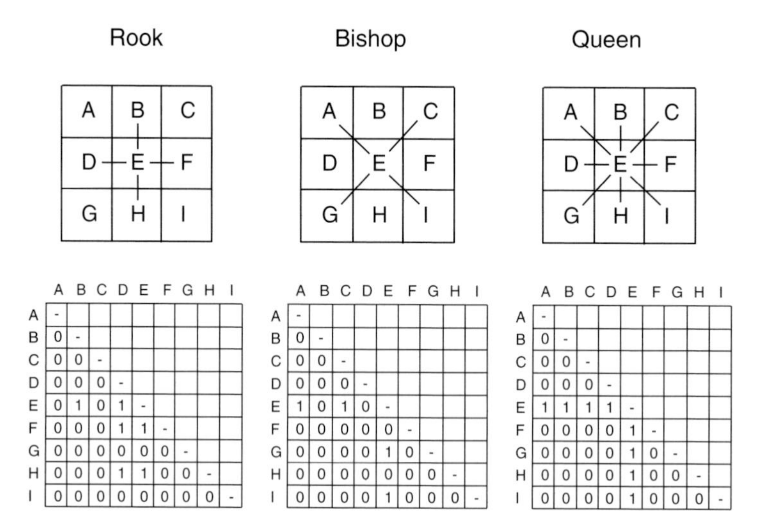


Figure 3.2 Connectivity based on chess moves among nine sampling locations on a lattice (A–I). The targeted sampling location E has four neighbours while using the rook and bishop moves and eight neighbours with the queen move as shown in the plots and translated in the binary connectivity matrices of first nearest neighbours (1, connected; 0, not connected).

connectivity matrix indicates which pairs of sampling units have first-order connections, by an entry of 1, and those that do not, by 0. The connectivity matrix can be extended to higher level neighbours such that spatial connectivity among the sampling units can be divided into $d$ neighbours, called a distance class connectivity matrix (Figure 3.1$c$), where $d = 1, 2, \ldots, m$, and where $m$ is the number of neighbours needed to link the two most distant sampling units. There may be cases in which the sampling units are contiguous squares, or form a regular spaced lattice. In such circumstances, the tessellation links between sampling units can be described as chess piece moves (Figure 3.2): the rook makes links in four cardinal directions, the bishop makes only diagonal links, and the queen makes links in all eight directions. By using topological space to establish the spatial relationship among sampling units, we are assuming implicitly that Euclidean distances among sampling units are not important, such that we are interested only in their relative positions. When absolute distances among sampling units are important, or their absolute positions in some frame of reference (e.g. the study area limits), then Euclidean space should be used.

Euclidean distances among sampling units can be computed using the $x$–$y$ coordinates of the centroid of the sampling units, resulting in a matrix of straight-line distances (Figure 3.3$a$). Based on these Euclidean distances, a distance class connectivity matrix can be created for a number of distance classes (Figure 3.3$b$). When other kinds of a priori knowledge of the point or surface pattern process

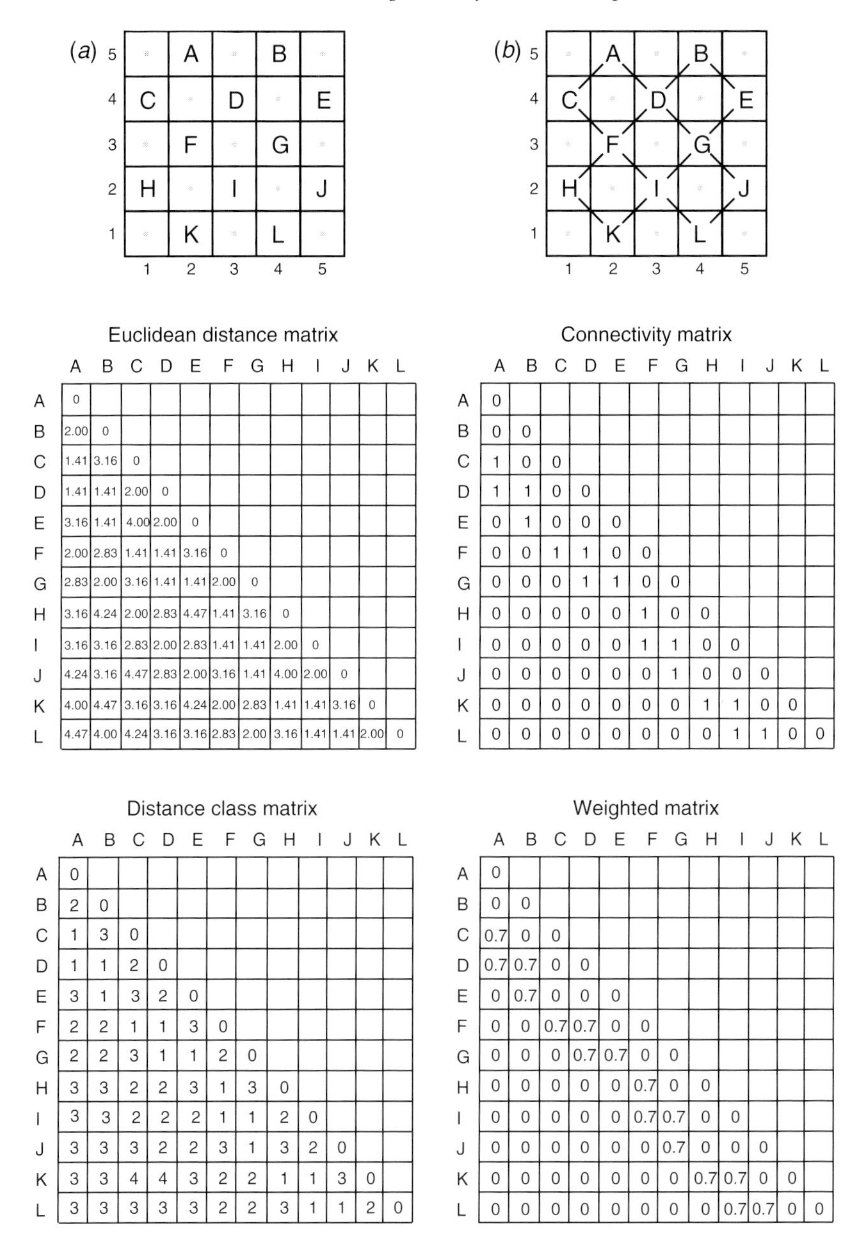


Euclidean distance matrix

| | A | B | C | D | E | F | G | H | I | J | K | L |
|---|---|---|---|---|---|---|---|---|---|---|---|---|
| A | 0 | | | | | | | | | | | |
| B | 2.00 | 0 | | | | | | | | | | |
| C | 1.41 | 3.16 | 0 | | | | | | | | | |
| D | 1.41 | 1.41 | 2.00 | 0 | | | | | | | | |
| E | 3.16 | 1.41 | 4.00 | 2.00 | 0 | | | | | | | |
| F | 2.00 | 2.83 | 1.41 | 1.41 | 3.16 | 0 | | | | | | |
| G | 2.83 | 2.00 | 3.16 | 1.41 | 1.41 | 2.00 | 0 | | | | | |
| H | 3.16 | 4.24 | 2.00 | 2.83 | 4.47 | 1.41 | 3.16 | 0 | | | | |
| I | 3.16 | 3.16 | 2.83 | 2.00 | 2.83 | 1.41 | 1.41 | 2.00 | 0 | | | |
| J | 4.24 | 3.16 | 4.47 | 2.83 | 2.00 | 3.16 | 1.41 | 4.00 | 2.00 | 0 | | |
| K | 4.00 | 4.47 | 3.16 | 3.16 | 4.24 | 2.00 | 2.83 | 1.41 | 1.41 | 3.16 | 0 | |
| L | 4.47 | 4.00 | 4.24 | 3.16 | 3.16 | 2.83 | 2.00 | 3.16 | 1.41 | 1.41 | 2.00 | 0 |

Connectivity matrix

| | A | B | C | D | E | F | G | H | I | J | K | L |
|---|---|---|---|---|---|---|---|---|---|---|---|---|
| A | 0 | | | | | | | | | | | |
| B | 0 | 0 | | | | | | | | | | |
| C | 1 | 0 | 0 | | | | | | | | | |
| D | 1 | 1 | 0 | 0 | | | | | | | | |
| E | 0 | 1 | 0 | 0 | 0 | | | | | | | |
| F | 0 | 0 | 1 | 1 | 0 | 0 | | | | | | |
| G | 0 | 0 | 0 | 1 | 1 | 0 | 0 | | | | | |
| H | 0 | 0 | 0 | 0 | 0 | 1 | 0 | 0 | | | | |
| I | 0 | 0 | 0 | 0 | 0 | 1 | 1 | 0 | 0 | | | |
| J | 0 | 0 | 0 | 0 | 0 | 0 | 1 | 0 | 0 | 0 | | |
| K | 0 | 0 | 0 | 0 | 0 | 0 | 0 | 1 | 1 | 0 | 0 | |
| L | 0 | 0 | 0 | 0 | 0 | 0 | 0 | 0 | 1 | 1 | 0 | 0 |

Distance class matrix

| | A | B | C | D | E | F | G | H | I | J | K | L |
|---|---|---|---|---|---|---|---|---|---|---|---|---|
| A | 0 | | | | | | | | | | | |
| B | 2 | 0 | | | | | | | | | | |
| C | 1 | 3 | 0 | | | | | | | | | |
| D | 1 | 1 | 2 | 0 | | | | | | | | |
| E | 3 | 1 | 3 | 2 | 0 | | | | | | | |
| F | 2 | 2 | 1 | 1 | 3 | 0 | | | | | | |
| G | 2 | 2 | 3 | 1 | 1 | 2 | 0 | | | | | |
| H | 3 | 3 | 2 | 2 | 3 | 1 | 3 | 0 | | | | |
| I | 3 | 3 | 2 | 2 | 2 | 1 | 1 | 2 | 0 | | | |
| J | 3 | 3 | 3 | 2 | 2 | 3 | 1 | 3 | 2 | 0 | | |
| K | 3 | 3 | 4 | 4 | 3 | 2 | 2 | 1 | 1 | 3 | 0 | |
| L | 3 | 3 | 3 | 3 | 3 | 2 | 2 | 3 | 1 | 1 | 2 | 0 |

Weighted matrix

| | A | B | C | D | E | F | G | H | I | J | K | L |
|---|---|---|---|---|---|---|---|---|---|---|---|---|
| A | 0 | | | | | | | | | | | |
| B | 0 | 0 | | | | | | | | | | |
| C | 0.7 | 0 | 0 | | | | | | | | | |
| D | 0.7 | 0.7 | 0 | 0 | | | | | | | | |
| E | 0 | 0.7 | 0 | 0 | 0 | | | | | | | |
| F | 0 | 0 | 0.7 | 0.7 | 0 | 0 | | | | | | |
| G | 0 | 0 | 0 | 0.7 | 0.7 | 0 | 0 | | | | | |
| H | 0 | 0 | 0 | 0 | 0 | 0.7 | 0 | 0 | | | | |
| I | 0 | 0 | 0 | 0 | 0 | 0.7 | 0.7 | 0 | 0 | | | |
| J | 0 | 0 | 0 | 0 | 0 | 0 | 0.7 | 0 | 0 | 0 | | |
| K | 0 | 0 | 0 | 0 | 0 | 0 | 0 | 0.7 | 0.7 | 0 | 0 | |
| L | 0 | 0 | 0 | 0 | 0 | 0 | 0 | 0 | 0.7 | 0.7 | 0 | 0 |

Figure 3.3 Connectivity based on Euclidean distance among 10 sampling locations (A–J). (*a*) The Euclidean distance matrix among the 10 sampling locations and the corresponding distance class matrix when using a distance interval class of 1.5 units. (*b*) The binary connectivity matrix based on first nearest neighbours and the corresponding weight matrix as a function of distance ($1/d = 1/1.41$ units, which gives a weight of 0.7).

are available, one may not just want to use a connectivity matrix (Figure 3.3*c*) that indicates in a simple binary fashion which sampling units are neighbours (1, neighbours; 0, otherwise), but a more sophisticated weight matrix where real values (usually between 0 and 1) stress that nearby neighbours are more important than ones located further away in the estimation of spatial structure (Figure 3.3*d*). The most common weight used is the inverse distance ($1/d$) or the inverse of distance squared ($1/d^2$).

There are some rules of thumb that can facilitate choosing the number of distance classes, based either on sample size or spatial lag. For example, Legendre & Legendre (1998; p. 717) suggested using Sturge's rule, based on sample size, to determine the number of histogram classes: for $n$ samples the recommended number of classes is:

$$D = 1 + 3.3 \times \log_{10}\left[\frac{n(n-1)}{2}\right]. \tag{3.1}$$

Once the number of distance classes is selected, there are two ways of distributing the Euclidean distances among them. The first view is that they should be divided by equal distance intervals (also referred to as spatial lag or the spatial interval, Figure 3.4*a*) and the second is that distance classes should have equal numbers of pairs of sampling units in them (as proposed by Sokal & Wartenberg 1983; Figure 3.4*b*). The equidistant approach is the most frequently used and is more intuitive, allowing comparison of results among different studies at a given distance interval. However, the number of pairs per distance class varies, which in turn affects the reliability of the estimation of spatial structure (Figure 3.4). The equal frequency approach provides a better estimation of spatial autocorrelation at each distance class because each distance class must contain the same number of pairs, thus minimizing the edge-effect problem presented in Chapter 1. This implies, however, that the values estimated to describe the spatial structure are not at regular distance intervals (Figure 3.4). The equal frequency distance class approach is less common, either because it is not widely known (software packages lack this option) or because the results are not as easy to compare from one study area to another when the distance intervals differ.

When using equidistant classes, one can select the distance interval instead of the number of distance classes. The minimum distance interval must be equal to or larger than the sampling unit length (or the longest side length in the case of a rectangular sampling unit). When the sampling units form a regular lattice, this minimum distance interval (i.e. the length of a sample unit, say 1 m) will make links equivalent to the rook's move, so that each sample unit has only four neighbours (Figure 3.4). In order to include all eight neighbours (equivalent to the queen's move), the minimum distance interval needs to include the diagonal,

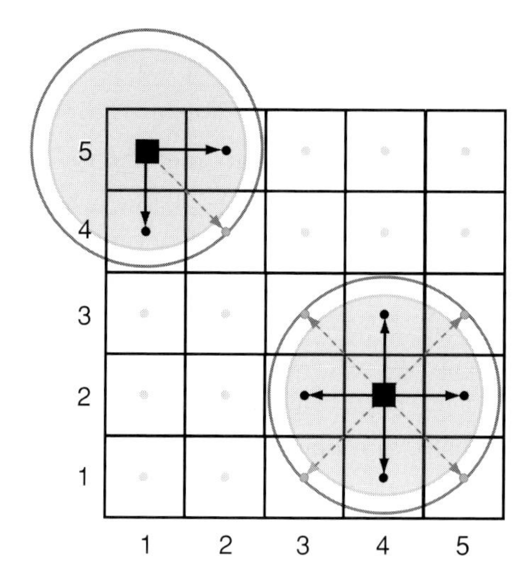


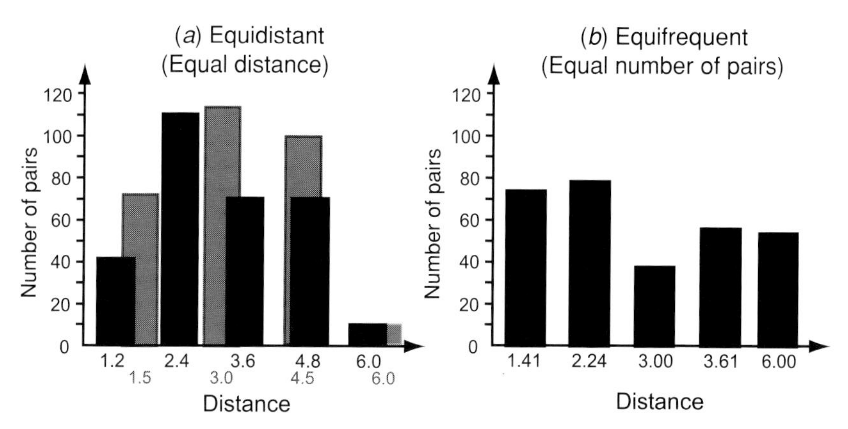


Figure 3.4 Distance class determination based on distance (*a*) or number of pairs (*b*) of sampling locations. (*a*) When defining the distance interval class at 1.2 units (in solid black), there are five equidistant classes (where the diagonal locations are not included in the first distance class – 1.41 units apart); while when the distance interval class is 1.5 units (in solid grey) there are only four equidistant classes (including the diagonal locations in the first distance class – 1.41 units apart) such that there are more pairs of sampling locations in the first distance class. (*b*) With the equifrequent distance class approach, each distance class has more or less 50 pairs of sampling locations, but the Euclidean distance from one distance class to another varies in interval (i.e. 1.41, 0.83, 0.76, 0.61 and 2.39).

i.e. $\sqrt{2} = 1.414$ (Figure 3.4). There are some trade-offs to consider in selecting a distance interval (Fortin 1999a): too small a distance interval will result in a very large number of distance classes and there will be fewer pairs of observations in each distance class making the estimation of spatial structure less reliable. Too large a distance interval will lead to too few distance classes where each distance class

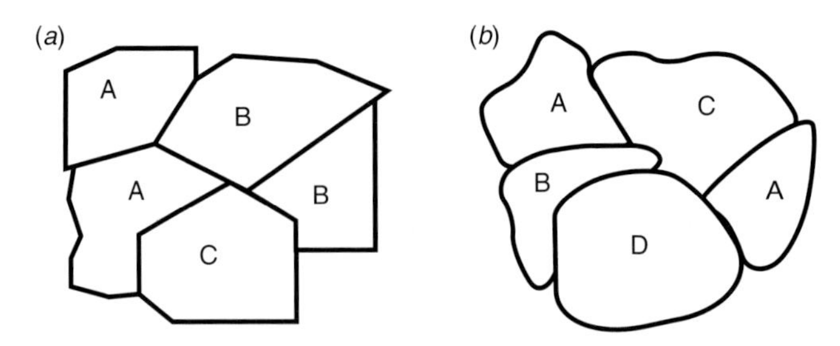


Figure 3.5 Spatial adjacency of polygons. (*a*) Arbitrary political boundaries among counties make it possible to have adjacent counties with the same category. (*b*) Natural forest stands have, by definition, different types of adjacent forest surrounding them.

will cover more area and the magnitude of spatial structure will be diminished. All these issues will be made more obvious throughout this chapter and will be revisited in Section 3.3.

## 3.2 Join count statistics

The characterization and testing of the spatial association of categories (e.g. nominal, qualitative, categorical, binary data) can be obtained using join count statistics (Moran 1948; Cliff & Ord 1981). In essence, and as the name of the method suggests, join count statistics test whether or not the occurrence of categorical attributes at spatially adjacent sampling locations can be accounted for by randomness alone. These statistics were first developed by human geographers to test whether or not adjacent counties (i.e. areas delimited arbitrarily; Figure 3.5*a*) showed a spatial pattern of disease. In such analyses, each county was given a binary nominal value (black indicating disease or white indicating no disease) and it was assumed that the nominal attribute prevailed for the entire county. Adjacency is defined by the 'join' between counties, or any data polygons (Figure 3.5*b*) that share boundaries. The join count statistics can also be used with lattice data (i.e. adjacent sampling units as illustrated in Figure 3.2). With lattice data, and as presented in Section 3.1, three types of neighbourhood rules can be used to determine which sampling units are joined to one another, as inspired by the chess board moves of the rook, the bishop and the queen (Figure 3.2). When sampling units are not spatially contiguous, adjacency can be established using any of the connectivity network algorithms presented in Chapter 2 (nearest neighbour, Delaunay triangulation, etc.) and the order of neighbourliness in that network (first order, second order and so on), or using Euclidean distances (see Section 3.1).

There are three join count statistics for binary data that count the numbers of pairs of adjacent sampling units having the same category ($J_{BB}$ and $J_{WW}$; B for black and W for white) or not ($J_{BW}$). The statistics $J_{BB}$ and $J_{WW}$ assess the presence of spatial association (positive spatial association) between adjacent sampling units, while $J_{BW}$ assesses the presence of spatial repulsion (negative spatial association). The total number of joins ($J$) among sampling units is determined by the spatial arrangement of the sampling units, as well as by the connectivity algorithm used, and is the sum of the three join count statistics:

$$J = J_{BB} + J_{WW} + J_{WB}. \tag{3.2}$$

Therefore, only two of the three statistics are independent, while the third one can be derived from the other two.

The observed join count statistic, $J_{BB}$, counts the number of joins in adjacent regions (or cells) having the same category (black–black):

$$J_{BB} = \frac{1}{2}\left(\sum_{\substack{i=1\\ i\neq j}}^{n}\sum_{\substack{j=1\\ j\neq i}}^{n}\delta_{ij}x_i x_j\right), \tag{3.3}$$

where $i$ and $j$ are the two sampling units compared, $x_i$ is the attribute of a sampling unit (black $= 1$ and white $= 0$) and $\delta_{ij}$ is the entry in the connectivity matrix indicating the adjacency of sampling units $i$ and $j$: 1, when $i$ and $j$ are adjacent; 0, otherwise. The statistic $J_{BW}$ counts the number of adjacent sampling units with unlike categories (black–white), and is computed as follows:

$$J_{BW} = \frac{1}{2}\left[\sum_{\substack{i=1\\ i\neq j}}^{n}\sum_{\substack{j=1\\ j\neq i}}^{n}\delta_{ij}(x_i - x_j)^2\right], \tag{3.4}$$

and $J_{WW}$ can be computed using the two previous statistics:

$$J_{WW} = \frac{1}{2}\left(\sum_{\substack{i=1\\ i\neq j}}^{n}\sum_{\substack{j=1\\ j\neq i}}^{n}\delta_{ij}\right) - (J_{BB} + J_{BW}). \tag{3.5}$$

The null hypothesis of complete spatial randomness (CSR) of categories can then be tested, assuming stationarity, by computing the expected values of joins based on the proportion of each category and the number of joins (links) in the study area. Then, using the observed and expected values, $z$ values can be calculated for comparison with the standard normal distribution, $N(0, 1)$. Normality is approximately ensured when the number of sampling units is at least 20 and the probability of neither category is less than 0.2 (Cliff & Ord 1973). One of two assumptions are required to establish the probability of each category:

(1) we assume that the assignment of the categories (here black and white) are independent for each sampling unit (free-sampling assumption, with replacement) such that $p = q = 0.5$; or
(2) we assume that they are dependent (i.e. the category in one county affects the category type in other counties, non-free-sampling assumption, without replacement) such that $p = n_b/n$ and $q = n_w/n = (1 - p)$, where $n_b$ is the number of black sampling units, $n_w$ is the number of white units and $n$ is the total number of sampling units.

Mathematical details about the equations for the expected values and variances are presented in Cliff & Ord (1973, 1981) and in Sokal & Oden (1978).

Using simulated data with known spatial patterns (random, uniform or regular and patchy), Figure 3.6 illustrates the effects of the three most commonly used connectivity algorithms for lattice data (rook, bishop and queen) on the detection of significant spatial associations. In this example, there are 100 sampling units (50 black and 50 white) and the total number of joins varies according to the connectivity criteria used (180, 162 and 342, respectively). For randomly distributed data, the join count statistics calculated using the rook directional joins indicate that the pattern is significantly different from random, while for the two other connectivity algorithms the statistics are not significant, as expected. For uniformly distributed data, both the rook and bishop connectivity algorithms result in significant statistics using the null hypothesis of the absence of spatial association. However, with the queen algorithm, the null hypothesis was not rejected. Finally, for patchy data, all three connectivity definitions produced join count statistics that were significantly different from a random distribution.

### 3.2.1 Considerations and other join count statistics

Join count statistics have been extended beyond the binary case to the analysis of the spatial association of $k$ categories (Cliff & Ord 1973, 1981; Sokal & Oden 1978). Also, join count statistics can be computed not only for the first neighbour but at several spatial distances defined either in terms of neighbour links, using neighbour network topology as a way to measure distances among sampling units (see Chapter 2 for connectivity network algorithms), or in terms of Euclidean distance (as illustrated in Figures 3.1, 3.2 and 3.3). Combining these two extensions, (3.3) can be rewritten as:

$$J_{rr}(d) = \frac{1}{2}\left[\sum_{\substack{i=1\\ i\neq j}}^{n}\sum_{\substack{j=1\\ j\neq i}}^{n}\delta_{ij}(d)x_{ri}x_{rj}\right], \tag{3.6}$$

Type of connectivity

| | Rook | Bishop | Queen |
|---|---|---|---|
| | 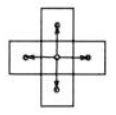 | 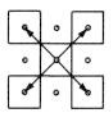 | 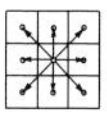 |
| $J$ | 180 | 162 | 342 |

Observed values ($z$ values)

Random

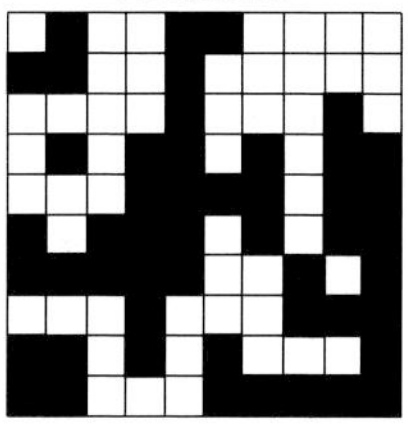

| | Rook | Bishop | Queen |
|---|---|---|---|
| $J_{BB}$ | 52 ($z$ = 2.06) | 35 ($z$ = −1.24) | 87 ($z$ = 0.39) |
| $J_{WW}$ | 54 ($z$ = 2.61) | 39 ($z$ = −0.26) | 93 ($z$ = 1.39) |
| $J_{WB}$ | 74 ($z$ = −2.54) | 88 ($z$ = 0.98) | 162 ($z$ = −1.19) |

Uniform

| | Rook | Bishop | Queen |
|---|---|---|---|
| $J_{BB}$ | 0 ($z$ = −12.32) | 81 ($z$ = 10.03) | 81 ($z$ = −0.60) |
| $J_{WW}$ | 0 ($z$ = −12.32) | 81 ($z$ = 10.03) | 81 ($z$ = −0.60) |
| $J_{WB}$ | 180 ($z$ = 13.40) | 0 ($z$ = −12.98) | 180 ($z$ = 0.81) |

Patch

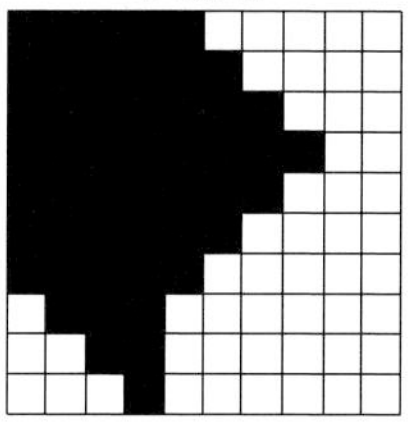

| | Rook | Bishop | Queen |
|---|---|---|---|
| $J_{BB}$ | 82 ($z$ = 10.36) | 76 ($z$ = 8.80) | 158 ($z$ = 12.20) |
| $J_{WW}$ | 75 ($z$ = 8.42) | 63 ($z$ = 5.61) | 138 ($z$ = 8.87) |
| $J_{WB}$ | 23 ($z$ = −10.21) | 23 ($z$ = −9.33) | 46 ($z$ = −14.11) |

Expected values

| | Rook | Bishop | Queen |
|---|---|---|---|
| $J_{BB}$ | 44.545 | 40.090 | 84.636 |
| $J_{WW}$ | 44.545 | 40.090 | 84.636 |
| $J_{WB}$ | 90.909 | 81.818 | 172.727 |

Figure 3.6 Join count statistics ($J_{BB}$, $J_{WW}$ and $J_{WB}$). The expected values are based on the number of joins (rook = 180, bishop = 162 and queen = 342) and the number of black and white values ($n_B = 50$; $n_W = 50$). The observed values of the three statistics (and corresponding $z$ values) are computed according to the three chess moves (rook, bishop and queen) based on binary data from a 10 × 10 lattice from three types of spatial patterns (random, uniform and patchy). This example stresses how the numbers of joins (i.e. the connectivity rule) affect the observed statistics and their significance. For example, in the case of the random pattern, the observed statistics based on the rook move are significant (values > $|-1.96|$), while these statistics are not significant with the bishop and the queen moves.

and (3.4) as:

$$J_{rs}(d) = \frac{1}{2}\left[\sum_{\substack{i=1\\ i\neq j}}^{n}\sum_{\substack{j=1\\ j\neq i}}^{n}\delta_{ij}(d)x_{ri}x_{sj}\right], \tag{3.7}$$

where $r$ indicates one category and $s$ another, ($d$) indicates the distance class either in terms of $d$ neighbours or $d$ Euclidean distance classes at which the sampling units are to be considered connected (1) or not (0); and $x_{ri}$ is an indicator function that takes the value 1 when unit $i$ belongs to category $r$, and 0 otherwise. Epperson (2003) presented a mathematical description of these $k$-category join count statistics and applied them to genetic data.

Up to this point, the spatial units we have considered were either arbitrarily determined a priori, as in the case of political counties, or as regularly or irregularly spaced sampling units. The implication of this is that adjacent counties (Figure 3.5*a*) or sampling units (Figure 3.6) can belong to the same category. There are cases, however, where this is not so. For example, while using forest inventory maps based on photographic interpretation, the delineation of each forest stand implies that it is different from neighbouring stands (Figure 3.5*b*). Lowell (1997) showed that with such forest inventory data, there are never any joins of the same category ($J_{rr}$) and he demonstrated, based on simulations, that when there is a minimum of five or so categories, the join count statistic ($J_{rs}$) is robust and provides unbiased results.

There are, however, circumstances where the join count statistics are not robust. This occurs when there is an obvious lack of stationarity over the entire study area (see Chapter 1), for example in the presence of a gradient or trend, implying a difference in the mean (known as first-order heterogeneity). The likelihood of using data showing properties of first-order heterogeneity is becoming more and more prevalent since ecologists are using remotely sensed data that cover large areas. To address the issue of first-order heterogeneity over a study area, and violation of the stationarity assumption, Kabos & Csillag (2002) developed an $H$ Moran statistic (or heterogeneous Moran) to analyse lattice data. This statistic assesses the probability of each category at each sample location given the values of neighbouring locations. By doing so, the significance of the estimated statistic at each sampling location is not influenced by the lack of stationarity over the entire study area.

## 3.3 Global spatial statistics

Ecologists are used to the notion of covariance and correlation between two variables $x$ and $y$ where one is testing whether the two variables covary positively, negatively or not (i.e. testing the null hypothesis of no relationship between two variables resulting in a correlation value of 0). Linear correlation between quantitative variables can be estimated using Pearson's product–moment correlation coefficient. This coefficient is the standardized covariance between two variables (e.g. $x$ and $y$) which measures the deviations of the variables from their respective averages ($\bar{x}$ and $\bar{y}$):

$$\hat{\rho}(x, y) = \frac{\sum_{i=1}^{n}(x_i - \bar{x})(y_i - \bar{y})}{\sqrt{\sum_{i=1}^{n}(x_i - \bar{x})^2 \sum_{i=1}^{n}(y_i - \bar{y})^2}}, \tag{3.8}$$

which estimates

$$\rho(x, y) = \frac{\mathrm{Cov}(x, y)}{\sqrt{\mathrm{Var}(x)\mathrm{Var}(y)}}. \tag{3.9}$$

Stemming from this notion of covariation between two variables, human geographers proposed estimating the covariance and the correlation of the values of a single variable with itself (auto) for all pairs of sampling units that are separated by a given spatial lag (spatial autocorrelation) or temporal interval (temporal autocorrelation). The spatial autocovariance, $C(d)$, of the variable $x$ can therefore be estimated by computing the product of the deviation of the value of the variable $x$ at the location $i$ with the expected value ($E(x_i)$) with the deviation of the value at the location $i + d$, where $d$ is a given distance:

$$C(d) = E\{[x_i - E(x_i)][x_{i+d} - E(x_i)]\}. \tag{3.10}$$

The spatial autocorrelation, $\rho(d)$, of the variable $x$ at a distance class $d$, is the autocovariance divided by the variance of the variable (i.e. when $C(d)$ is at $d = 0$):

$$\rho(d) = \frac{C(d)}{C(0)}, \tag{3.11}$$

where

$$C(0) = \mathrm{Var}(x) = \sigma^2. \tag{3.12}$$

The estimation of spatial structure of a variable can be computed using various statistics derived from Pearson's product–moment correlation (Eqn (3.8)) as presented in this section. These statistics are based on the assumption of stationarity. We refer to these statistics as 'global' because they estimate the intensity of spatial dependence for the entire study area and summarize it with a single value.

As stressed in Chapter 1, spatial patterns (i.e. lack of independence among data at nearby locations) are often the result of several processes. Consequently, the observed spatial structure is a mix of both induced spatial dependence (i.e. variable response to the spatial structure of exogenous process) and inherent spatial autocorrelation (i.e. inherent in the ecological process of the variable of interest). In fact, most of the time, it is difficult to disentangle them. The first spatial statistics were developed to estimate the degree of spatial autocorrelation in the data. Through the years it became more and more obvious that spatial autocorrelation coefficients, presented in this section, actually estimate the degree of spatial dependence (i.e. induced spatial dependence and inherent spatial autocorrelation). Hence, although we are going to use the term 'spatial autocorrelation' in this section, keep in mind that spatial autocorrelation coefficients cannot discriminate between the spatial structure induced by spatial dependence and that inherent to the variable itself.

### *3.3.1 Spatial autocorrelation coefficients for one variable*

The first coefficient of spatial autocorrelation was proposed by Moran (1948) and can be computed at each $d$ distance class:

$$I(d) = \left[\frac{1}{W(d)}\right] \frac{\sum_{\substack{i=1\\ i\neq j}}^{n} \sum_{\substack{j=1\\ j\neq i}}^{n} w_{ij}(d)(x_i - \bar{x})(x_j - \bar{x})}{\frac{1}{n}\sqrt{\sum_{i=1}^{n}(x_i - \bar{x})^2}}, \tag{3.13}$$

where $w_{ij}(d)$ is the distance class connectivity matrix (also called the weight matrix) that indicates whether a pair of sampling locations are in distance class $d$ (Figure 3.3*b*); $x_i$ and $x_j$ are the values of the variable $x$ at sampling location $i$ and $j$; and $W(d)$ is the sum of $w_{ij}(d)$, here being the number of pairs of sampling locations per distance class. As with Pearson's correlation coefficient, positive autocorrelation is indicated by positive values (usually ranging from 0 to 1), negative autocorrelation by negative values (usually ranging from 0 to $-1$) and the expected value for the absence of spatial autocorrelation is close to 0: $E(I) = -(n - 1)^{-1}$ (Cliff & Ord 1973, 1981). When there are too few pairs of sampling locations in a given distance class $d$, and the spatial layout of the data looks non-stationary, the estimated value is unstable and can fall outside the expected bounded range of $-1$ to $+1$ (usually this occurs at the largest distances where there are fewer pairs of distances, as illustrated in Figures 1.4 and 3.4).

When estimating spatial autocorrelation using (3.13), the researcher needs to understand that the estimation of spatial autocorrelation is computed by first

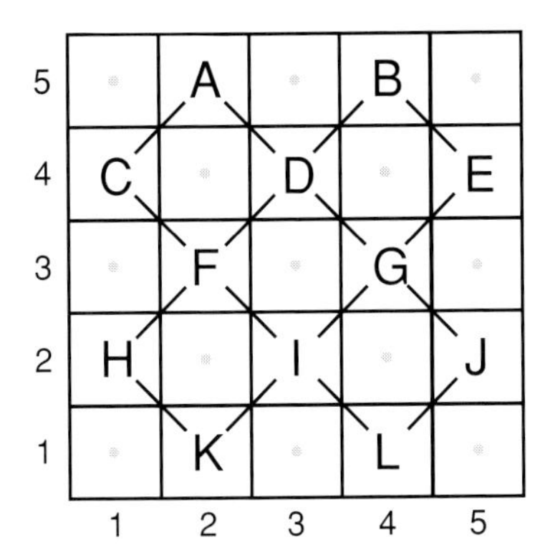


$$I_{(1)} = I_{(A,1)} + I_{(B,1)} + I_{(C,1)} + I_{(D,1)} + I_{(E,1)} + I_{(F,1)} + I_{(G,1)} + I_{(H,1)} + I_{(I,1)} + I_{(J,1)} + I_{(K,1)} + I_{(L,1)}$$

Figure 3.7 Example of how to compute Moran's $I$ for the first distance class $I_{(1)}$ (where the distance interval class corresponds to 1.5 units as in Figure 3.3*a*) among 10 sampling locations A–J. $I_{(1)}$ is the sum of the $I_{(1)}$ values computed at each of the 10 sampling locations (e.g. $I_{(A,1)}$ at location A to $I_{(J,1)}$ at location J); $I_{(A,1)}$ is the sum of deviations of the $Z_A$ with the values of its two first nearest neighbours C and D, respectively, divided by the number of neighbours, here 2; $Z_A$ indicates the quantitative value of the variable $Z$ at the sampling location A (and so on for all the other sampling locations); $\bar{Z}$ is the average value of the variable Z. Consequently, to estimate $I_{(1)}$, it is necessary to compute $I_{(1)}$ at each sampling location first.

summing the covariation between sampling units a given distance apart, *d*, as illustrated in Figure 3.7. Then, this sum of covariations from each sampling location of the entire study at a given distance class *d*, is divided by the actual number of pairs of locations, *W(d)*, in that distance class, *d*. Hence, the spatial autocorrelation coefficient, for a distance class *d*, is the average value of spatial autocorrelation at that distance (in all directions) for the entire study area: a *global average isotropic* estimated value of spatial autocorrelation.

Note that, in some circumstances, the spatial layout of the data looks non-stationary as the estimated value is based on fewer pairs and can reflect an extreme spatial pattern of high or low spatial autocorrelation resulting in estimated values higher than 1 (or lower than −1). Furthermore, by computing the deviation of each value from the average of the variable, the estimation of spatial autocorrelation can be biased when the data are not normally distributed. Indeed, a skewed distribution resulting from the presence of a few outliers (either extremely low or high values) may bias the estimate of the average, which in turn will result in an under- or overestimation of the degree of spatial autocorrelation because the deviations are computed using a biased average. This is why some researchers (mostly geostatisticians) do not like to use Moran's $I$ coefficient. It is favoured, however, by ecologists as it is easier to relate to the notion of correlation and it is more intuitive given that it behaves like a Pearson's correlation coefficient.

To avoid the use of average, another spatial autocorrelation index was proposed by Geary (1954). Geary's coefficient, $c$, measures the difference between values of a variable at nearby locations, so that the degree of spatial autocorrelation is based on differences as a function of distance:

$$c(d) = \left[\frac{1}{2W(d)}\right] \frac{\sum_{\substack{i=1 \\ i \neq j}}^{n} \sum_{\substack{j=1 \\ i \neq j}}^{n} w_{ij}(d)(x_i - x_j)^2}{\frac{1}{n-1}\sqrt{\sum_{i=1}^{n}(x_i - \bar{x})^2}}. \tag{3.14}$$

Geary's $c$ behaves like a distance measure and varies from 0 (highest value of positive autocorrelation) to 2 (strong negative autocorrelation). The expected value, $E(c)$, is 1, indicating the absence of spatial autocorrelation. As for Moran's $I$, the estimated values of Geary's $c$ based on too few pairs of sampling locations will result in strange values, here values greater than 2. Also, the estimated values of spatial autocorrelation with Geary's $c$ will be biased in the presence of skewed data (as for Moran's $I$, but for another reason) because the differences between adjacent locations are squared (see Eqn (3.14)). The squared difference between an outlier and other values will have more effect on the index and may distort the estimation of spatial autocorrelation.

The significance of each individual value of spatial autocorrelation estimated at a given distance class can be tested using either a randomization procedure or a normal distribution approximation test. Mathematical details about the respective variance equations used to test the significance of these coefficients can be found in Cliff & Ord (1973, 1981). The significance testing of the coefficients estimated at several distance classes is, however, problematic due to the lack of independence in the data. Indeed, the same data are used to estimate the coefficient values at different distance classes. This problem is inherent to all multiple testing analyses such as the quadrat variance methods presented in Chapter 2. Several procedures have been proposed to address this issue. The most widely used because of its simplicity and robustness (Oden 1984) is the Bonferroni correction that adjusts the probability level, at which to test the significance, by dividing $\alpha$ by the number of distance classes, $k$:

$$\alpha' = \frac{\alpha}{k}. \tag{3.15}$$

For example, when using a probability level of $\alpha = 0.05$ and a number of distance classes of $k = 10$, $\alpha' = 0.005$. Thus in order to consider all the coefficient values as significant, given the number of distance classes (here 10), the probability value of each coefficient needs to be smaller than $\alpha' = 0.005$. The Bonferroni-corrected level value is directly related to the number of distance classes selected, however. As usual, the selection of the number of distance classes is more or less arbitrary. This can affect our ability to detect significant spatially autocorrelated pattern. To diminish this effect, we can use a progressive (sequential) Bonferroni correction (Legendre & Legendre 1998), where the Bonferroni-corrected level is computed for each distance class separately. This is done by using for each distance class the number of tests actually performed up to that distance class. For example, using the same values as above ($\alpha = 0.05$ and $k = 10$), the progressive Bonferroni levels for each distance class are: $\alpha'(d = 1) = 0.05$, $\alpha'(d = 2) = 0.025$ and so on up to $\alpha'(d = 10) = 0.005$ (the Bonferroni corrected level $\alpha'$). Equation (3.15) can therefore be rewritten as:

$$\alpha'(d) = \frac{\alpha}{d}, \tag{3.16}$$

where $d$ is the distance class of interest ($d = 1$ to 10).

Significance testing allows us to determine which coefficient values can be used to interpret the spatial structure of the data. Characterization of the spatial structure includes its intensity (magnitude, degree), its spatial range and its shape (isotropic or not). Determination of the pattern's characteristics can be facilitated by plotting the values of spatial autocorrelation against the distance classes $d$. This plot is called a spatial correlogram (e.g. Figures 3.8 and 3.9), where the significance of each individual coefficient value can be indicated by a specific symbol (solid circles for significant values at $\alpha = 0.05$ and open circles for non-significant ones). Note that when distance $d$ equals zero (i.e. correlation of the variable with itself), the value of Moran's $I$ is 1 and of Geary's $c$ is 0. Given that most ecological data show some degree of positive autocorrelation, at short distances, the values of autocorrelation are generally positive. In fact, most of the time, the strongest value of spatial autocorrelation is within the first distance class. As the strength of the process decreases with distance (e.g. in the case of seed dispersal, most of the seeds are found close to the seed tree), the values of spatial autocorrelation decrease with distance. The way in which this occurs and is displayed by the spatial correlogram can be used to characterize the spatial pattern: thus a trend in the correlogram, from positive through zero to negative spatial autocorrelation with increasing distance, is indicative of a gradient in the data. A plot (Figure 3.8*b*) showing levelling off of values around the expected value of no autocorrelation ($E(I)$) is characteristic of the absence, or non-detection, of spatial pattern. When the values at short distances are positive and then, at a given distance, show fluctuation around the expected value, this can be interpreted as an indication of patchiness (Figure 3.8*c*–*f*). Repeated

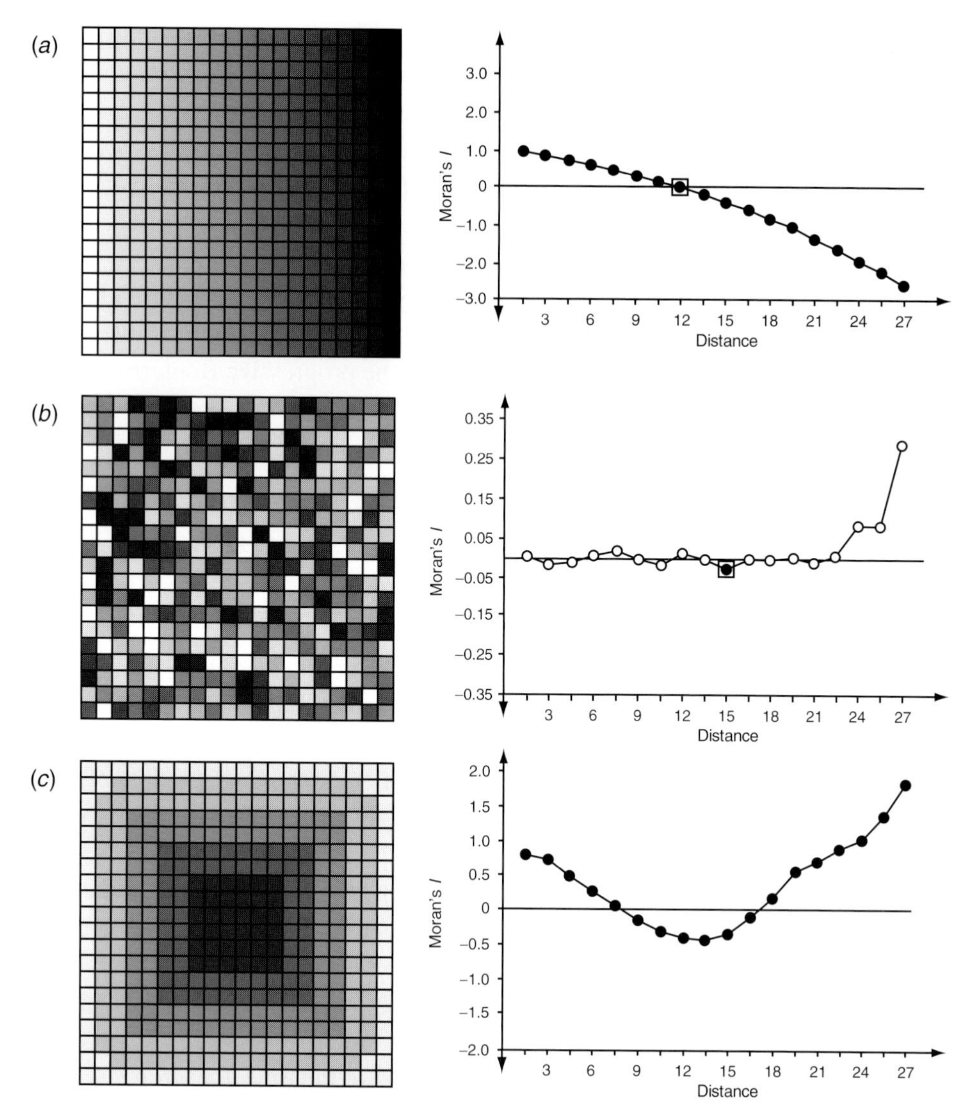


Figure 3.8 Moran's *I* spatial correlograms corresponding to simulated spatial patterns (a 20 × 20 lattice where the values increase from 0 to 10 from white to black). (*a*) Gradient: the correlogram shows the corresponding characteristic trend of significant positive values at short distances to negative ones at large distances. (*b*) Random: the values are oscillating along the zero value (i.e. the absence of significant spatial autocorrelation). (*c*) One big patch: the values are all significant and positive at short and large distances, while negative at intermediate. The spatial range (zone of influence, patch size) is around 7.5 units, a distance at which the sign of the values changes from positive to negative. (*d*) 16 patches: a first change of sign from positive to negative values occurs around 2.0 units, which corresponds to the spatial range of the patches. Then the correlogram repeats this oscillation in decreasing amplitude with distance revealing a repetitive spatial pattern of patches. Note here that the patches all have the same size and distance among them.

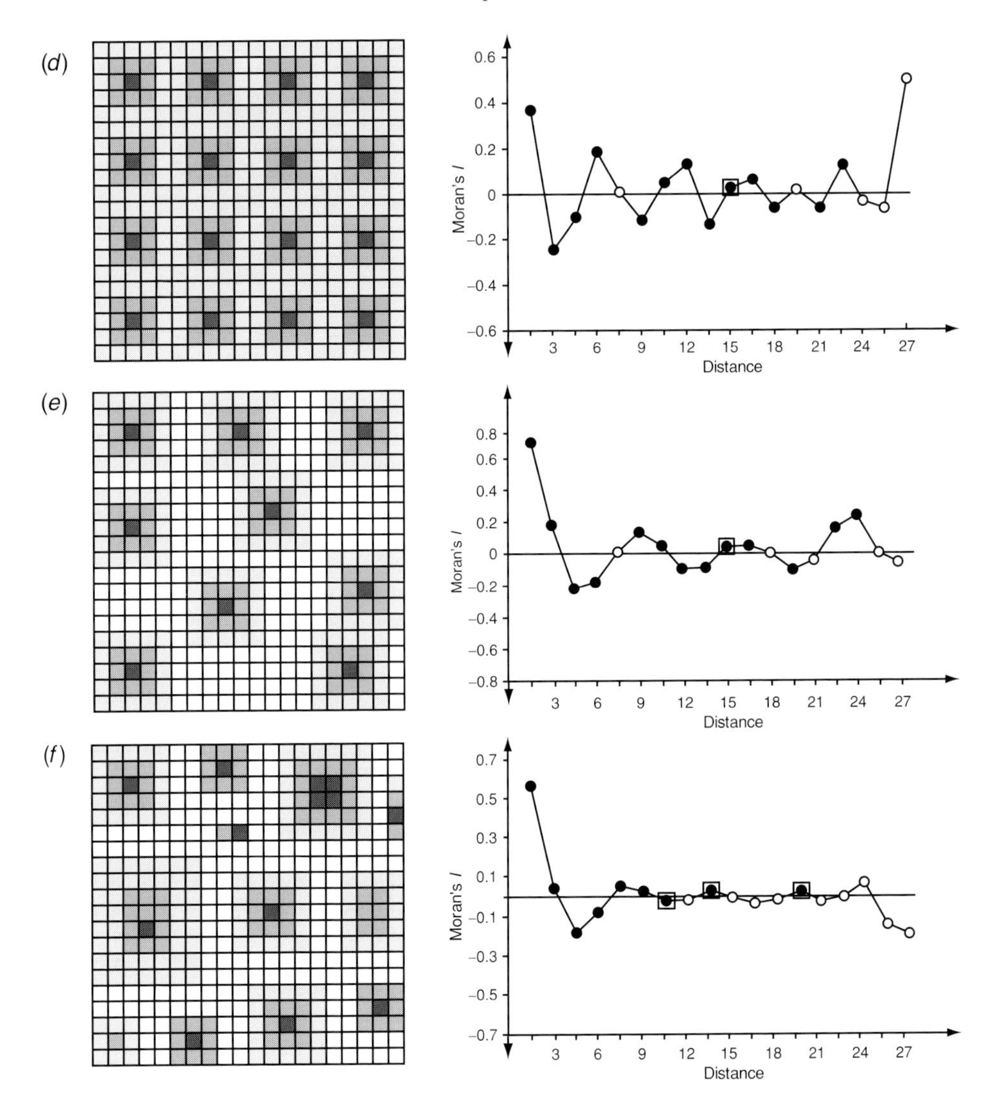


Figure 3.8 (*cont.*) (*e*) 9 patches: as described previously, the sign change of the values indicates that the patch size is around 3.0 units and that there is a repetitive pattern of patches. Note here that the patches have the same size but not the same distance among them. (*f*) 12 patches: the sign change of the values indicates that the patch size is around 2.5 units. Here the repetitive pattern of patches is not detected by the correlogram as both the patch size and distance among the patches vary. Solid circles indicate significant coefficient values at $\alpha = 0.05$; open circles indicate non-significant coefficient values; open squares indicate coefficient values that are non-significant after progressive Bonferroni correction.

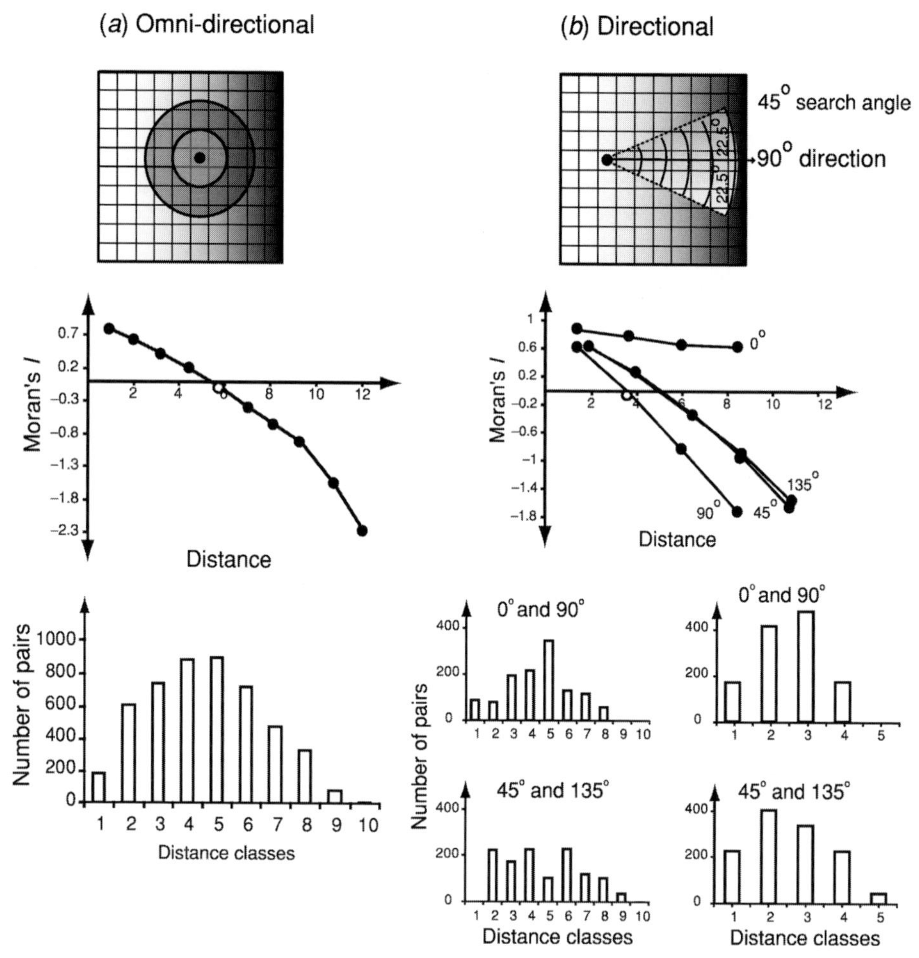


Figure 3.9 Moran's $I$ spatial correlograms. (*a*) Omnidirectional correlogram based on 10 isotropic equidistant classes. The trend in the 10 × 10 lattice data is detected by the characteristic shape of the correlogram (see Figure 3.8*a*). The histogram of the pairs of sampling locations per distance class shows well the edge effects at short and large distance classes (as mentioned in Chapter 1), but, with the exception of the last distance class, there are many more than 20 pairs. (*b*) Directional correlograms computed in four directions (0°, 45°, 90° and 135°) using a search angle of 45° (22.5° on each side of the direction). The directional correlograms are based on five equidistant classes; because with 10 equidistant classes, the number of pairs of locations per distance and angle class lower than in the isotropic case. By using fewer equidistant classes, here five, instead of 10, the number of pairs per distance and angle class increases such that the coefficients based on them are more reliable and comparisons among them can be carried out. Here, the four directional correlograms allow us to detect the presence of an anisotropic spatial pattern because the values of Moran's $I$ do not overlap one another (only the 45° and 135° ones are perfectly overlapping) and the correlograms have different spatial ranges (0°: no spatial range, 45° and 135°: around 5.0 units and 90°: 3.8 units).

alternation of values, from positive to negative, is indicative of patch structure in the study area (Figure 3.8*d*, *e*). The distance at which the value of autocorrelation reaches, or crosses, the expected value is considered the 'spatial range', the 'zone of influence' or the 'patch size' of the spatial pattern under study.

To interpret the shape of the correlogram as an indicator of the type of spatial pattern (intensity, spatial range, shape, etc.), the significance of the entire correlogram needs to be considered. This can be determined using either the Bonferroni correction, where at least one value needs to be smaller than the corrected level $\alpha'$, or the progressive Bonferroni correction. In the former type of correction, it is recommended that we do not consider the values estimated at large distances because these estimations are based on very few pairs of observations. Use of the progressive Bonferroni correction avoids the arbitrary selection of non-interpretable values such as these (see Figure 3.8). Consequently, the interpretation of the spatial pattern is facilitated if non-significant values at the progressive corrected $\alpha'(d)$ levels are not considered.

Another approach that minimizes the effect of multiple tests is to compute a partial spatial correlogram instead of a spatial correlogram. This is the spatial equivalent of the partial time series analysis using Durbin's autoregressive procedure. The mathematical details are briefly presented in Cliff & Ord (1981).

Up to now, we have only considered the case where the values of spatial autocorrelation are estimated solely by comparing the distance between sampling locations. This therefore assumes that the process of interest is isotropic (i.e. the same in all directions, see Chapter 1). This is why these spatial correlograms are often referred to as 'omnidirectional', 'all directions' or simply 'one-dimensional' (Figure 3.9*a*). There are, however, several ecological processes (e.g. species responses to environmental conditions, downwind seed dispersal, etc.) that can generate spatial patterns that vary with direction, such as along a gradient, resulting in elongated patches. These spatial patterns are called anisotropic (i.e. the pattern varies with direction, see Chapter 1). To determine the degree of anisotropy in a spatial pattern, spatial autocorrelation can be estimated by considering both the distances among the sampling locations (using a distance class matrix) and their orientation using an angle class matrix (Oden & Sokal 1986). As for the distance class matrix, we can select the number of angles, directions, in which we want to compute the degree of spatial autocorrelation. The angle class matrix indicates, therefore, which sampling locations are in a given orientation with respect to one another given a specified search angle (this is equivalent to spatial lag). Hence, for example, we could be interested in determining the spatial structure downwind in the 90° direction using a search angle of 45° that is divided equally on both sides of the selected direction (i.e. 22.5° on each side of the selected direction as illustrated in Figure 3.9*b*). Spatial autocorrelation can be computed in several directions and the resulting spatial correlograms are called 'directional' correlograms (Figure 3.9*b*). The detection of spatial anisotropy is then established by comparing whether or not the shape and spatial range of the directional correlograms coincide. For example, in Figure 3.9*b*, while two directional correlograms (45° and 135°) coincide perfectly, both in shape and spatial range, this is not true for all four directional correlograms because the shapes and spatial ranges differ (e.g. the 0° correlogram does not have a spatial range). Note that the estimation of spatial autocorrelation based on both distance and direction is computed with fewer pairs of sampling locations than in the omnidirectional case, and so we recommend using fewer distance classes (see Figure 3.9*b*) and only four angle classes (usually the four directions: 0°, 45°, 90° and 135°).

To avoid having to compute spatial autocorrelation at fixed distances, by means of distance classes, Bjørnstad & Falck (2001) proposed a continuous non-parametric estimator of spatial covariance resulting in a spline correlogram where the significance is based on a bootstrap confidence interval envelope (this is comparable to significance testing for Ripley's $K$ as presented in Chapter 2).

### 3.3.2 Variography

Parallel to the development of spatial statistics by human geographers, mining engineers (Matheron 1970) built a family of spatial statistics, known as geostatistics. Geostatistics use the spatial structure estimated, from sampled data by computing the spatial variance (as presented in this section) to predict values at unsampled locations by modelling the spatial structure using kriging techniques (as described in Section 3.5). In this section, we provide a brief overview of the notions related to the estimation of spatial variance, referred to as variography. More information about the wide range of geostatistics techniques can be found in numerous textbooks (among others, Journel & Huijbregts 1978; Isaaks & Srivastava 1989; Haining 1990, 2003; Cressie 1993; Goovaerts 1997; Chilès & Delfiner 1999; Webster & Oliver 2001).

Geostatistics are based on 'regionalized variable theory', which assumes that the value of a variable $z$ at a given location $x$ is a particular realization of a random function $Z(x)$. This value, $z(x)$, is composed of three components:

$$z(x) = m(x) + \varepsilon(x) + \varepsilon', \tag{3.17}$$

where $m(x)$ is the deterministic structural function of the variable at location $x$; $\varepsilon(x)$ is the spatially dependent residual from $m(x)$, i.e. the spatial variance component; and $\varepsilon'$ is the spatially independent normally distributed residual component. When the data are stationary, $m(x)$ is the average value of the variable within the study area. Estimation of the spatial structure of a variable from sampled data is based on the assumption that the regionalized variable respects the intrinsic hypothesis of stationarity. That is, the expected difference between the values at two sampling locations at a given $h$ distance apart is 0, and that the variance of this difference varies only according to $h$. Given that it is rare to have a process that is stationary over the entire study area, the spatial variance can be estimated only when quasi-stationarity (also called weak stationarity) prevails. Weak stationarity implies that the first-order moment, $E(Z(x))$, and second-order moment, $\mathrm{Var}(Z(x))$, are assumed to be stationary only within the range of a relatively small neighbourhood (i.e. within a moving window).

The notation used in geostatistics differs slightly, however, from the notation used in spatial statistics. For example, in geostatistics, the spatial lag parameter is '$h$' (when it in bold, **h** denotes a vector of both distance and direction) rather than '$d$'. Hence (3.11) for spatial autocorrelation can be rewritten using $h$ instead of $d$:

$$\rho(h) = \frac{C(h)}{\sigma^2}. \tag{3.18}$$

Then, there is a direct relationship between spatial covariance, $C(h)$, and spatial variance, $\gamma(h)$, where the spatial variance is the variance, $\sigma^2$, of the variable minus the autocovariance function:

$$\gamma(h) = \sigma^2 - C(h). \tag{3.19}$$

When the assumption of stationarity is met, spatial autocovariance and spatial variance are mirror images of one another. Similarly, the relationship between spatial autocorrelation and spatial variance is:

$$1 - \rho(h) = \frac{\gamma(h)}{\sigma^2}. \tag{3.20}$$

The spatial variance of a quantitative variable is then estimated using the semi-variance function, $\hat{\gamma}(h)$:

$$\hat{\gamma}(h) = \frac{1}{2n(h)} \sum_{i=1}^{n} [z(x_i) - z(x_i + h)]^2, \tag{3.21}$$

where $z$ is the value of the variable $x$ at the sampling location $i$, and $n(h)$ is the number of pairs of sampling locations located at distance $h$ from one another. Some geostatistics books use **h** instead of $h$ to indicate the more general case where the spatial variance is computed according to both spatial lag, $h$, and direction, $\theta$. Since the summation is from 1 to $n$ (the number of sampling locations), each pair of sampling locations is considered twice in the calculation. This is why the value

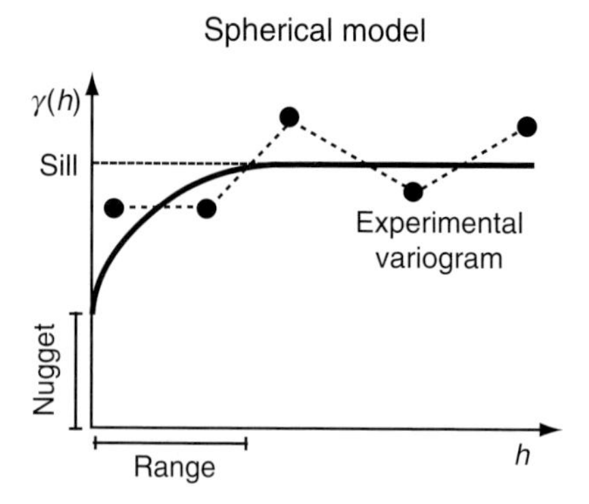


Figure 3.10 Variogram. Experimental variogram (solid circles linked by the dashed line) and corresponding theoretical spherical variogram (solid line) and its parameters: the range, the nugget and the sill ($h$ indicates the spatial lag; $\gamma(h)$ is the semi-variance).

is divided by 2 and the function is called semi-variance. Technically, the plot of the semi-variance values against the spatial lag $h$ is a semi-variogram. However, for simplicity, we will refer to it as only a 'variogram' as the majority of the literature on the subject does. When a variogram is computed from sampled data, it is called an experimental variogram (also called a sample or observed variogram); when it is modelled to fit the experimental variogram, it is called a theoretical variogram or a variogram model.

The equation for the semi-variance function (3.21) is quite comparable to the one of Geary's $c$ (3.14) except that it lacks the division by the standard deviation in the denominator, which standardizes the spatial autocorrelation value. Hence, the semi-variance function is in the same units as the data, and is not bounded as are Moran's $I$ ($-1$ to 1) or Geary's $c$ (0 to 2) values. At short distance lags, the values of semi-variance are also small (close to zero) indicating that the spatial structure is at its strongest intensity. As the distance lags increase, the semi-variance values rise to level off at a plateau called the sill. Three key parameters are estimated from an experimental variogram to fit a theoretical variogram (Figure 3.10): the nugget effect, $c_0$, the spatial range, $a$, and the sill, $c_1$. The nugget is the intercept at the origin that is greater than zero. Theoretically, at $h = 0$, the semi-variance is also equal to 0. However, based on the shape of the experimental variogram, it can be unrealistic sometimes to force the theoretical variogram to go through 0. The nugget parameter is therefore used to account for the observed variability at short distances due to local random effects or measurement errors (e.g. accuracy of measurements, inappropriate sampling unit size, etc.). The spatial range indicates the distance up to which the spatial structure varies. In other words, the range indicates the maximal distance at which the variable is spatially autocorrelated. Beyond the range, the distance among sampling locations does not affect the spatial structure of the data and the semi-variance values level off, forming the sill. There are cases where the experimental variogram does not have a sill. This also implies that the range is undetermined (Figure 3.11*a*). This can happen when the extent of the study area is smaller than the spatial pattern, i.e. the spatial range, of the variable of interest.

As Geary's $c$, the semi-variance is affected by outlier values from skewed data because their estimation is based on the squared differences among the values of a given variable. This is why geostatisticians recommend that the data should be:

(1) plotted at different spatial lags to identify outliers using $h$ scattergrams (Rossi *et al.* 1992); and
(2) transformed to reduce the degree of skewness in the data.

Furthermore, as for spatial correlograms, the experimental variogram is a plot of semi-variance versus spatial lag $h$, where the number of pairs of distances decreases as distance increases. It is a rule of thumb to interpret only the first two-thirds of a variogram in order to determine the spatial structure of the sampled data.

As mentioned above, **h** is used to denote a vector including both spatial lag and direction. Anisotropic pattern can therefore be determined by computing directional variograms. In geostatistics, three types of anisotropy are distinguished:

(1) when the ranges differ but the sills are the same (geometric anisotropy),
(2) when the ranges are the same but the sills differ (zonal anisotropy), and
(3) when both the ranges and the sills differ.

Geometric anisotropy can be modelled, using a linear relationship, to predict the values of a variable at unsampled locations. It is not as straightforward for the two other types of anisotropy where more than one variogram model is needed in order to model spatial structure. This can be achieved by determining different models as a function of distance, as illustrated in Figure 3.11*b*, where up to the range a linear model is used and beyond the range, the sill is used.

Ecological data are rarely stationary and often show some trend or gradient called drift in geostatistics. The presence of drift in the data implies that $m(x)$ is not a good estimate of the average value of the variable over the entire study area. Drift can be detected by comparing the values of $m(x)$ computed using only the values at the sampling locations at the beginning of vector **h** (i.e. at $z(x_i)$, called 'head' locations) with those computed with sampling locations at the end of vector **h** (i.e. at $z(x_i + \mathbf{h})$, called 'tail' locations). When these average values differ, the covariance is said to be non-ergodic (where ergodic covariance implies that $m = m_{-\mathrm{h(head)}} = m_{+\mathrm{h(tail)}}$; Isaaks & Srivastava 1989; Rossi *et al.* 1992). In the presence of drift, the use of generalized random intrinsic functions of order $k$ is recommended to estimate the spatial variance rather than the experimental

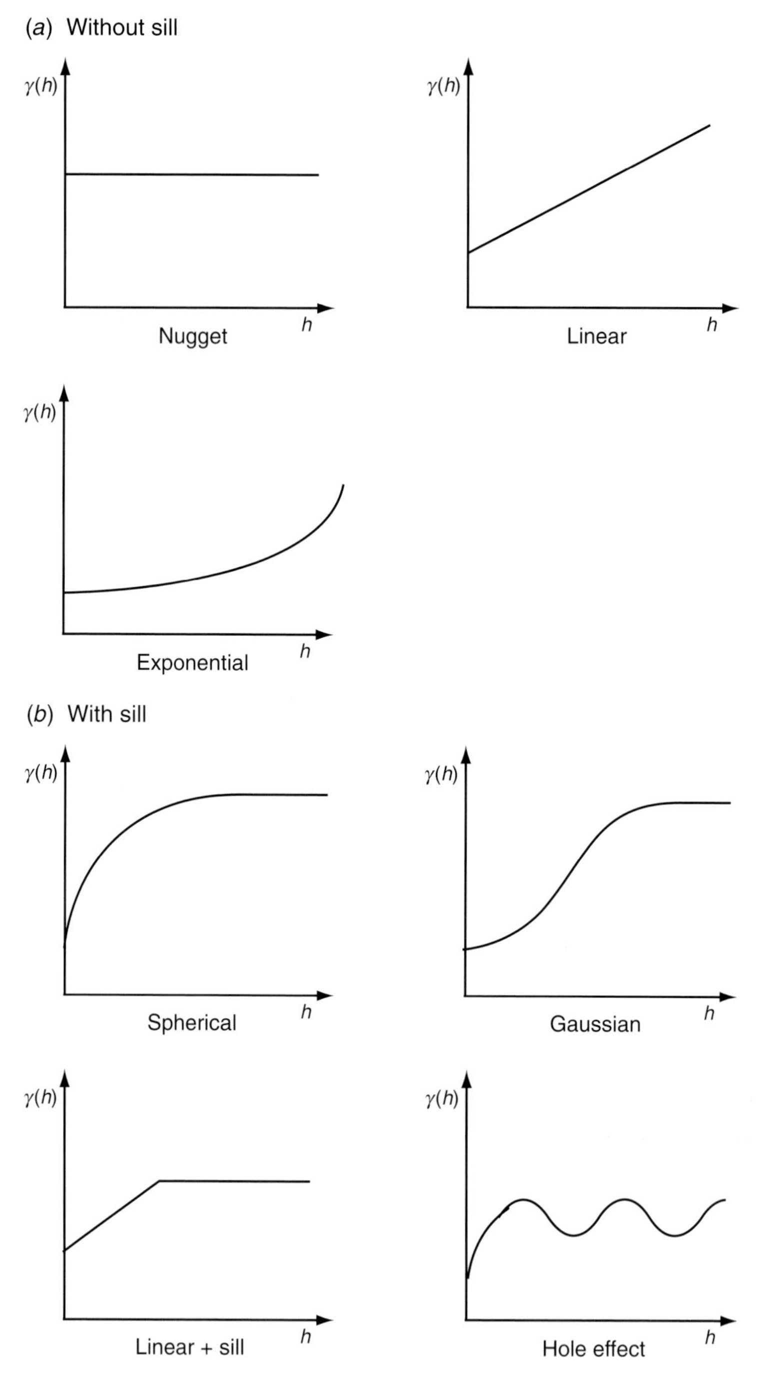


Figure 3.11 Theoretical variograms: (*a*) without a sill: showing the pure nugget effect, linear and exponential models; (*b*) with a sill: showing spherical, Gaussian and nested models (linear and a sill), and the hole effect. See text for equations describing these conditions.

variogram (see more advanced geostatistical textbooks for mathematical details, Journel & Huijbregts 1978; Goovaerts 1997; Chilès & Delfiner 1999; Webster & Oliver 2001).

It is often the case that ecological data are binary (e.g. presence : absence) or that threshold values are used to indicate state responses that are of interest. Spatial variance of such qualitative data can be estimated using indicator functions, where $k$ is the threshold value at which continuous data are considered as the response of interest (e.g. presence) and are set to 1 or otherwise are set to 0. Continuous variables can therefore be transformed into binary indicator variables $I(x; z)$, where the average of the indicator variable is the proportion of the variable in the study area (Rossi *et al.* 1992). The indicator function is computed as in (3.21):

$$\hat{\gamma}_{I(x;z)}(h) = \frac{1}{2n(h)} \sum_{i=1}^{n} [I(x_i; z) - I(x_i; z + h)]^2. \tag{3.22}$$

The indicator function can be extended to several thresholds, $k$ values, corresponding to multiple-state variables, or to explore the sensitivity of the selected thresholds (Webster & Oliver 2001).

Determination of the experimental variogram is only the first step in a series leading to the prediction of values at unsampled locations using kriging (i.e. a family of spatial interpolation techniques as presented in Section 3.5). The next step involves fitting the best (derived analytically) theoretical variogram to the experimental variogram. The most commonly used unbounded theoretical variograms (without a sill) are (Figure 3.11*a*):

- the nugget model:

$$\gamma(h) = c_0, \tag{3.23}$$

- the linear model:

$$\gamma(h) = c_0 + bh, \tag{3.24}$$

  where $b$ is the parameter of the slope,
- the exponential model:

$$\gamma(h) = c_0 + c_1 \left[ 1 - \exp\left( \frac{h}{a} \right) \right], \tag{3.25}$$

and bounded variograms (with a sill, Figure 3.11*b*):

- the spherical model:

$$\begin{aligned} \gamma(h) &= c_0 + c_1 \left[ \frac{3h}{2a} - \frac{1}{2} \left( \frac{h}{a} \right)^3 \right], \quad \text{for } 0 < h < a, \\ \gamma(h) &= c_0 + c_1, \quad \text{for } h \geq a, \end{aligned} \tag{3.26}$$

- the Gaussian model:

$$\gamma(h) = c_0 + c_1 \left[1 - \exp\left(-3\frac{h^2}{a^2}\right)\right], \tag{3.27}$$

- the linear model with a sill:

$$\begin{aligned} \gamma(h) &= c_0 + bh \quad \text{for } 0 < h < a, \\ \gamma(h) &= c_0 + c_1 \quad \text{for } h \geq a, \end{aligned} \tag{3.28}$$

- the hole-effect model:

$$\gamma(h) = c_0 + c_1 \left[1 - \frac{a \sin\left(\frac{h}{a}\right)}{h}\right]. \tag{3.29}$$

Given that the strongest spatial variance occurs, usually, at short distances (i.e. up to the spatial range), it is important (as much as possible) to fit the theoretical variogram model at short distances. Adjustments of the theoretical variogram, and its parameters, are needed for the next step, which is spatial interpolation of the data using kriging techniques (see Section 3.5). Inappropriate parameter estimations can result in very different predicted values. For example, when the spatial range is too small, the spatial structure will be modelled for only small distances generating small patches, and when the spatial range is too large, the resulting spatial structure will be overly smoothed.

Selection of the best variogram model, and its parameters, used to be determined by eye. The reliability of these parameters can be evaluated using cross-validation techniques that require withholding some sampling locations and then comparing the variogram models' performance in predicting these values using kriging. Nowadays, generalized least-squares and maximum likelihood (e.g. Akaike information criteria, AIC) can be used to select the parameters making the process more objective (Cressie 1993; Goovaerts 1997; Chilès & Delfiner 1999).

To conclude this brief overview of variography, it is important to stress that variograms aim to determine the parameters used to model spatial pattern. Hence, to keep the same measurement units as the data, semi-variance values are not standardized. Furthermore, while no significance tests have been developed for semi-variance function values, randomization and bootstrap procedures can be used to assess their significance. Finally, there is a much wider range of geostatistics techniques available than those presented here but further details are beyond the scope of this book. We refer the reader to more advanced textbooks for greater detail (see among others, Deutsch & Journel 1992; Cressie 1993; Goovaerts 1997; Chilès & Delfiner 1999).

### 3.3.3 Fractal dimension

When dealing with landscape features and entities that can be delineated, such as rivers, islands, patches, etc., we can be interested in characterizing and quantifying their spatial structure. The most fundamental property of these landscape entities is their geometrical dimension. In fact, objects can lie either in Euclidean space, with an integer number of dimensions (0 for points, 1 for lines, 2 for surfaces and 3 for volumes), or in a fractal dimension (fractal for fractional), introducing a new way of characterizing the occupancy of space by objects (between 0 and 1 for clusters of points, 1 and 2 for curves, 2 and 3 for surfaces and 3 and 4 for volumes). The fractal dimension is therefore a mathematical coefficient that measure the fractal geometry (i.e. non-integer dimension) of objects in physical space (Mandelbrot 1983). In theory, fractal objects have several interesting properties such as having the same fractal dimension at all scales of observation: the property of self-similarity. Furthermore, the fractal dimension is independent of units of measurement (e.g. metres, kilometres), allowing comparisons among objects, organisms, community structures and so on. The fractal dimension value can also be seen as a measure of the degree of occupancy of a low-dimensional object in a higher-dimensional space. The concept of fractal dimension has since been extended to include quantification of the spatial structure of continuous data using statistical fractals. Ecologists have been exploring the potential usefulness of the concept of fractal dimension for spatial characterization of landscape patches (Gustafson 1998), for modelling of an organism's geometric growth such as exhibited by plants and shells, as well as for modelling of landscape spatial structures (Milne 1992; Palmer 1992; Kenkel & Walker 1993; Hargrove *et al.* 2002).

One of the simplest ways of computing fractal dimension is to measure the length of the object using several different sizes of a divider (e.g. a ruler). Hence, the geometric fractal dimension of objects, $D$, can be estimated using a simple relationship between the number of units, $N$, needed to measure either the length (divider dimension, Figure 3.12*a*) or area occupied (grid dimension, Figure 3.12*b*), and the size of the 'divider' or the 'box', $n$, used. Using the box method:

$$N(r) \propto r^{-D}, \tag{3.30}$$

where the slope of the log of $N(r)$ against the log of $r$, is an estimate of $-D$, the fractal dimension.

Similarly, the statistical fractal dimension of ecological patterns from continuous data can be computed as the slope of a log variogram (assuming that the variogram is isotropic, linear and without a sill):

$$2\hat{\gamma}(h) = h^{4-2D}, \tag{3.31}$$

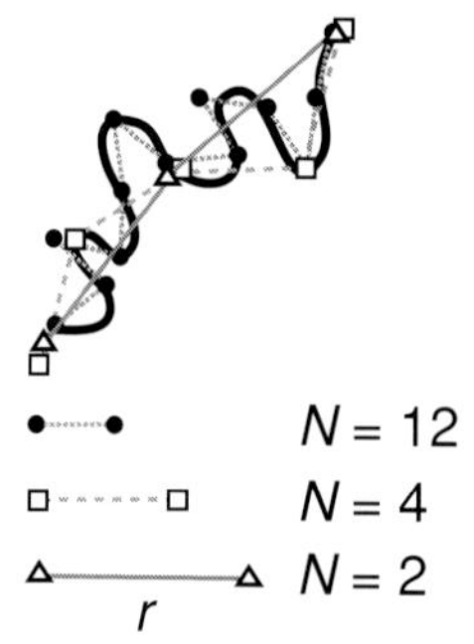




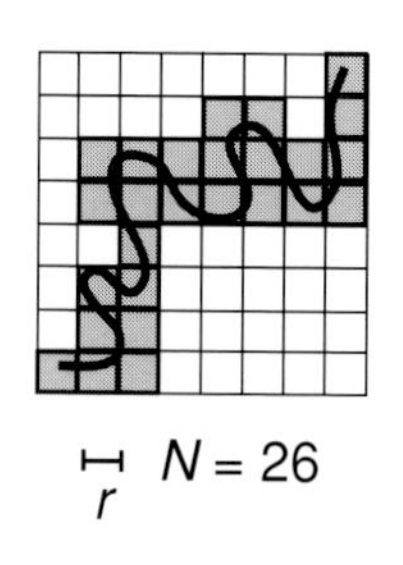


Figure 3.12 Fractal dimension templates. (*a*) Divider method: three sizes of sticks, *r*, and the respective number of sticks needed to measure the shaped line. (*b*) Grid method: the grey shaded boxes indicate those that include the shaped line and their number.

where the slope is $(4-2D)$. Consequently, there is a direct relationship between both spatial variance (variogram) and fractal dimension as defined in (3.31), as well as between spatial autocorrelation (correlogram) and fractal dimension as illustrated in Figure 3.13. This figure shows, by comparing the fractal dimensions estimated from a variogram based on eight spatial lags (none are linear) with those based on a variogram with only four spatial lags (more linear), how fractal dimension values can be distorted when the variogram is not linear. We recommend plotting the variogram before estimating the fractal dimension with continuous data to evaluate the relevance of using this method.

While examining the geometric fractal of objects, what is perhaps most interesting is to find scales of observation at which the fractal dimension values change (Burrough 1981; Allen & Hoekstra 1992, Figure 3.14). Indeed, this is because at these critical scales, ecological processes that act upon the variable of interest may also be changing (Dungan *et al.* 2002). Thus fractal dimensions may be used to distinguish the range of influence of a process, implying different levels of

Figure 3.13 Moran's *I* correlogram: (*a*) omnidirectional and (*b*) directional. Solid circles indicate significant values at 0.05 (open circles indicate non-significant values). The experimental variogram (solid circles) and fractal dimension (based on a linear variogram with eight semi-variance values (solid and dashed lines) or four values (solid line only) record sassafras abundance data ($n = 84$ sampling locations from a woodlot sampled on Long Island, NY (Fortin 1992)). Key features to notice are: (1) Moran's *I* correlograms show significant values but the variogram does not; (2) the shapes of the correlograms and variograms are more or less comparable; (3) by computing the fractal dimensions using a variogram, it is important first to determine whether or not the variogram is linear. Here most of the variograms are non-linear when eight semi-variance values are considered, but are more linear when only four values are used.

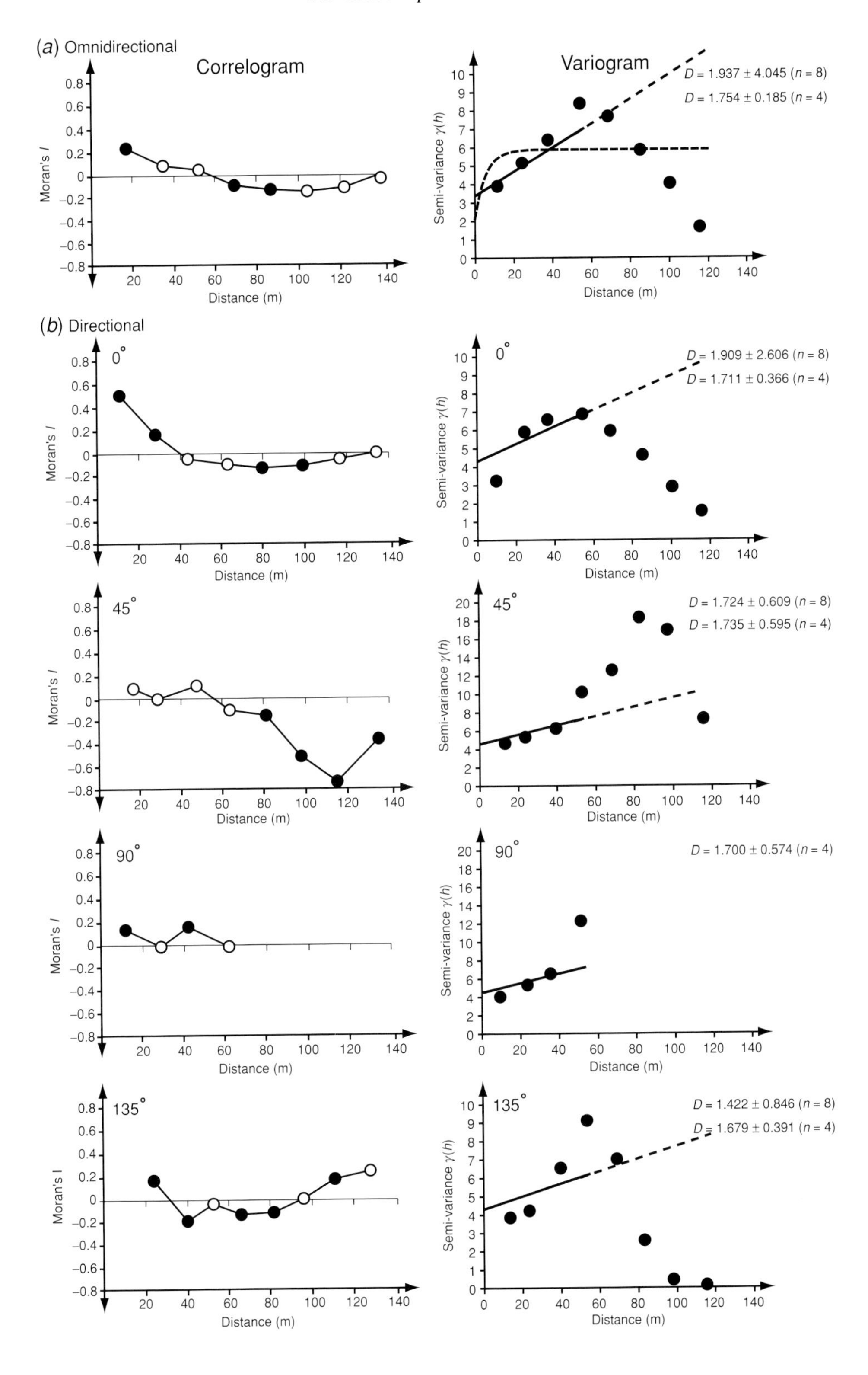
(a) Omnidirectional
Correlogram
Moran's I
Distance (m)
Variogram
Semi-variance γ(h)
D = 1.937 ± 4.045 (n = 8)
D = 1.754 ± 0.185 (n = 4)
(b) Directional
0°
D = 1.909 ± 2.606 (n = 8)
D = 1.711 ± 0.366 (n = 4)
45°
D = 1.724 ± 0.609 (n = 8)
D = 1.735 ± 0.595 (n = 4)
90°
D = 1.700 ± 0.574 (n = 4)
135°
D = 1.422 ± 0.846 (n = 8)
D = 1.679 ± 0.391 (n = 4)

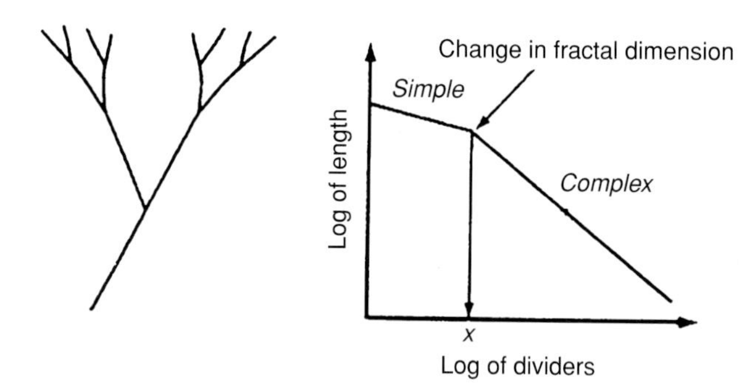


Figure 3.14 Change in fractal dimension values as divider size increases. In the left-hand panel, the shape of the aquatic network is smooth (no branching) at short distances as indicated by the middle slope (i.e. low fractal dimension), and complex (branching) at intermediate distances as indicated by the steeper slope (i.e. higher fractal dimension). The scale at which the change in fractal dimension (i.e. pattern) occurs can be used as an indicator of change in ecological processes.

organization of organisms in space; furthermore, in moving from one scale to another, we may be sure that we are changing between levels of organization. Therefore a shift in the fractal dimension could be interpreted as a change in ecological processes, reflected by a re-organization in the structural scale of the parameters measured from the data.

### *3.3.4 Sampling design effects on the estimation of spatial pattern*

As presented in Chapter 1, regardless of the type of spatial analysis used, the sampling design (including the determination of three components: sample size, sampling unit size and their layout in space) determines the power to detect significant signal in the sample data. Furthermore, determination of one of the components of the sampling design affects the other two (Dungan *et al.* 2002). The first component to be determined is usually the sample size as it is directly related to our sampling effort or ability (usually limited by cost and time). In time series analysis, it is recommended that at least 100 data points be sampled in order to detect temporal patterns. In geostatistics, the recording of at least 100 sampling locations is also recommended. In ecology, it is quite rare to have such large sample sizes when field work is involved. In fact, when a spatial signal is strong, fewer sampling locations, say around 50, can detect it. Moreover, if the spacing among sampling units is such that it is within the spatial range of the pattern, then even fewer sampling locations, say between 20 and 30, may be able to capture the essence of the spatial pattern. We stress that the spatial pattern needs to be clear in order to detect it with so few sampling locations (Fortin 1999a). Indeed, this does not hold when the pattern is weak to start with, as illustrated by Figure 3.15: the spatial pattern is detected

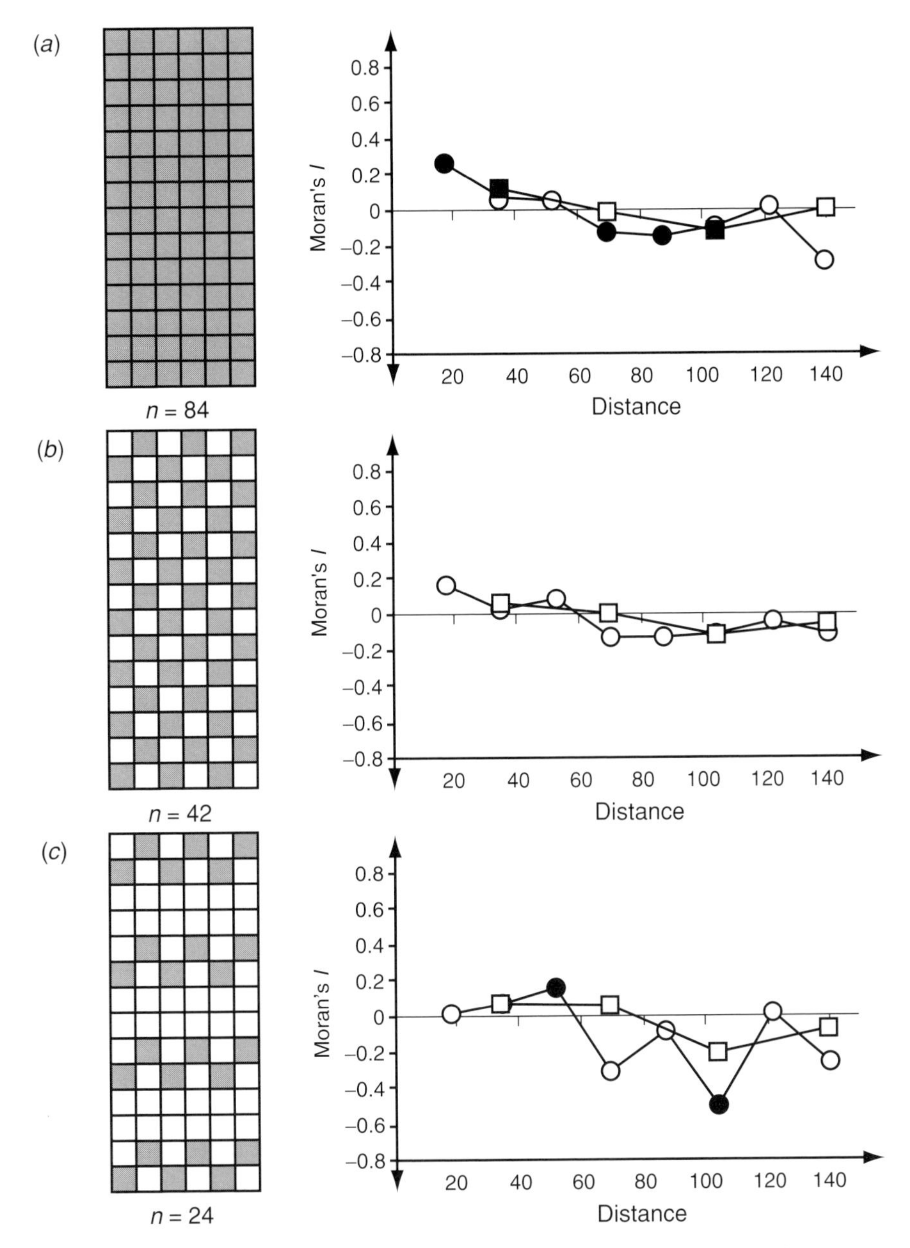


Figure 3.15 Sample size effect on the estimation of spatial autocorrelation. Moran's $I$ omnidirectional correlograms of sassafras abundance data using (*a*) 84 sampling locations (indicated in grey), (*b*) 42 sampling locations (indicated in grey) and (*c*) 24 sampling locations (indicated in grey). Solid symbols indicate significant values at 0.05; open symbols indicate non-significant values. Correlograms were computed using eight equidistant classes (distance interval = 17.4 m; circles) and four equidistant classes (distance interval = 34.8 m; squares). Significant values only occur when 84 sampling locations with eight equidistant classes are used for the analysis. All other subsampling combinations ($n = 42$ or $n = 24$) and numbers of equidistant classes do not detect the patchy structure of the sassafras data.

only when 84 sampling locations are used and not when the sample size is reduced to 42 or 24 (while keeping the sampling unit size constant). This holds true, even when different equidistant classes are used (here 17.4 and 34.8 m; Figure 3.15). Therefore, failure to detect significant spatial pattern does not necessarily imply that there is no spatial pattern and a revised sampling design may be able to detect it (Fortin *et al.* 1989, Fortin 1999a).

There are two ways to revise the sampling design: (1) by changing the spacing among sampling units and (2) by changing the size of the sampling unit. The spacing of the sample locations can be arranged to maximize, as much as possible given the sample size, the number of locations within the spatial range of the variable (see Figure 1.12). Spatial layouts where there is more than one spatial lag seem to be more efficient at capturing spatial pattern with relatively small sample sizes (Fortin *et al.* 1989; Webster & Oliver 2001). Interestingly, a random sampling design is able to detect significant spatial pattern because there is a wide range of spatial distances among the sampling locations (Fortin *et al.* 1989).

The other aspect to consider is whether the choice of sampling unit size is the most appropriate one. For example, when spatial autocorrelation at the first distance class is not significant (high nugget effect in the variogram) or negative, this can indicate that the sampling unit size is larger than the spatial range of the spatial pattern, or that it includes more than one spatial pattern (Fortin 1999a). In such circumstances, a smaller sampling unit size should be used. The question is: how small? When no prior knowledge is available about a system, we recommend using the smallest sampling unit size that is large enough to include more than one object of interest (as presented in Chapter 1). Then, with this relatively fine spatial resolution, the data can be aggregated into a coarser resolution (Figure 3.16; Qi & Wu 1996; Fortin 1999a). By doing so, the spatial domain of a pattern (and the underlying process(es)) can be identified, as well as the size (distance) at which it changes. This is comparable to searching for a change in fractal dimension value (Figure 3.14). Indeed, for a range of sampling unit sizes, estimation of spatial autocorrelation will be comparable up to a given size at which the intensity of spatial autocorrelation will change (Fortin 1999a). For example, in Figure 3.16, as the sampling unit size increases from $5 \times 5$ to $10 \times 10$ m, the autocorrelation value in the first distance class increases from 0.113 (significant) to 0.250 (significant); then at a sampling unit size of $15 \times 15$ m, it reaches its maximum value of 0.303 (significant), and begins to decrease thereafter to 0.154 with a $20 \times 20$ m sampling unit (not significant). Although the values in the first distance class using $5 \times 5$ and $20 \times 20$ m units are quite alike, the reasons for obtaining these values are different: the $5 \times 5$ m unit is too small and includes only one or two tree stems, so several sample units are empty, resulting in a weak spatial autocorrelation structure because we are almost at a random spacing

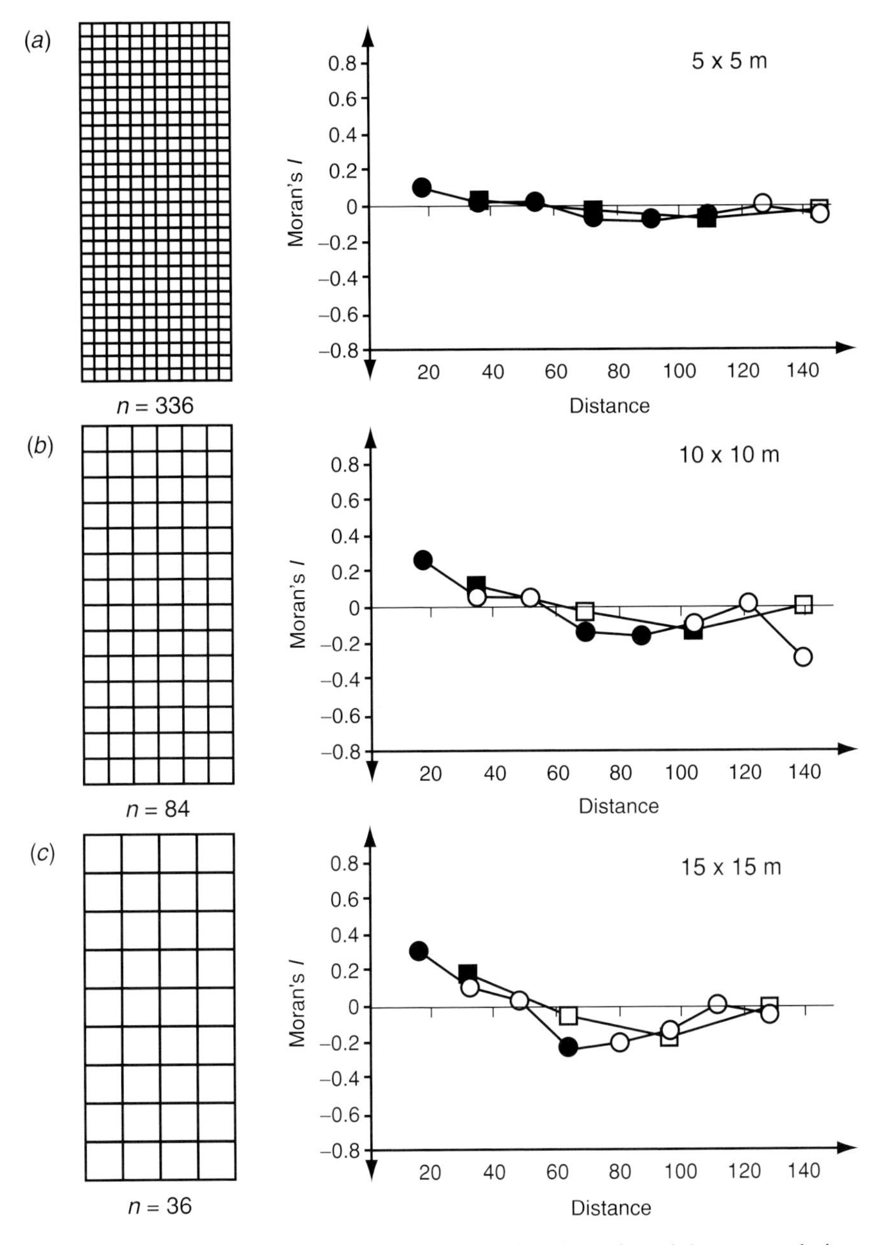


Figure 3.16 Sampling unit size effect on the estimation of spatial autocorrelation. Moran's $I$ omnidirectional correlograms of sassafras abundance data using (*a*) a $5 \times 5$ m sampling unit size ($n = 336$), (*b*) a $10 \times 10$ m sampling unit size ($n = 84$), (*c*) a $15 \times 15$ m sampling unit size ($n = 36$) and (*d*) a $20 \times 20$ m sampling unit size ($n = 21$). Solid symbols indicate significant values at 0.05; open symbols indicate non-significant values. Correlograms were computed using eight (circles) and four (squares) equidistant classes. Overall, with two equidistant intervals, the estimation of spatial autocorrelation increases (from 0.113 to 0.250 to 0.303) with increasing sample unit size (5, 10 and 15 m, respectively) but decreases to 0.154 at the larger sampling unit size (20 m). Comparison among these correlograms can be used to select the optimal sampling unit size needed to analyse the spatial structure. Here, the sampling unit of 15 m estimated the highest values of spatial autocorrelation.

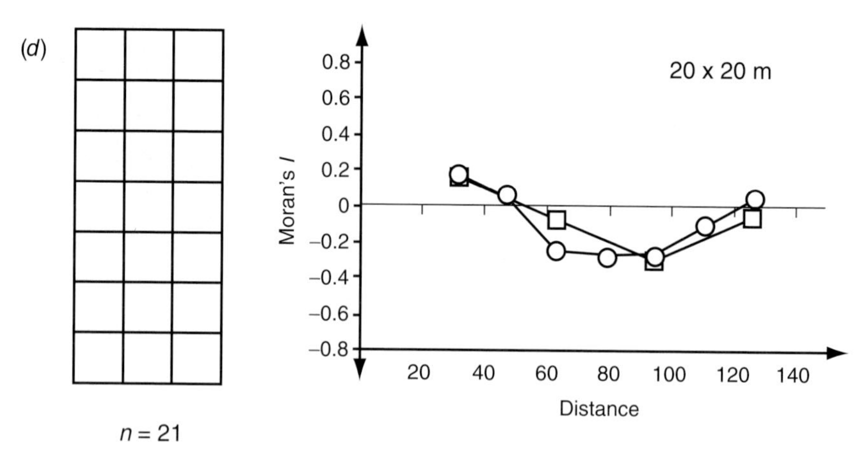


Figure 3.16 (*cont.*)

level; with the 20 × 20 m unit, the sampling unit size is large enough to include tree stems but may also include more than one process. It may also be larger than the spatial range, such that there is random noise as well. By comparing the results from Figure 3.15 with those of Figure 3.16, using more or less comparable sample sizes of 42 and 36, respectively, we see that the spatial autocorrelation estimated was strongest with 36 sampling locations. Thus, this sampling unit size was more appropriate because it could contain more than one tree. For comparable sample sizes, 24 and 21, respectively, the spatial pattern was too weak to be detected regardless of the fact that with 21 sampling units, the entire study area was sampled.

The fact that values of spatial autocorrelation, or spatial variance, change with the sample unit size is a known issue in spatial analysis (Wu & Qi 2000; Dungan *et al.* 2002) as well as in plant ecology (Greig-Smith 1961, Chapter 2). It is referred to as the 'modifiable area unit problem' (MAUP; Openshaw 1984) or as the 'change of support' in geostatistics (Journel & Huijbregts 1978). This is related to the 'ecological fallacy' problem that inferences about individual objects can be made based on aggregated information (e.g. using data recorded at the sampling unit level to make conclusions about individual tree stems while no information about individual trees was in fact recorded). These are inherent problems with any kind of spatial data and are an active area of research in geography, especially with the advent of geographical information systems (GIS; O'Sullivan & Unwin 2003) and satellite imagery (Woodcock & Strahler 1987; Arbia *et al.* 1996; Bradshaw & Fortin 2000; Dungan 2001), in geostatistics (Cressie 1996; Bellehumeur & Legendre 1997, Bellehumeur *et al.* 1997), as well as in landscape ecology (Wu *et al.* 2004).

### 3.3.5 Spatial relationship between two variables

So far, we have presented only spatial methods that quantify the spatial structure of one variable at a time. There are obvious cases where it is of interest to analyse the spatial interaction between two variables as presented in Chapter 2 (e.g. bivariate Ripley's $K$) or to predict the spatial pattern of one variable based on the known correlation between two variables (e.g. co-kriging, as presented in Section 3.5). To do so, spatial autocorrelation coefficients can be modified to estimate cross correlation between two variables:

$$I_{xy}(d) = \left[\frac{1}{W(d)}\right] \frac{\sum_{\substack{i=1\\ i\neq j}}^{n} \sum_{\substack{j=1\\ j\neq i}}^{n} w_{ij}(d)(x_i - \bar{x})(y_j - \bar{y})}{\left[\frac{1}{n}\sqrt{\sum_{i=1}^{n}(x_i - \bar{x})^2}\right]\left[\frac{1}{n}\sqrt{\sum_{i=1}^{n}(y_i - \bar{y})^2}\right]}. \tag{3.32}$$

Similarly, the semi-variance function can be modified to estimate the cross-covariance function:

$$\hat{\gamma}_{uv}(\mathbf{h}) = \frac{1}{2n(\mathbf{h})} \sum_{i=1}^{n} [z_u(x_i) - z_u(x_i + \mathbf{h})][z_v(x_i) - z_v(x_i + \mathbf{h})]. \tag{3.33}$$

While the mathematics is quite straightforward, very few software packages offer the option of computing cross correlation. The cross-variance function, being a step in co-kriging analysis, is more widely available in geostatistical packages.

### 3.3.6 Spatial relationships among several variables

In ecology, many studies are interested in the spatial structure of species assemblages (i.e. multivariate data). In the same way that the spatial autocorrelation coefficients and semi-variance functions were extended to deal with bivariate data (Section 3.3.5), they can be extended to analyse the spatial pattern of multivariate data (Wartenberg 1985; Wackernagel 2003). When we are interested in summarizing the spatial pattern of species assemblages or other multivariate data, with one number, the spatial autocorrelation coefficients, and semi-variance functions, can be written as a cross product between two matrices (Getis 1991). This approach was first proposed by Mantel (1967) to quantify the spatial ($x_i$) and temporal ($t_i$) relationships of multivariate data sampled at the same locations (Mantel 1967). It was extended to study the linear relationship between two symmetric matrices:

$$Z_M = \sum_{\substack{i=1\\ i\neq j}}^{n} \sum_{\substack{j=1\\ j\neq i}}^{n} w_{ij} x_{ij}, \tag{3.34}$$

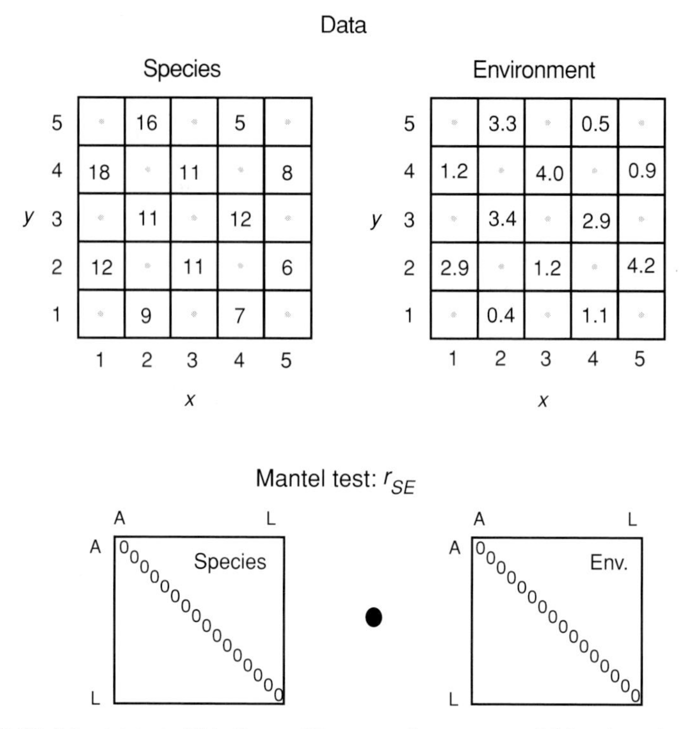


Figure 3.17 Mantel test. This figure illustrates how two variables (species and environment) sampled at the same 10 locations need to be converted into symmetrical distance matrices (10 × 10) before the standardized Mantel statistic, $r_{SE}$, can be computed.

where $Z_M$ is the Mantel statistic, $w_{ij}$ is either a connectivity matrix (**A**) or a Euclidean distance matrix, among the $n$ sampling locations and $x_{ij}$ is a dissimilarity or distance matrix (**B**) of the values of all the variables at each sampling location. The advantage of a matrix approach is that any dissimilarity or distance coefficient can be used (see Legendre & Legendre 1998 for a full list); this may reflect better the community structure of the variables of interest. Given that multivariate data are first transformed into a symmetric matrix, (3.34) can be rewritten using matrix notation:

$$Z_{\mathbf{AB}} = \mathbf{A} \cdot \mathbf{B}, \tag{3.35}$$

where, $Z_{\mathbf{AB}}$ is the sum of all of the element-by-element products (as indicated by '•') between the distance matrices **A** and **B** (Figure 3.17). $Z_{\mathbf{AB}}$ is unbounded, and, to facilitate its interpretation, each matrix can be standardized to obtain a bounded $r_M$ statistic ranging from $-1$ to 1. The Mantel statistic therefore estimates the linear relationship between the distance values of two matrices. However, given that this computation is performed using distance measures rather than the raw data themselves, the values of $r_M$ ($r_{\mathbf{AB}}$) are not directly comparable to those of a Pearson's correlation coefficient (but see Dutilleul *et al.* 2000). Dietz (1983) suggested ranking the distance measures before computing the cross-product between the two matrices. Thus, a monotonic relationship between the two matrices can be estimated and this is equivalent to performing a Spearman correlation analysis.

The null hypothesis for a Mantel test is that the variables in matrix **A**, as summarized by a distance measure into a synthetic value per pair of sampling locations, are independent (no relationship) of variables in matrix **B**. This can be tested either by an approximate *t*-test when the sample size is large or by a randomization test when the sample size is small. The randomization test is restricted to retain the relationships between sampling locations by randomly shuffling the rows and columns of one of the matrices (Legendre & Legendre 1998; Fortin & Gurevitch 2001). When the null hypothesis is true, the observed Mantel's statistic is expected to have a value located near the middle of the reference distribution. In the presence of a significant relationship, positive or negative, between the two matrices, the observed Mantel statistic is expected to be more extreme, either higher or lower, than most of the reference distribution values. The precision of the probability value is directly related to the number of randomizations used. We recommend generating as many as 10,000 randomizations (Manly 1997; Fortin *et al.* 2002), where the observed statistic is included in the reference distribution (Hope 1968).

Regardless of the type of relationship computed, linear or monotonic, Mantel tests are based on distance values rather than on the raw data, and the magnitude of the effect is often weaker than with the raw data. But what is really computed? It is the relationship between distance measures. Hence, pairs of sampling locations can have the same degree of dissimilarity because they both have either high or low values. Consequently, the magnitude of $r_{\mathbf{AB}}$ should not be used as a correlation coefficient, Pearson or Spearman, but rather more in a comparative way with other $r_{\mathbf{AB}}$ values. Furthermore, given that the significance of the observed value is tested against the re-arrangement of the sample data, it can be significant even if the value is very small (close to zero).

The Mantel statistics can be used to test the relationship between two sets of variables recorded at the same sampling locations. For example, we could be interested in the relationship between the abundances of 14 species of trees and their relative elevation, as described in Fortin (1992). In that example, the $r_{\mathbf{AB}}$ between the tree species matrix, **A** (Euclidean distances among tree abundance data of 14 species, see Fortin 1992 for detail), and the topography matrix, **B** (Euclidean distances among relative elevation in metres), is positive (0.255) and significant (Table 3.1). The Mantel test can also be computed between the same tree species matrix, **A**, and a geographical distance matrix, **B** (Euclidean distances among sampling locations).

Table 3.1 *Standardized Mantel ($r_{\mathbf{AB}}$) and partial Mantel ($r_{\mathbf{AB.C}}$) values between the tree community data and the relative topography controlling the x–y coordinates of sampling units (n = 84) using Euclidean distances*

| A | B | C | $r_M$ | Prob($r_M$)[a] |
|---|---|---|---|---|
| Tree | Topography | | 0.255 | 0.0001 |
| Tree | $x$–$y$ | | 0.232 | 0.0001 |
| Topography | $x$–$y$ | | 0.839 | 0.0001 |
| Tree | Topography | $x$–$y$ | 0.113 | 0.0391 |

[a] The probability is based on a reference distribution having 9,999 randomizations plus the observed statistic.

In that case, the $r_{\mathbf{AB}}$ is also positive and significant (0.232) and corresponds to an averaged isotropic intensity of spatial structure for the tree species community for the entire study area (Table 3.1). Note that there is a strong positive relationship between the relative topography and the sample locations: 0.839. We will come back to this issue below.

The **B** matrix can be converted into a connectivity matrix, such as a distance class matrix (as with Moran's $I$, the estimation of spatial autocorrelation is done at several distance classes), to compute a Mantel (multivariate) correlogram (Figure 3.18, Oden & Sokal 1986; Legendre & Fortin 1989). By doing so, a trend is identified in our example with a spatial range of about 60 m and where $r_M$ for the first distance class is 0.090 ($p < 0.001$).

The two-matrix approach was extended to a three-matrix method, called the partial Mantel test (Smouse *et al.* 1986), aiming to quantify the relationship between two matrices, **A** and **B**, while controlling for the effects of a third one, **C** (Figure 3.19, Legendre & Legendre 1998; Fortin & Gurevitch 2001). The partial Mantel test, $r_{\mathbf{AB.C}}$, can be computed by 'detrending' the effects of the variables in matrix **C** on those of matrix **A** and then matrix **B**, using a linear regression. Then with the residuals from both regressions, $\mathrm{Res}_{\mathbf{A}|\mathbf{C}}$ and $\mathrm{Res}_{\mathbf{B}|\mathbf{C}}$, a Mantel test is performed as in (3.35). Another way to compute the partial Mantel is to compute a partial correlation using the three matrices (Legendre & Legendre 1998):

$$r_{\mathbf{AB.C}} = \frac{r_{\mathbf{AB}} - r_{\mathbf{AC}} r_{\mathbf{BC}}}{\sqrt{1 - r_{\mathbf{AC}}^2}\sqrt{1 - r_{\mathbf{BC}}^2}}. \tag{3.36}$$

By comparing $r_M$ values, computed with two (Mantel test, $r_{\mathbf{AB}}$) and three matrices (partial Mantel test, $r_{\mathbf{AB.C}}$), we can test alternative causal relationships among

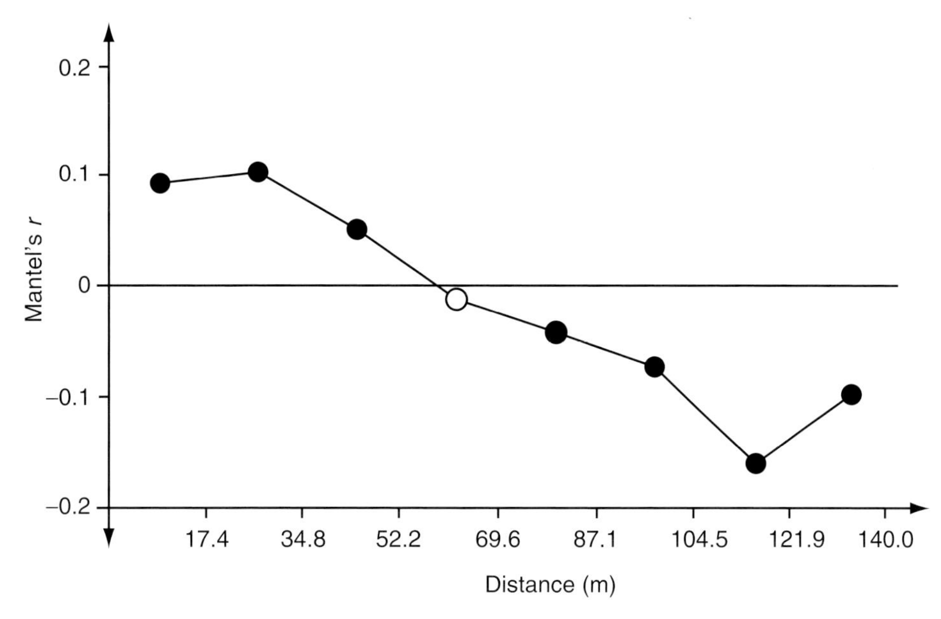


Figure 3.18 Mantel correlogram based on the abundance of 14 tree species where the standardized Mantel statistic, $r_M$, is plotted against distance in metres. The overall spatial structure of the tree community is a trend with a spatial range of around 60 m.

the three matrices (Legendre & Fortin 1989, Leduc *et al.* 1992; Legendre & Legendre 1998). Furthermore, the third matrix, **C**, can be a design matrix with dummy variables corresponding to treatments and control locations (see Fortin & Gurevitch 2001) or a set of covariables, allowing us to test specific hypotheses by coding it as a contrast matrix in ANOVA (Sokal *et al.* 1993) or by using geographic locations as a surrogate for variables not measured (Fortin & Payette 2002). For example, it could be that the topography, being related to the sampling locations, is not influencing tree species abundance that much. A partial Mantel test can be performed (Table 3.1) to test this hypothesis. In doing so, the relationship between tree abundance and topography decreases to 0.113 and is barely significant at $\alpha = 0.05$. Consequently, the topography influences tree abundances, but less so than we first thought by computing only the Mantel test.

When we control for the effects of a third matrix, **C**, such as the Euclidean distances matrix among the sampling locations, we are not controlling for the degree of spatial autocorrelation of the variables but only for the relative distance among the sampling locations (Fortin & Payette 2002). Furthermore, when the variables are strongly spatially autocorrelated, the restricted randomization (by rows and columns of the matrices) are no longer equally likely, so that the significance of the

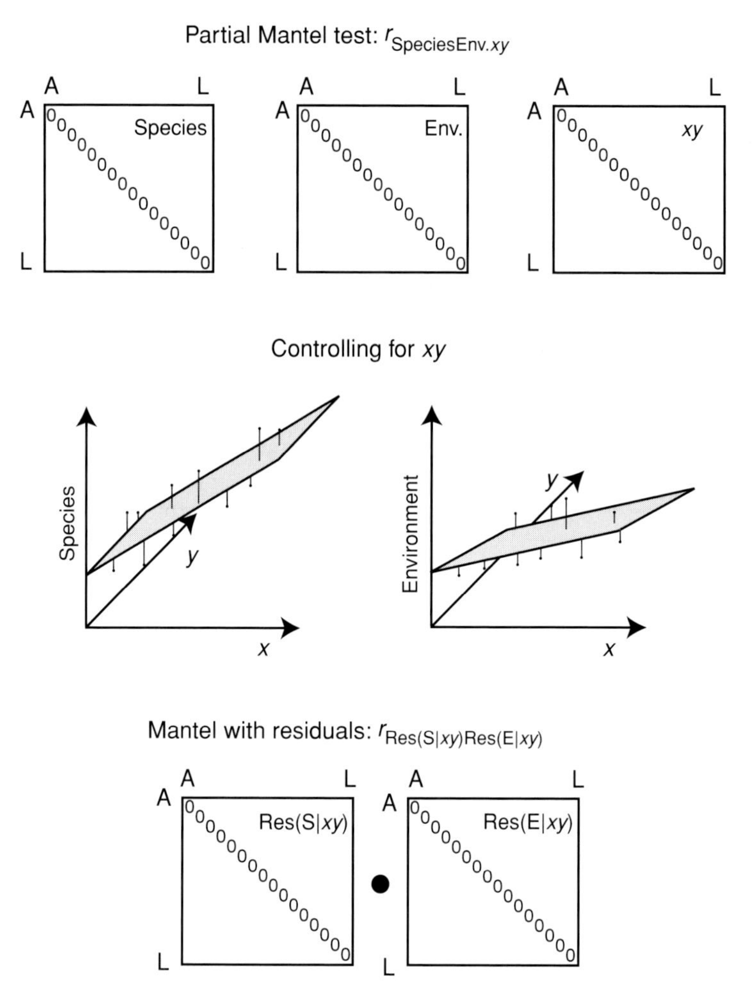


Figure 3.19 Partial Mantel test. This figure illustrates how the relationship between two variables (species and environment) sampled at the same 10 locations can be computed by first factoring out the effect of a third matrix (here the $x$–$y$ coordinates of the sampling locations) using linear regression, and then computing a standardized Mantel statistic, $r_{\text{Res}(S|xy)\text{Res}(E|xy)}$, with the residuals of these two linear regressions. See text for details.

partial Mantel test is not adequately evaluated (Oden & Sokal 1992). This problem has been acknowledged recently and different ways to restrict the randomization procedure have been proposed (Dutilleul *et al.* 2000; Legendre 2000).

Another problem with the Mantel and partial Mantel tests is that information about several variables is summarized into a single distance or dissimilarity value such that the resulting relationship is a global outcome for all of the variables and it is not possible to identify which variable(s) contribute the most to its intensity.

Hence, to be able to identify which variables contribute most to the relationship among sets of variables, canonical ordination techniques, CCA or RDA, can be used (Legendre & Legendre 1998). In these analyses, species ordination axes are constrained, using multiple regression, to be linear combinations of the environmental variables in maximizing species variance. The best linear fit between the species and environmental ordination axes is obtained by an iterative procedure. Often the relationship between species and environmental variables is induced by other underlying factors such as climate, topography or historical events. One way to control for the effects of these other variables is to use partial canonical ordination techniques such as partial CCA or partial RDA (Borcard *et al.* 1992; Legendre & Legendre 1998). The advantage of ordination, and partial ordination, techniques over Mantel and partial Mantel tests is that the relationship is computed on the raw data values rather than on distance measures. Also, these techniques permit us to disentangle the relative contributions of each variable to the relationship between the two matrices **A** and **B**.

One important issue is how to translate the information about the effects of space solely using sampling location *x–y* coordinates. At first, Borcard *et al.* (1992) used polynomials up to the third order of the *x–y* coordinates (i.e. a trend surface analysis approach, as presented in Section 3.5), which corresponds to a global, large-scale, trend surface response. Hence, such a polynomial trend surface analysis could not reflect local patch structure. To address this, Pelletier *et al.* (1999) suggested using a neighbourhood matrix based on local adjacency or connectivity among the sampling locations. Recently, Borcard & Legendre (2002) developed spectral decomposition of the spatial arrangement of sampling locations based on principal coordinates of neighbour matrices (PCNM), which they called the scalogram approach. This scalogram approach provides orthogonal sine waves with decreasing periods such that both large and local spatial layout of the sampling locations can be analysed as a **C** matrix in a partial canonical ordination analysis or as independent variables in a multiple regression.

## 3.4 Local spatial statistics

As more and more ecologists are designing ecological studies with large spatial extents, the likelihood that their data sets violate the assumption of stationarity is very high. When using global spatial statistics with such data sets, local and small areas of spatial heterogeneity are masked by the fact that they summarize the spatial pattern for the entire study area into a single average value of spatial autocorrelation (or a series of average values calculated at different distances as in a correlogram or variogram). In such cases, a global assessment of spatial dependence may be misleading because these average values of spatial autocorrelation provide information about neither the range of variability in intensity of spatial dependence nor the exact localization of patterns. For example, in a study area on a slope, tree abundance may vary from the top (say high) to the bottom (say low) of the slope, with some small tree gaps here and there creating localized patches with lower tree abundance. While global statistics may detect the large-scale trend in abundance values, they may miss local patterns because, by the construct of their algorithm, they lump all the local deviations together by summing and then averaging. While this average value of spatial dependence is meaningful where only one process occurs (either induced or inherent), it is misleading when several processes act at various intensities in different parts of the study area.

These limitations and the misapplication of global spatial statistics have been acknowledged for more than a decade (Fortin 1992; Getis & Ord 1992; Anselin 1995; Sokal *et al.* 1998; Fotheringham *et al.* 2000; Boots 2002). Also, the recognition that local estimation of the intensity of spatial autocorrelation may reveal interesting insights about local spatial processes resulted in the development of local spatial statistics (Getis & Ord 1992; Anselin 1995; Kabos & Csillag 2002; Boots 2003). Anselin (1995) proposed 'LISA' (local indicator of spatial association) as an acronym for these local spatial statistics. Here, we will be using the acronym LISA in a more inclusive way to refer to local statistics quantifying spatial dependence in general.

As presented in Section 3.3 and illustrated in Figure 3.7, the calculation of global spatial statistics, such as Moran's $I$ and Geary's $c$, requires the computation of spatial deviation at each sampling location (from the mean or neighbouring locations, respectively). Then, these deviations, which are local deviations, are lumped together. Hence, in essence, global spatial statistics are averages of the local spatial variations in a study area. It is therefore not too surprising that the first local spatial statistics proposed were modified global spatial statistics, calculated at each sampling location $i$ based only on its $j$ local neighbours. Consequently, the local Moran's $I_i$ can be estimated as follows:

$$I_i(d) = \frac{(x_i - \bar{x})}{\frac{1}{n}\sum_{i=1}^{n}(x_i - \bar{x})^2} \sum_{\substack{j=1 \\ j \neq i}}^{n} w_{ij}(d)(x_j - \bar{x}), \tag{3.37}$$

where $w_{ij}(d)$ is the weight matrix given a local neighbourhood search of radius $d$. Here the weight matrix can be binary stressing only the connectivity among sampling locations (see Figure 3.3) or actual weights (e.g. inverse distance weighting function) to emphasize further the local neighbourhood effect on local spatial pattern.

Under the assumption of complete randomization (i.e. that all randomizations are equally likely), the expected value of $I_i$ is:

$$E(I_i) = \frac{-1}{n-1} \sum_{j=1}^{n} w_{ij}. \tag{3.38}$$

Consequently, unlike global Moran's $I$, which has the same expected value for the entire study area and the different distance classes, the expected value of local Moran's $I$ varies for each sampling location because it is proportional to the number of neighbours. The significance of local $I_i$ can be tested based on a normal distribution where $I_i$ is transformed into $z(I_i)$:

$$z(I_i) = \frac{[I_i - E(I_i)]}{\sqrt{\mathrm{Var}(I_i)}}. \tag{3.39}$$

Then, the equation for $\mathrm{Var}(I_i)$ can be derived assuming either complete randomization or conditional randomization (mathematical details of these equations can be found in Boots (2002)). As for global Moran's $I$, local Moran's $I_i$ can be computed with different neighbour search distances, $d$. Any type of Bonferroni correction used to adjust for both the number of sampling locations and the number of neighbour search distances used, which results in a very large number of multiple comparisons, would be too conservative (i.e. only rarely would some $I_i$ be considered as significant). In general, we recommend using local spatial statistics as spatial exploratory tools to detect localized spatial structures that could indicate lack of stationarity.

As with global Moran's $I$, the local $I_i$ still computes the deviation from the average value of the variable, $\bar{x}$, for the entire study area. Positive $I_i$ values occur when the value at location $i$ is similar to those of its neighbours in their deviation from the average ($\bar{x}$). In other words, positive values of $I_i$ indicate that the values around $i$ and at $i$ are either all larger (positive deviation) or smaller (negative deviation) than the average. Negative values of $I_i$ also indicate that deviation from the average is either larger or smaller than the average but where the value at location $i$ is of a different sign from its neighbours. When the value of $I_i$ is close to zero, this means that deviation from the average is very small and no local spatial structure can be detected. This can occur either because there is no spatial pattern or because there is a subtle pattern but we cannot detect it because the local values are too similar to the overall average. Figure 3.20*a* shows significant local $I_i$ at $\alpha = 0.05$ based on the same simulated data used to compute the omnidirectional correlograms in Figure 3.8*d* (16 regularly spaced patches of equal size), Figure 3.8*e* (9 irregularly spaced patches of equal size) and Figure 3.8*f* (12 regularly spaced patches of different size). These maps of $I_i$ help to identify the position, size, shape

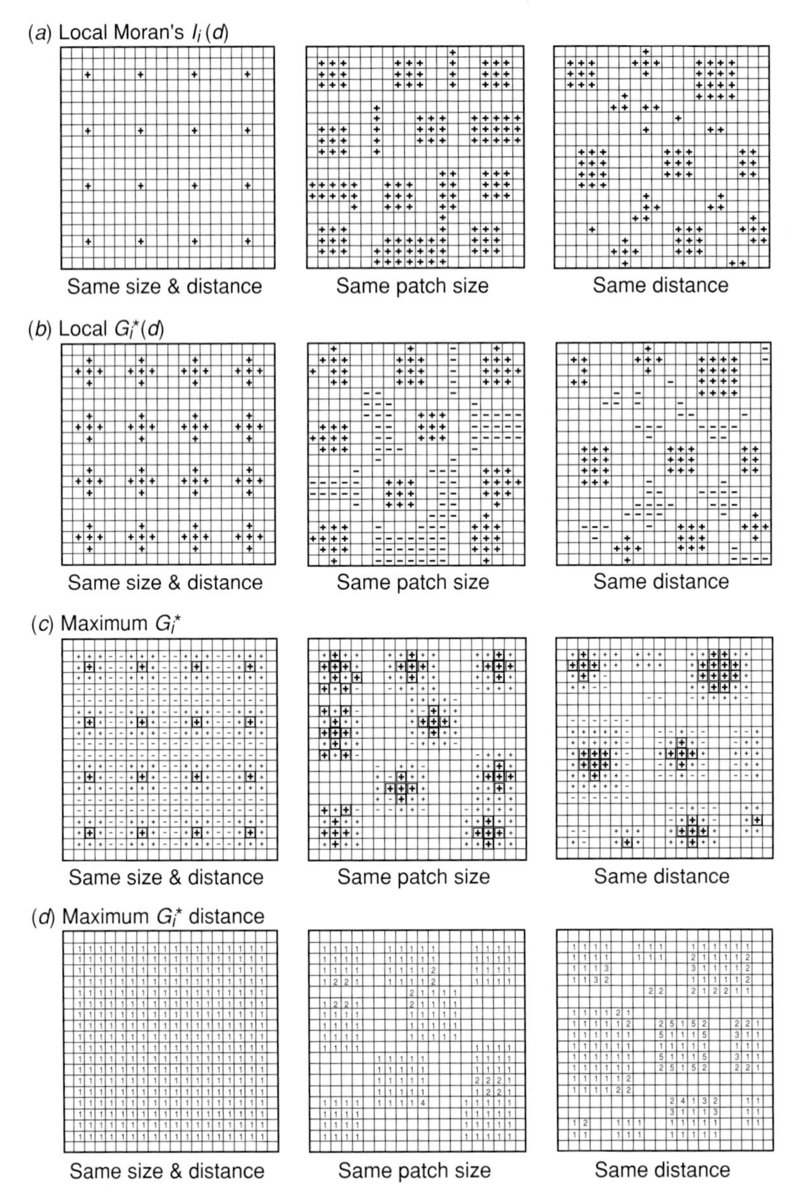


Figure 3.20 LISA (Local Indicators of Spatial Association) using the same simulated data shown in Figure 3.8: 16 patches (Figure 3.8*d*); 9 patches (Figure 3.8*e*) and 12 patches (Figure 3.8*f*). (*a*) Local Moran's $I_i$ estimated using a distance interval class of 1.5 units (+ indicates significant values at 0.05, positive or negative). (*b*) Local $G_i^*$ estimated using a distance interval class of 1.5 units (+ indicates positive and – negative significant values at 0.05). (*c*) Maximum local $G_i^*$ estimated using five distance classes of 1.5 units each (+ in a square indicates positive and – negative significant values at 0.05; + indicates positive values but non-significant). (*d*) Maximum local $G_i^*$ distance (number indicates the distance class at which the maximum local $G_i^*$ was estimated). See text for more detail.

and layout of local spatial structures. For example, in the case of the 16 regularly spaced patches (Figure 3.8*d*), the centroid of each patch is identified as having the most positive spatial autocorrelation with its neighbours (Figure 3.20*a*). The spatial maps of local $I_i$ values for the two other cases (Figure 3.20*a*) are not as informative, however, because we cannot discriminate between clusters of high and low spatial structure as significant $r$ because positive values of $I_i$ (as indicated by +) can result from either clusters of high or low values. Indeed, without looking at the maps of the raw data (Figure 3.8*e*, *f*), we cannot determine which clusters have high or low values. Other local spatial statistics may be more appropriate in such cases, as described below.

Global Geary's $c$ can also be modified to obtain a local spatial statistic, the local $c_i$:

$$c_i(d) = \frac{1}{\frac{1}{n}\sum_{i=1}^{n}(x_i - \bar{x})^2} \sum_{\substack{j=1 \\ j \neq i}}^{n} w_{ij}(d)(x_i - x_j)^2. \tag{3.40}$$

The difference between the local $I_i$ and the local $c_i$ is in the numerator, where for local $I_i$ it is the deviation from the value at location $i$ and the overall average of the variable, while for local $c_i$ it is the value of the variable at location $i$. Similarly, under the assumption of complete randomness, the expected value of $c_i$ is proportional to the number of local neighbours:

$$E(c_i) = \frac{2n}{n-1} \sum_{j=1}^{n} w_{ij}. \tag{3.41}$$

The equations for complete and conditional randomness for Var($c_i$) can also be found in Boots (2002). Here, positive values of local Geary's $c_i$ correspond to cases where the value at location $i$ is similar to its neighbours, while negative values indicate a difference in sign from its neighbours.

Getis & Ord (1992) proposed two new local spatial statistics: local $G_i$, where the value at location $i$ is excluded from the computation; and local $G_i^*$ where the value at location $i$ is included. Local $G_i$ is computed as follows:

$$G_i(d) = \frac{\sum_{\substack{j=1 \\ j \neq i}}^{n} w_{ij}(d)x_j}{\sum_{\substack{j=1 \\ j \neq i}}^{n} x_j}, \tag{3.42}$$

where the expected value, under the assumption of complete randomness, depends on the number of local neighbours:

$$E(G_i) = \frac{1}{n-1} \sum_{\substack{j=1 \\ j \neq i}}^{n} w_{ij}. \tag{3.43}$$

Similarly, the local Getis $G_i^*$ is computed as:

$$G_i^*(d) = \frac{\sum_{j=1}^{n} w_{ij}(d) x_j}{\sum_{j=1}^{n} x_j}, \tag{3.44}$$

where the expected value, under the assumption of complete randomness, also depends on the number of local neighbours:

$$E(G_i^*) = \frac{1}{n} \sum_{j=1}^{n} w_{ij}. \tag{3.45}$$

The expected values of the variance of the two local Getis statistics are described in Boots (2002).

These two local Getis statistics are, in essence, the ratio of local averages (around the location $i$ and at the location $i$) over the global average of the variable of interest for the entire study area, i.e. local spatial moving averages. These local spatial statistics detect clusters of either high or low values, which are often referred to as 'hot spots', or 'cold spots', respectively. As with local Moran's $I_i$, the local $G$ statistics cannot differentiate between cases where there is an absence of spatial structure and cases where the local average equals the global average.

As the number of neighbours increases with neighbour search distance, the local $G$ statistics are asymptotically normally distributed and can be standardized to facilitate their interpretation (Getis & Ord 1996): positive local $G$ values indicate clusters with high values (hot spots), and negative ones indicate clusters with low values (cold spots). Figure 3.20*b* illustrates how, unlike local Moran's $I_i$, the local $G_i^*$ can discriminate between locations of significant hot spots, indicated by $+$, and cold spots, indicated by $-$, in the three different spatial arrangements of patches.

Wulder & Boots (1998) proposed that the local $G_i^*$ statistic could be computed using different neighbour search distances, $d$, identifying:

(1) the maximum $G_i^*$ value calculated at each location $i$ regardless of $d$, and
(2) the distance $d$ at which the maximum $G_i^*$ value occurred.

Maps of these two values provide, respectively, information about the maximum intensity of spatial dependence at a location *i*, and its local spatial extent. Following Wulder & Boots' (1998) suggestion, we mapped the maximum $G_i^*$ (Figure 3.20*c*) and the maximum $G_i^*$ distance (Figure 3.20*d*) based on the three spatial arrangements of patches. The maximum $G_i^*$ map allows us to identify the centre of the patches, as indicated by positive values (the + in squares), while the maximum $G_i^*$ distance map shows the local spatial range (extent) of each patch.

Despite the fact that local *G* statistics, $G_i$ and $G_i^*$, are more informative than local Moran's $I_i$, they are still estimated relative to the global average of the data and are therefore sensitive to the presence of overall global spatial structure in the data (e.g. a trend). The *H* Moran statistic (as mentioned in Section 3.2) proposed by Kabos & Csillag (2002) addresses this issue by computing a local statistic for qualitative data that is not affected by the presence of a global pattern both in its computation and in its significance tests. In the spirit of assessing significance locally, Boots (2003) proposed a local join count approach that he called the local indicator for categorical data (LICD). In short, the LICD assesses the significance of the local spatial arrangement (configuration) of categorical data by adjusting for the proportion (composition) of each category within a local neighbourhood search window. So far, there are no equivalent local statistics for quantitative data that can account for the presence of global patterns in the data and detect local spatial patterns.

## 3.5 Interpolation and spatial models

Most of the time, we collect data to obtain information about a target population. In ecological studies, such information could be related to species abundance, species behaviour and so on. In a spatial context, we are often interested in the estimation and prediction of the values of a variable at unsampled locations. This can be achieved by modelling spatial pattern using spatial interpolation techniques. As in any regression and modelling context, a spatial model attempts to summarize the spatial pattern using as few parameters as possible. In fitting these parameters to the data, we are modelling the main spatial signal and trying to minimize the error. Hence, any interpolation techniques produce smoothed estimated values at locations not sampled. Several interpolation methods are available, each having various advantages and limitations in the way the spatial structure is modelled. Interpolation techniques can be classified according to the following broad properties:

- Global: A single interpolation function is used to interpolate the values for the entire study area. The resulting map of the interpolated data is usually a smooth surface (e.g. trend only). Change in one value will affect the function and thus the predicted values (trend surface analysis).

- Local: The interpolation function is applied locally for a limited number of locations. The resulting map of the interpolated data is still smooth but includes both global and local patterns. Change in one value will affect only neighbouring locations (proximity polygons, inverse distance weighting, kriging).
- Approximate: At the sampling locations the predicted values will not be the same as the observed ones (trend surface analysis).
- Exact: At the sampling locations the predicted values will be exactly the observed values (proximity polygons, inverse distance weighting, kriging).

Other properties can also be used to characterize these interpolation techniques: deterministic (i.e. there is only one possible predicted value at a given location – all methods except kriging, which can be stochastic when used in a conditional annealing procedure as described in Section 3.5.4), point interpolators (again all methods except the proximity polygons method) and areal (only the proximity polygons and kriging). Here, we provide only a brief overview of these methods as there are several text books that present them in great detail (Journel & Huijbregts 1978; Isaaks & Srivastava 1989; Haining 1990, 2003; Cressie 1993; Goovaerts 1997; Chilès & Delfiner 1999; Webster & Oliver 2001; O'Sullivan & Unwin 2003).

There are other spatial models (moving average, MA; simultaneous autoregressive, SAR; conditional autoregressive, CAR) that simulate data with a known degree of spatial dependence. These spatial autoregression models are mostly used to simulate data rather than to interpolate and therefore we will present them in Chapter 5.

### *3.5.1 Proximity polygons*

We interpolate from data daily without even realizing it. Indeed, by using sampling units, we assume that the sampled quantity does not vary over the entire sampling unit. This concept can be extended to point data (which we refer to as events in Chapter 2) where polygons (also called Dirichlet, Thiessen or Voronoi) can be determined based on the $x$–$y$ coordinates of each point given their proximity to each other (Okabe *et al.* 1992). The notion of proximity refers here both to the spatial arrangement of all neighbouring points and to how these points can potentially interact with one another. The value at the $x$–$y$ coordinates is then assigned to all the area within the polygon (Voronoi polygons): when the value is qualitative this creates a categorical attribute; when it is quantitative, this numerical value is assumed to be uniform over the entire polygon. Therefore, these Voronoi polygons can be used to define the spatial area of influence around each point (Figure 3.21). This simple technique can result in abrupt changes from one polygon to another.

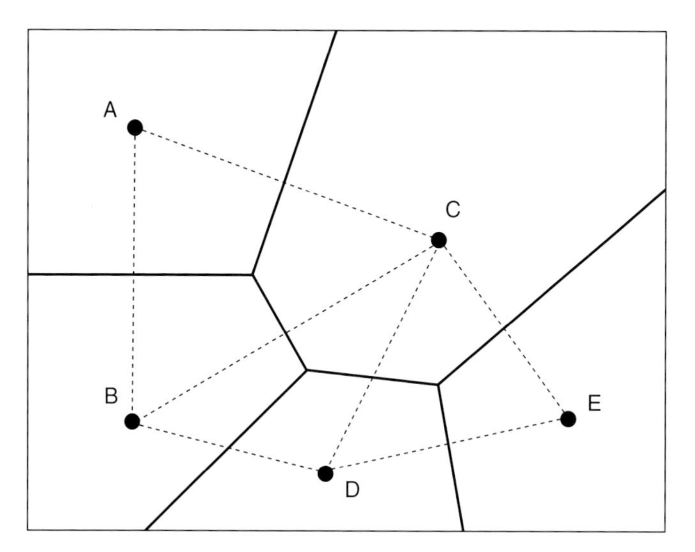


Figure 3.21 Proximity polygon interpolation using Voronoi polygons. The five sampling locations (A–E) were first linked using a Delaunay tessellation (dashed lines) algorithm and then the Voronoi polygons (solid lines) were determined. Each polygon indicates the zone of influence of each sampling location.

### *3.5.2 Trend surface analysis*

In an aspatial context, and when only information about a sampled quantitative variable is available, the best predictor at an unsampled location is the average value of the variable over the entire sample. When data and knowledge of the relationship between an independent variable and the variable of interest exists, the most commonly used interpolation method is regression. In a spatial context, the $x$–$y$ coordinates of sampling units can be used as independent variables in a regression. Hence, when the spatial pattern is a linear trend, values can be interpolated using multiple linear regression:

$$\hat{z}(x_0) = b_0 + b_1 x + b_2 y, \tag{3.46}$$

where $\hat{z}(x_0)$ is the predicted value at location $x_0$, $b_0$ is the intercept, and $b_1$ and $b_2$ are coefficients of the slope of the surface (Figure 3.22*a*). This multiple regression approach is simple and useful because a general equation can be used over the entire study area to model large-scale spatial patterns.

When the overall spatial pattern of the study area is a non-linear trend, the values can be approximated using polynomial regression of various orders such as quadratic ($(x, y)^2$), cubic ($(x, y)^3$) or higher orders. Hence when the pattern is a relatively smooth, monotonic, curved surface (e.g. hill or valley shape),

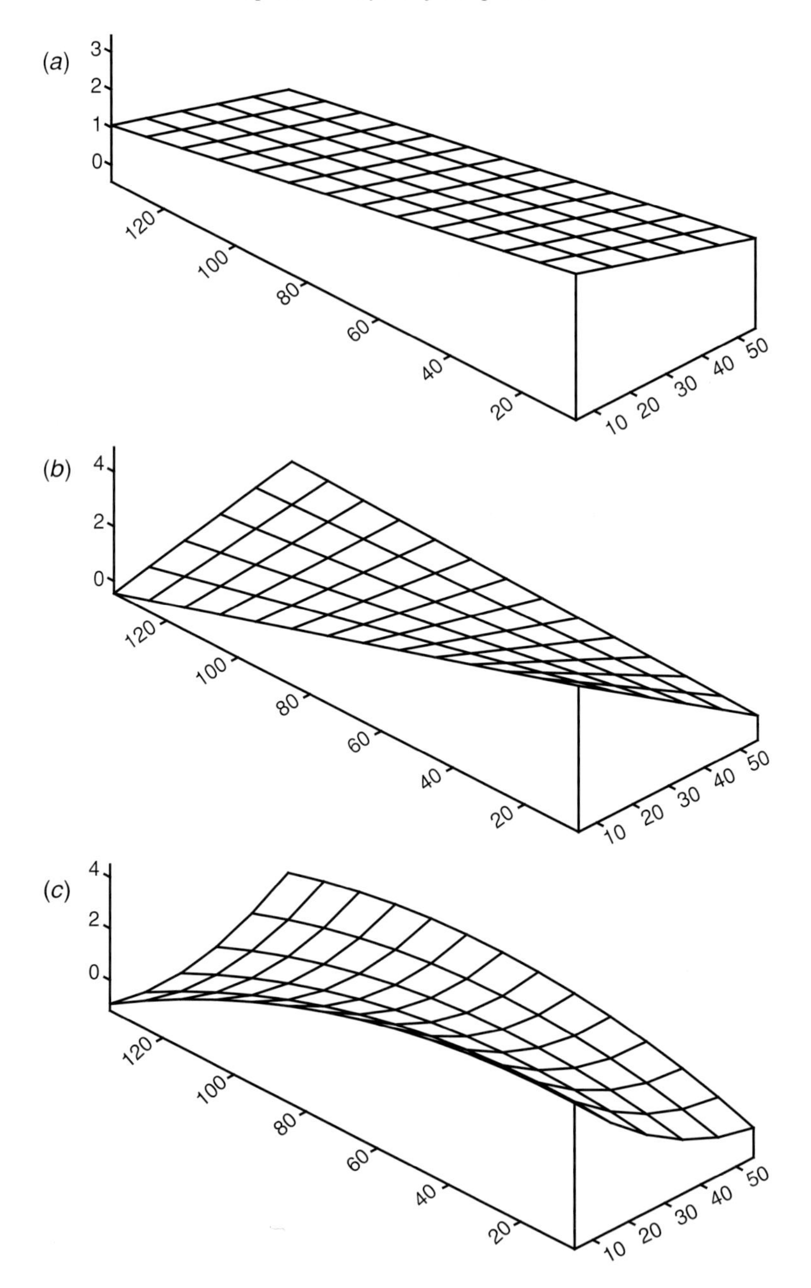


Figure 3.22 Trend surface analysis: (*a*) first-order polynomial (linear surface), (*b*) second-order polynomial (non-linear surface: valley type), (*c*) third-order polynomial (non-linear surface: saddle surface) and (*d*) fourth-order polynomial (non-linear surface: wavy surface).

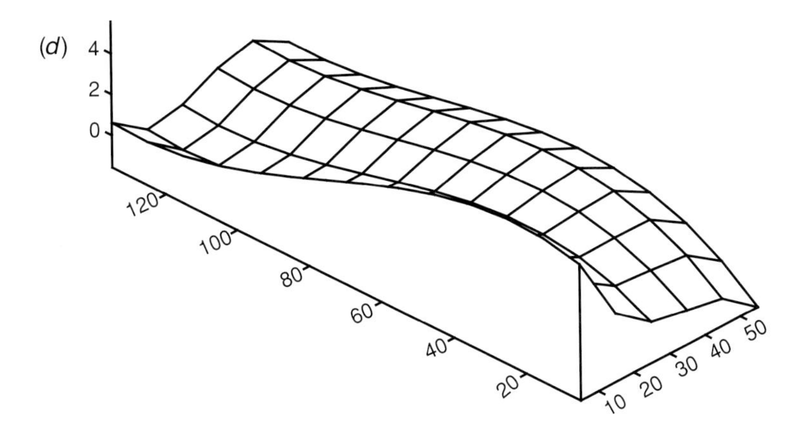


Figure 3.22 (*cont.*)

a second-order polynomial (quadratic) can be used to interpolate the data (Figure 3.22*b*):

$$\hat{z}(x_0) = b_0 + b_1 x_i + b_2 y_i + b_3 x_i^2 + b_4 x_i y_i + b_5 y_i^2. \qquad (3.47)$$

When the pattern shows some saddle shape, a third-order polynomial (cubic) could be used (Figure 3.22*c*):

$$\hat{z}(x_0) = b_0 + b_1 x_i + b_2 y_i + b_3 x_i^2 + b_4 x_i y_i + b_5 y_i^2 + b_6 x_i^3 + b_7 x_i^2 y_i + b_8 x_i^2 y_i + b_9 y_i^3. \qquad (3.48)$$

We could continue to fit the sample data using higher and higher order polynomials (e.g. Figure 3.22*d*). This is not recommended, however, because the gain in simplicity acquired with polynomial regression is lost due to the necessity of estimating several regression coefficients. More importantly, by improving the fit in some areas, others may start to fit less well, becoming distorted. In fact, trend surface analysis is a global interpolator and should not be used to model local spatial pattern. So, in a nutshell, the advantages of using a trend surface analysis method to interpolate data are:

(1) no a priori knowledge of the spatial pattern is needed as the interpolation is based on empirical data,
(2) both multiple and polynomial regression are available in most statistical software packages which also offer the possibility of testing the significance of regression coefficients using *F*-test. This can be used as a way to find out which order of polynomial is best to use (linear, quadratic or higher).

The disadvantages are, as mentioned, that it should not be used when there are several small patches and that it is not an exact interpolator but rather an approximate one, such that at the sampled locations the predicted values are not equal to the observed ones.

### 3.5.3 Inverse distance weighting

In the spirit of trend surface analysis, linear interpolation can be used to interpolate data based on the data from sampled locations given a restricted neighbourhood search area. The underlying premise is that nearby locations are more likely to have similar values and the linear interpolator weights the interpolated data, $\hat{z}(x_0)$, at unsampled locations $x_0$, according to the proximity of known sampled data as follows:

$$\hat{z}(x_0) = \sum_{j=1}^{m} w_j z(x_j). \tag{3.49}$$

Here $z(x_j)$ is the value of variable $z$ at the sampled location $j$, $m$ is the number of neighbouring sampling locations based on some definition such as being within a search radius, and $w_j$ is the weight according to the distance among the unsampled and sampled locations such that:

$$\sum_{j=1}^{m} w_j = 1. \tag{3.50}$$

The most common form of the inverse distance weight is:

$$\hat{z}(x_0) = \frac{\sum_{j=1}^{m} z(x_j) d_{ij}^{-k}}{\sum_{j=1}^{m} d_{ij}^{-k}}, \tag{3.51}$$

where $k$ is a real value from 0 to 1, and $d_{ij}$ is the distance between the unsampled location $i$ and sampled location $j$. More weight can be put to nearby locations by varying the value of the exponent $k$. When the distance between the sampled and the unsampled locations is zero, the interpolated value will be the observed one. The advantage in this type of linear interpolator is that it is weighted locally around each sampling location therefore preserving more of the complexity of local spatial patterns (Figure 3.23) than trend surface analysis. It is also very easy to use and does not require prior knowledge about the data. When a map of the study area is needed for illustration purposes alone, this linear interpolation technique is quite useful. It does not, however, provide any information about the discrepancies between the interpolated values and the 'real' spatial pattern at the unsampled locations. This is why kriging is often preferred over the inverse distance weighting method.

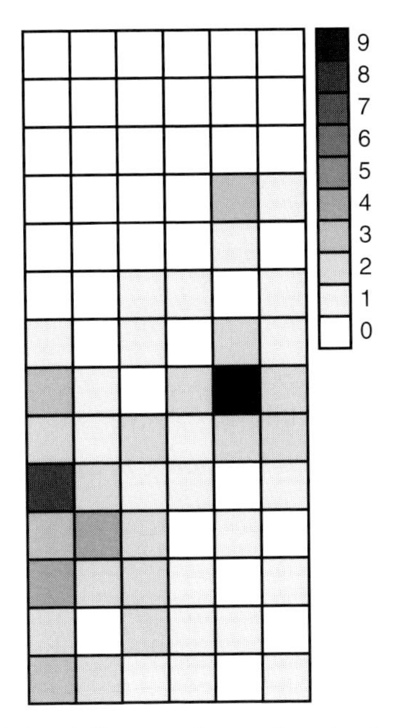


Figure 3.23 Inverse weighted distance. Interpolated values of sassafras abundance data ($n = 84$) based on 42 sampling locations (as shown in Figure 3.15*d*).

### 3.5.4 Kriging

Stemming from the two previous interpolation methods, Krige (1966) proposed an interpolation method using a system of linear equations, based on prior knowledge of the degree of spatial dependence in the data. This spatial interpolation technique is, in essence, a weighted moving average technique and it is called kriging. Kriging is one of the geostatistical techniques (Journel & Huijbregts 1978) and it uses the spatial parameters (spatial range, nugget and sill) estimated by the experimental variogram as described in Section 3.3.2.

Having its origin in application to mining questions, kriging was first developed to address specific needs in the spatial prediction of ore resources, such as interpolating from either punctual (point) or block (area) samples over a two-dimensional region, as well as predicting the values of the ore for a given volume (three-dimensional). This last feature made this technique known quite rapidly in oceanography (see Simard *et al.* 2003 for a recent example) and meteorology, but it took longer to become widespread in ecology (Legendre & Fortin 1989; Rossi *et al.* 1992).

Kriging is a set of linear regressions that determine the best combination of weights to interpolate the data as in the inverse weight distance method by minimizing the variance as derived from the spatial covariance in the data. Here, the weights, $w_i$, are based on the spatial parameters of the variogram model, which are derived from an experimental variogram such that sampling locations within the spatial range (distance) have more influence on the predicted value. To solve this system of linear algebraic equations, the sum of the weights is constrained to equal 1 ($\sum_{i=1}^{n} w_i = 1$), such that there are more equations than unknown parameters to estimate:

$$\hat{z}(x_0) = \sum_{j=1}^{m} w_j z(x_i), \tag{3.52}$$

where $m$ is the number of neighbouring sampling locations within a given search neighbourhood. To minimize the estimator error, the Lagrangian multiplier ($\lambda$) is added as a constant (for details see Journel & Huijbregts 1978):

$$\sigma_E^2(x_0) = \sum_{i=1}^{m} w_i \gamma(x_i, x_0) + \lambda. \tag{3.53}$$

The estimation error is also called the kriging variance or kriging error. Equation (3.52) can be written in matrix notation:

$$\mathbf{Cw} = \mathbf{c}, \tag{3.54}$$

where the predicted value at unsampled location $x_0$, is a vector $\mathbf{c}$ that is obtained by multiplying the variance–covariance matrix, $\mathbf{C}$, between known locations, $i$ and $j$, as estimated by the theoretical variogram model selected; by the vector of weights, $\mathbf{w}$, which are to be determined; $\mathbf{c}$ is the vector of covariances between the sampled locations $i$ and the unsampled location 0. The covariance matrix and vector values are given by the variogram model, where the vector of weights is to be estimated:

$$\begin{bmatrix} \gamma(d_{11}) & \gamma(d_{12}) & \cdots & \gamma(d_{1m}) & 1 \\ \vdots & \vdots & \ddots & \vdots & \vdots \\ \gamma(d_{m1}) & \gamma(d_{m2}) & \cdots & \gamma(d_{mm}) & 1 \\ 1 & 1 & \cdots & 1 & 0 \end{bmatrix} \cdot \begin{bmatrix} w_1 \\ \vdots \\ w_m \\ \lambda \end{bmatrix} = \begin{bmatrix} \gamma(d_{10}) \\ \vdots \\ \gamma(d_{m0}) \\ 1 \end{bmatrix}. \tag{3.55}$$

This is achieved by multiplying both sides of (3.54) with the inverse covariance matrix, $\mathbf{C}^{-1}$, such that the vector $\mathbf{w}$ can be solved:

$$\begin{aligned} &\mathbf{Cw} = \mathbf{c} \\ &\mathbf{C}^{-1}\mathbf{Cw} = \mathbf{C}^{-1}\mathbf{c} \\ &\mathbf{Iw} = \mathbf{C}^{-1}\mathbf{c} \\ &\mathbf{w} = \mathbf{C}^{-1}\mathbf{c}, \end{aligned}$$

where ($\mathbf{C}^{-1}\,\mathbf{C}$) is the identity matrix $\mathbf{I}$ (equivalent to a multiplication by 1 in matrix algebra). The determination of weights is therefore related to both the variogram model and the number of sampling locations considered. Most of the geostatistical software packages offer two types of rules to determine the search neighbourhood, i.e. advice on how to select the number of sampled locations to consider in an interpolation procedure. The first method is to define a search distance. Given that most of the spatial dependence occurs within the spatial range, the search distance should not exceed the spatial range of the variogram. There are cases, however, when the spatial layout of the sampling locations is such that there are very few locations within the range. It is then recommended that the search radius be increased until a minimum number of locations is reached. Usually, a minimum of 15–20 neighbouring points is used. It is important to realize, however, that in these circumstances beyond the spatial range, the values at sampled locations contribute little to the kriged values and the resulting kriged values will show a smooth spatial pattern. Also, since the best estimation of the weights necessitates the inversion of the covariance matrix at each interpolated location, it is faster to perform if this matrix is not too large. The search neighbourhood does not need to be isotropic and can be more elliptic or even a volume in the case of three-dimensional kriging.

Kriging has some of the same features as trend surface analysis in that only one model is used for the entire study area. It is also similar to the inverse weighted distance method in that the interpolation is performed locally. Thus, given that the weights are proportional both to the spatial variance of the data and to the distance among sampling locations, kriging is an exact interpolator because the kriged values at sampling locations are equal to the observed ones. That is not to say that the selected variogram model and its parameters are the best ones. This is why cross-validation was proposed in order to evaluate the overall robustness of the model. Cross-validation consists of removing each sampled location one at a time and then kriging at that location; and by then comparing how reliably this kriged value matches the observed one. This procedure was very important early on when maximum likelihood methods did not exist to facilitate model selection. The effectiveness of the kriging depends on how well the selected model fits the data.

Another way to evaluate the plausibility of kriged values is to map both the kriged values and their associated estimation error (Figure 3.24). Indeed, given that the kriging variance is in the same units as the kriged data, areas where the errors are higher than others can be identified. These areas of high variance can be due to:

(1) too few sampled locations within the spatial range in those areas, or
(2) the selection of an inappropriate variogram model.

Note that the kriged errors are relative to the variogram model used. If the selected theoretical model seems to be the best one, these areas of high errors can be used

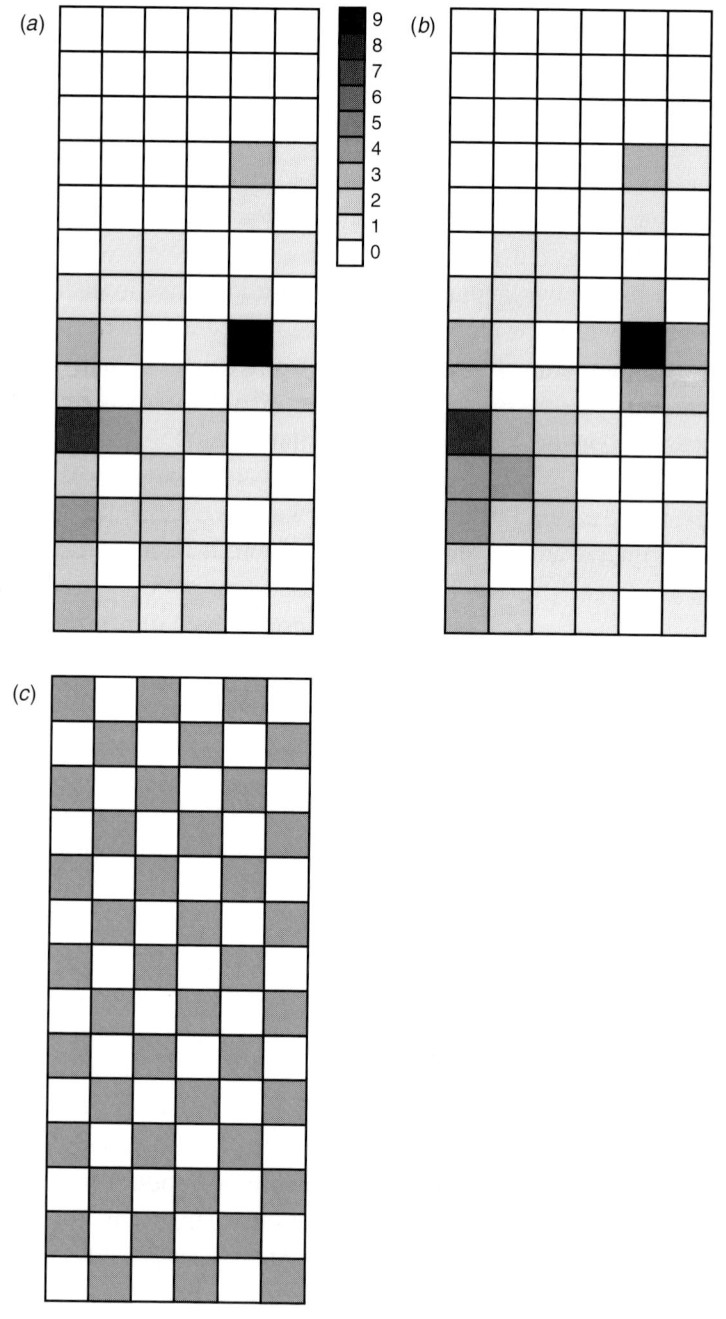


Figure 3.24 Kriged values of sassafras abundance data ($n = 84$) based on 42 sampling locations (as shown in Figure 3.15*d*) using an isotropic linear model (*a*) and an isotropic spherical model (*b*). The associated kriging errors with the kriged values based on the isotropic spherical model are shown (*c*). As grey values indicate, the kriged errors are higher at the unsampled locations than at the sampled ones, but given the uniform spatial sampling design, the errors are within the same range (4 to 5).

to detect sectors requiring more sampling effort. This procedure is often used to determine the optimal spatial sampling design (Webster & Oliver 2001).

Building from this system of linear equations, several kriging variants have been developed to account for the particular characteristics of environmental data such that the property of unbiased predictors can hold (Goovaerts 1997; Chilès & Delfiner 1999; Webster & Oliver 2001). First, interpolation can be performed for specific $x$–$y$ coordinates (*punctual* kriging) or for an area (*blocked* kriging). When the mean is known, i.e. when second-order stationarity applies, this system of linear equations is referred to as *simple* kriging. In most cases, the mean is not known and only the weak stationarity assumption prevails, thus *ordinary* kriging should be used. However, especially with environmental data, there could be a large-scale trend as well as local spatial patterns in the data. This implies that there is a 'drift' in the mean and that the simple system of linear equations cannot be used. *Universal* kriging was developed to model such large-scale trends and then the residuals are kriged after the trend is removed. As with trend surface analysis, the trend may not necessarily be linear, so an 'intrinsic random function of order $k$' should be used. These functions are the equivalent of the $k$-order polynomial functions in trend surface analysis and account for non-linear trends in the data before kriging. When the spatial pattern shows non-linearity that cannot be fitted by a polynomial function, *non-linear* kriging can be used.

As described earlier in this chapter, spatial pattern can be anisotropic where the spatial variance changes according to direction. When anisotropy is such that the sill is the same and only the range varies (i.e. geometric anisotropy), the distance matrix can be adjusted directly to account for it. When the anisotropy is 'zonal' (i.e. the range stays the same but the sill differs) then the adjustment needs to be made in the covariance matrix by adding more terms in a nested way (see Deutsch & Journel 1992). A nested procedure can also be implemented by using different variogram models as a function of distance. Stemming from this nested property of adding more terms in the system of linear equations, stratified kriging offers a way to interpolate over regions that have different spatial variances due to some change in strata types. For example, in forested landscapes, it is inappropriate to krige deciduous and coniferous stands using the same variogram. When information about the forest-stand type is available, stratified kriging, using different variograms for each forest type, should be used (Wallerman *et al.* 2002).

Too often in ecological and environmental studies, the variable of interest, $z_1$, is too costly to sample. If, however, knowledge of how this costly variable correlates with another variable, $z_2$, that can be sampled more easily (more cheaply or available from other sources), *co-kriging* can be used to interpolate $z_1$ given the spatial variance of $z_2$ as estimated by a cross-variogram between the two variables (for a recent example see Hudak *et al.* 2002). This is an appealing method, but it assumes that the linear relationship between the two variables holds even at locations where $z_2$ is sampled but $z_1$ is not. Often, the resulting kriged map of $z_1$ looks like a mirror image of $z_2$, which may or may not reflect the real spatial pattern of $z_1$.

Sometimes, in ecology, we are interested in mapping the spatial structure of an assemblage of species as a community rather than individually. This can be done using *multivariate* kriging (see the textbook by Wackernagel (2003) on the subject). Finally, given measurement errors, thresholding responses of variables or availability of only presence : absence data, we may want to use our quantitative variables as qualitative ones but would still like to determine their spatial structure and krige them. In such circumstances, an indicator variogram can be estimated (as presented above) and *indicator* kriging can be used (for a recent example see Todd *et al.* 2003).

There are many more things to know about geostatistics as described in more depth in advanced geostatistical textbooks (Goovaerts 1997; Chilès & Delfiner 1999; Webster & Oliver 2001; Wackernagel 2003). Here we have attempted to present a succinct overview, but have only lightly touched upon this very active research field. Before concluding, however, there is one more aspect of geostatistics that is very useful to ecological studies and deserves to be mentioned: stochastic simulation based on conditional annealing (Deutsch & Journel 1992). Spatial stochastic simulations are used more and more to generate a series of spatial data that have a given degree of spatial dependence in order to evaluate whether or not observed sample data show significant spatial patterns (Fortin *et al.* 2003). Here, the parameters of the variogram model are derived from an experimental variogram and can be used to generate stochastic simulations having the same degree of spatial variance as the observed data. This approach was first proposed in geostatistics to generate maps having more spatial variability than the kriged ones and hence looking more realistic in comparison to the observed map. Such simulated data are generated by an iteration process where the values at the sampled locations are kept as anchor locations from which the annealing algorithm iteratively spreads data values around them while ensuring that the overall degree of spatial variance is maintained (i.e. the range, nugget and sill values). This spatial stochastic simulation approach, based on a theoretical variogram, is computer intensive but allows us to address significance testing of spatially autocorrelated data.

## 3.6 Concluding remarks

As sketched in this chapter, several spatial statistics can estimate spatial dependence for sample data (Table 3.2). They share a common root in the determination of

Table 3.2 *Summary of the spatial analysis methods presented in Chapter 2*

| Spatial analysis method | Template |
|---|---|
| Join count statistics (Topology: network) | 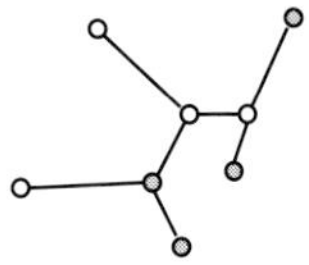 |
| Join count statistics (Lattice: chess moves) | 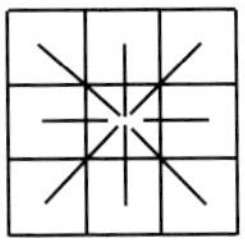 |
| Global spatial statistics (Isotropic spatial lag) | 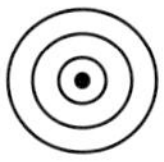 |
| Global spatial statistics (Anisotropic spatial lag) | 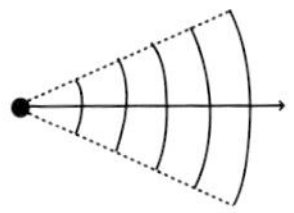 |
| Fractal dimension (Dividers) | 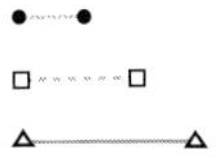 |
| Fractal dimension (Boxes) | 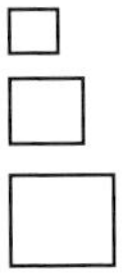 |
| Local spatial statistics (Topology: network) | 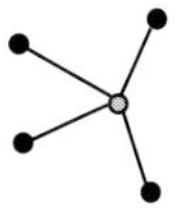 |
| Local spatial statistics (Lattice: chess moves) | 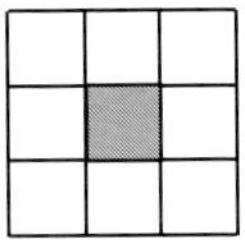 |

spatial covariance among the values of variable(s) of interest at different sampling locations. Hence the question becomes: which one should be used?

The first issue to decide upon is whether spatial analysis of the data should be global (i.e. should it summarize the spatial structure for the entire study area) or local (i.e. a measure of the local spatial structure at each sampling location). The choice between these two levels of analysis should be guided by both the goal(s) of the study and knowledge (or lack of knowledge) about the stationarity of the processes of interest. When no prior information is available, we recommend that both types of methods are performed and their results compared to identify whether or not the stationarity property holds.

Then, while using global spatial statistics, Moran's $I$ is favoured only because of its direct correspondence of meaning with Pearson's linear correlation (Eqn (3.8)). Users should be aware of its sensitivity to extreme outlier values as they will influence the value of the average which is used to estimate the spatial deviation (see Eqn (3.13)). This is why most would prefer Geary's $c$ or semi-variance $\gamma$ because outlier values affect only the spatial deviations (differences) computed with them. Unfortunately, these differences will have more weight because they are squared (see Eqns (3.14) and (3.21)).

Similarly, as local spatial statistics exist, it is important to realize that some are affected by the presence of a global pattern which may result in biased estimation of the spatial pattern. Two newly developed methods, $H$ Moran and LICD, are to be favoured when a large trend in the data is present. In the absence of overall structure, local $G_i^*$ in their standardized versions are easier to interpret as they indicate local areas of high (hot spots) or low (cold spots) values.

Both global and local spatial statistics provide information only about the spatial structure. When insight about the underlying causal processes that may have generated these spatial patterns is of interest, Mantel tests and partial Mantel tests can be used. In a multivariate context, partial ordinations (e.g. partial RDA, partial CCA) are more informative than the partial Mantel test about the relative contribution of each variable to the overall spatial structure.

Interpolation techniques can be used for illustration purposes, e.g. to map spatial pattern, where the simplest method (inverse weighted distance) provides good results. Such maps can also be modified by using various smoothing algorithms, such as a spline, to make it even more visually pleasing. On the other hand, when information about the actual values at unsampled locations, as well as an estimation of their associated errors, is needed, then one of the various kriging techniques should be used. There is no magical recipe, however, for which methods, and their respective parameter values, should be favoured. The ability to perform a meaningful interpolation using kriging comes with experience (i.e. by trial and error). The general rule of thumb is to capture the intensity and range of the spatial variance at short distances. Last, keep in mind that kriging errors associated with kriged values are a function of the theoretical model selected and the parameter values provided, not of the data themselves. The results depend on the appropriateness of the model on which kriging is based.

Finally, as in Chapter 2, the unresolved issue is that of significance testing while estimating spatial dependence at several distances based on the same data. Although spatial statistics traditionally have significance tests, and even progressive Bonferroni corrections, applied to them, these do not fully account for the dependence of values from one distance class to another. We will revisit this problem in Chapter 7.

# 4

# Spatial partitioning of regions: patch and boundary

## Introduction

In order to understand ecological processes at multiple scales, ecological studies are often carried out over large regions. In doing so, most study areas include more than one ecological process that can act at different spatial and temporal scales (Dungan *et al.* 2002). In such cases, it is unlikely that the assumption of stationarity of process (i.e. the same mean, variance and isotropy; see Chapter 1) is true. To analyse ecological data from large regions in a meaningful way, it is recommended that the region be partitioned into smaller, more spatially homogeneous areas (i.e. patches), that are more likely to be governed by the same ecological process. Stratifying a region can also be useful for monitoring and managing resources. There are two main approaches to spatial partitioning:

(1) grouping adjacent locations that have similar values of the variable(s) under study by generating spatial clusters (Figure 4.1), or
(2) dividing areas, based on their degree of dissimilarity, by delineating boundaries (Figure 4.2).

In theory, the outcomes of these two types of approaches should provide the same partitioning. In practice, however, there can be more or less pronounced spatial mismatches between the two methods. This is because the majority of the partitioning methods are descriptive and somewhat subjective in their use and interpretation. In this chapter, we are going to present the spatial partitioning methods, both spatial clustering and boundary detection, which are most relevant to ecologists (Figure 4.3). Indeed, the field of edge detection is currently growing in both computer vision and image analysis techniques with remote sensing and medical applications. Here we will concentrate on the analytic tools that are appropriate for ecological data and questions.

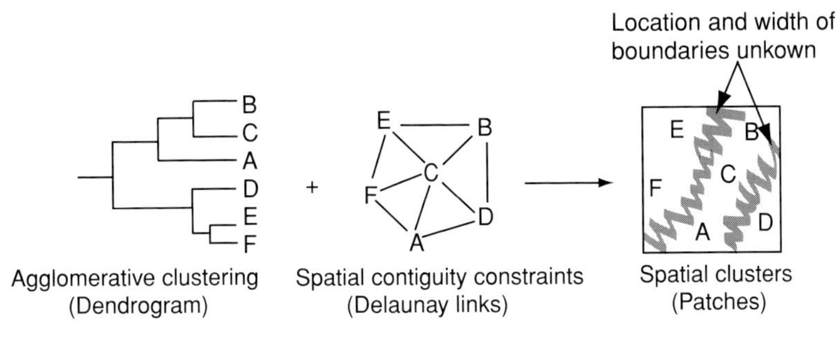


Figure 4.1 Spatial clustering method where spatial clusters are formed only when both the degree of similarity between sampling locations (A, B, C, D and E), based on a clustering algorithm (here an agglomerative one), is high and these sampling locations are adjacent to one another (e.g. A is spatially adjacent to F, C and D) based on spatial connectivity (here Delaunay links). Note that the exact location and width of the boundaries of the spatial clusters are not determined by this method as illustrated by the grey zigzags. In fact, the spatial clustering algorithm only identifies the membership of each sampling location to a spatial cluster.

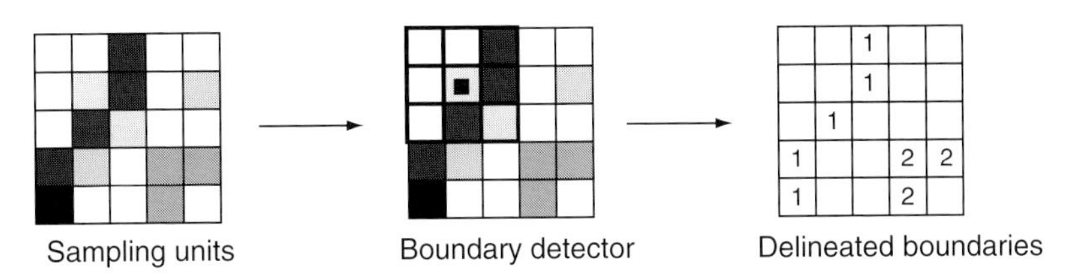


Figure 4.2 Boundary detection using a kernel (see Section 4.2) where the 1s indicate the locations where the boundaries are the most pronounced (sharp) and the 2s the second most pronounced boundaries. The grey shades correspond to quantitative values of a variable (low, black; high, white).

## 4.1 Patch identification

### *4.1.1 Patch properties*

A patch can be defined as a spatially homogeneous area where at least one variable has similar attributes either of category (e.g. forested area) or of quantitative value (e.g. tree age). Consequently, adjacent patches are different from one another in at least one variable. The juxtaposition of patches creates a mosaic at the regional level where each patch can be characterized by its structural properties such as area (e.g. small, large), shape (circle, elliptic, square, sinuous, with peninsulas, etc.), boundary properties (such as sharpness; see Section 4.2.1 below) and contrast between adjacent patches (e.g. low contrast between deciduous and mixed forest; high contrast between forest and agriculture land). Patch properties can be computed using landscape metrics (Li & Reynolds 1995; Gustafson 1998; Tischendorf 2001;

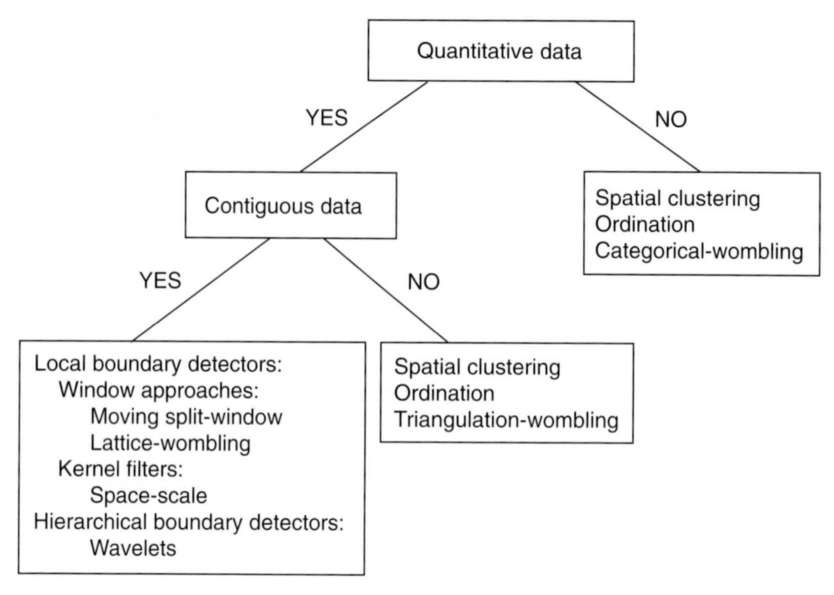


Figure 4.3 Decision tree to guide the selection of boundary detection methods for ecological data.

Fortin *et al.* 2003; Turner *et al.* 2003). The above definition of patches is data driven and implies that the spatial distribution of the values of targeted variable(s) is more-or-less uniform. This does not imply that all the variables measured in a given patch have a spatially uniform distribution. Similarly, when patches are arbitrarily delineated, as political or administrative units often are, the within-patch spatial structure of the data can range from a weak monotonic trend to strong spatial autocorrelation. The presence of within-patch spatial structure will reduce our ability to delineate boundaries accurately (Burrough & Frank 1996; Csillag *et al.* 2001; Edwards & Fortin 2001).

### *4.1.2 Spatial clustering*

Patches are, in effect, clusters of sampling locations that are spatially adjacent (i.e. spatial clusters) and have similar values. Clusters are usually based on the degree of similarity between data at different sampling locations. There are several clustering algorithms (see Legendre & Legendre (1998) for a complete review and mathematical details) that can be employed to perform spatial clustering. The ones used most often are the hierarchical agglomerative methods (e.g. single, intermediate and complete linkage, as well as centroid described below) which start with all sampling locations separated and then merged into larger groups based on the degree of similarity among them and the clusters already formed. To address the particular nature of ecological data better (e.g. presence : absence data, rare species, presence of double-zero), several similarity and dissimilarity metrics have been developed (Legendre & Legendre (1998). The researcher can define a priori the degree of similarity at which sampling locations can merge into an existing cluster. The linkage family of methods can be described as a gradient of criteria that need to be met before a sampling location can merge into a group, for example, in the single-linkage two sampling locations or clusters merge based on the minimum distance (i.e. maximum similarity) between them; in intermediate linkage, the similarity of a sampling location or cluster is compared to all possible pairs in the existing cluster before merging, and in complete linkage, the merging of clusters occurs when the maximum distance (i.e. minimum similarity) between clusters is reached. The centroid algorithm is comparable to the intermediate-linkage example where a sampling location merges with a cluster when its similarity value is comparable to the centroid value of the cluster. The advantage of agglomerative clustering procedures is that they create non-overlapping clusters. Several papers and books report the subtleties of clustering methods (e.g. Legendre & Legendre 1998). The most important drawback is the subjectivity involved in the selection of the degree of similarity for the creation of clusters.

The method of *k*-means partitioning has also been employed to determine spatial clusters (Legendre & Fortin 1989). With this clustering approach, the user needs to determine a priori the number of clusters to be obtained. The *k*-means algorithm optimizes, using an iterative process that minimizes the within-cluster sum of squares error term (SSE), i.e. the similarity from each sampling location data set to the centroid of the cluster to which it belongs. In this method, a certain amount of subjectivity comes in when choosing the number of clusters.

Spatial constraints among the sampling locations can be determined from any of the connectivity network algorithms or neighbour networks presented in Chapter 2. These networks identify which sampling locations are adjacent to one another (Figure 4.1) and can be used as spatial constraints while merging sampling locations. Hence, sampling locations can merge into a group only when adjacent sampling locations have comparable values (Figure 4.1). A by-product of forming spatial clusters, or patches, is that there are 'boundaries' between them (Figure 4.1), but their exact location and width are unknown. This is one of the possible weaknesses of spatial clustering methods: only the sampling locations have a known membership; the position of the boundaries among patches is arbitrary. On the other hand, the strength of spatial clustering approaches is that spatial clusters can be obtained from any data type (qualitative or quantitative, univariate or multivariate) sampled with any design (contiguous sampling units or not; Figure 4.3).

There are two major problems with spatial clustering methods. First, in the absence of a priori knowledge, or independent information about the ecological data,

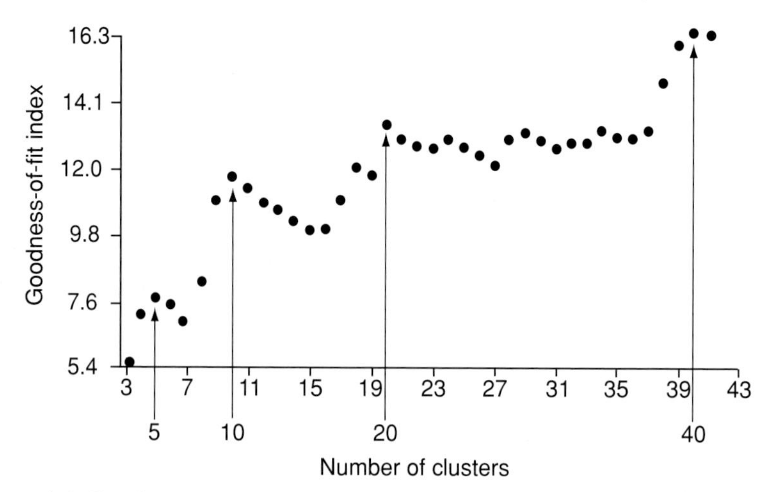


Figure 4.4 Goodness-of-fit index values according to the number of spatial clusters. The higher the index, the higher the contrast between the within-patch and among-patch sum of errors. Here, we are interested in the best number of spatial clusters to partition a study area having 84 sampling locations (6×14) based on tree abundance data (Fortin 1992). For comparison purposes, we selected four numbers of clusters: 5, 10, 20 and 40 (see Figure 4.6).

the researcher needs to select either the level of similarity for the agglomerative algorithms or the number of spatial clusters for $k$-means partitioning at which the spatial clusters will be obtained and subsequently interpreted ecologically. To achieve this, Gordon (1999) developed a goodness-of-fit index that indicates how the number of clusters selected contributes to minimize the sum of squares error of the between-cluster variability, $B$, when compared to the within-clusters example, $W$:

$$\text{Goodness-of-fit index} = \left[\frac{B/(k-1)}{W/(n-k)}\right], \tag{4.1}$$

where $k$ is the number of clusters and $n$ is the number of sampling locations. The value of this goodness-of-fit index can therefore be used as a guideline to determine the number of spatial clusters. It is important to have an underlying ecological question to provide an upper limit to the number of clusters requested. The choice of the appropriate number of clusters to select should therefore be guided by the goal of the research and knowledge about the study area.

For example, when looking at the 'best partition' of a study area where a known demarcation in forest canopy composition occurred (consider the data of the abundance of 26 tree species measured using 10×10 m sampling units used in Chapter 3, Fortin 1997), the highest value of goodness-of-fit was obtained with 40 spatial clusters (Figure 4.4). This is quite a large number of spatial clusters given that there are only 84 sampling locations. Based on our knowledge of the study area (Fortin 1992), three to five spatial clusters should be able to describe the spatial

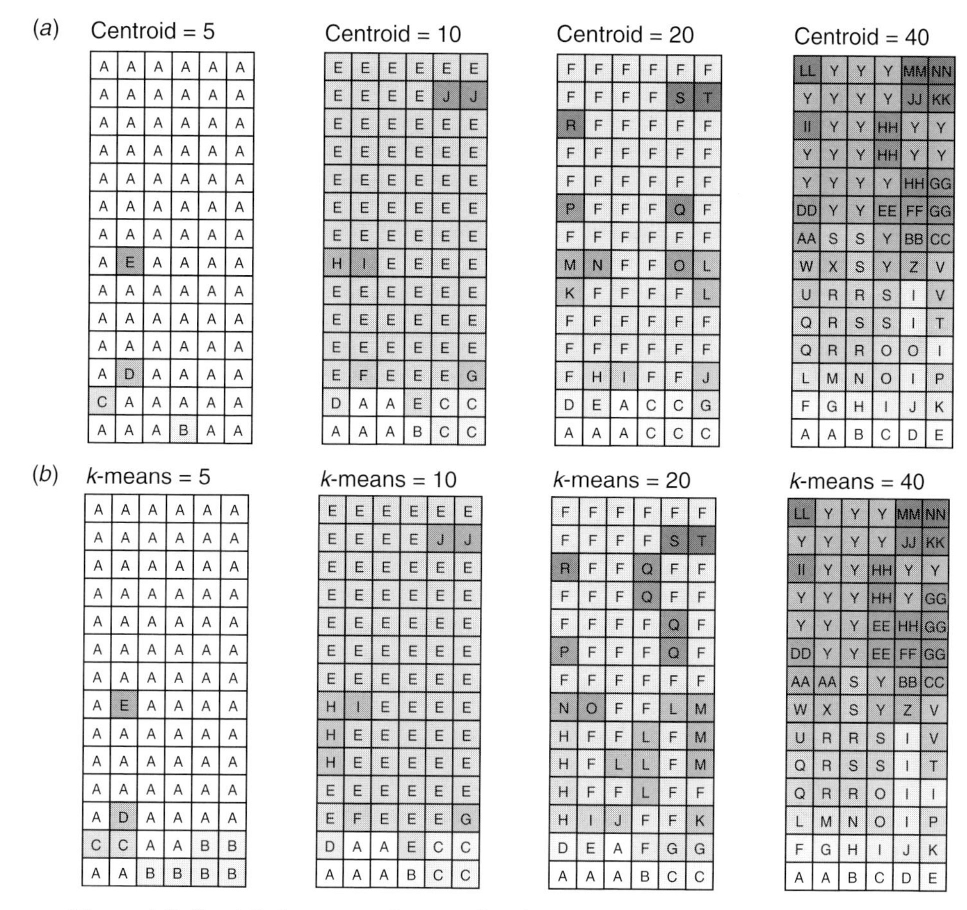


Figure 4.5 Spatial clusters on the tree abundance data of 84 sampling locations based on Delaunay connection links and (*a*) the centroid agglomerative algorithm and (*b*) the *k*-means method. Spatial partitions based on five spatial clusters identify that most of the spatial heterogeneity is concentrated in the lower left part of the study area and there is some heterogeneity in the middle of the plot. Spatial partitions based on 10 and 20 spatial clusters divide the large spatial cluster A, based on five spatial clusters, into smaller spatial clusters stressing much spatial heterogeneity over the entire plot. For spatial partitions based on 40 spatial clusters, although the number of clusters provided the highest goodness-of-fit index value, several spatial clusters were created having only one or two sampling locations. Overall, the *k*-means clustering algorithm creates larger spatial clusters than the agglomerative one.

arrangement of the woodlot tree canopy. For comparison purposes, however, we selected four values where the numbers of clusters show a local maximum in the goodness-of-fit index value: 5, 10, 20 and 40. In Figure 4.5, the spatial layout of the clusters based on different clustering algorithms (hierarchical agglomerative centroid and *k*-means) and number of clusters (5, 10, 20 and 40) can be visually compared. In doing so, we observe that both types of clustering algorithms provide more or less comparable spatial clusters but that the agglomerative algorithm creates more spatial clusters of only one sampling location (4 out of 5; 6 out of 10; 15 out of 20; 30 out of 40 clusters) than with the $k$-means method (2 out of 5; 5 out of 10; 12 out of 20 and 26 out of 40 clusters). Here, the spatial partition based on 20 clusters divides the study area into patches that characterize the differential spatial heterogeneity of the lower part of the study area and the relatively greater spatial homogeneity of the upper part. The advantage of the hierarchical agglomerative centroid algorithm is that it is hierarchical across the partitioning when increasing the number of clusters, i.e. the spatial clusters found with five clusters are maintained as such when the number of clusters increases. This is not the case with the $k$-means algorithm, where at each increase in the number of clusters, the new partition does not contain the spatial clusters of the previous partition with fewer clusters. This is because the $k$-means algorithm is an iterative procedure starting with a random assignment of the sampling locations to a spatial cluster. To reduce this problem, spatial partitioning based on a hierarchical clustering algorithm, or the $k$-means algorithm, could be used as a starting assignment input for the $k$-means procedure.

The second problem, which is characteristic of both spatial clustering and boundary detection, is that these algorithms are designed to provide clusters or boundaries even when none are present. An obvious example is the case where there is a gentle gradient across a region: the spatial clusters, or boundaries, identified will not reflect any true discontinuities in ecological process but will respond to local noise in the data. Furthermore, in some circumstances, spatial clusters may include sampling locations that have a high degree of similarity among adjacent sites, but other spatial clusters may have a high degree of dissimilarity. Finally, applying spatial constraints in clustering sampling locations may not necessarily create spatial clusters that have the strongest degree of similarity among sampling locations. For example, there could be situations where two sampling locations are not spatially adjacent but have comparable values because they are under the same climatic regime (Figure 4.6) or are influenced by the same underlying topography in terrestrial ecosystems or bathymetry in aquatic ecosystems (Figure 4.7). In such circumstances, it may be appropriate to customize the spatial constraints among the sampling locations by adding or removing links accordingly, not only to the topology of the connectivity network, but also according to physical and environmental conditions.

### 4.1.3 *Fuzzy classification*

A clear membership dichotomy by which every candidate is either a member, or otherwise, of any given class, such as presented in Section 4.1.2, may not be

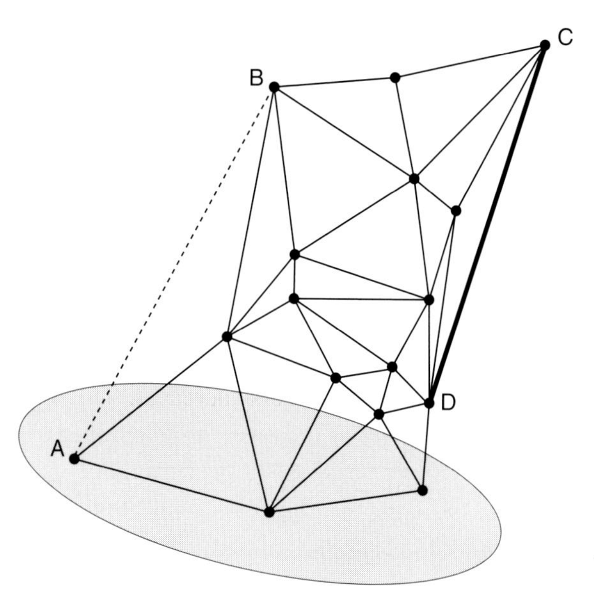


Figure 4.6 Delaunay connection links (solid lines) among 16 sampling locations. In a terrestrial context, e.g. forest, where only local neighbourhood effects prevail, this Delaunay tessellation network corresponds well to the ecological and environmental processes among sampling locations. However, if large-scale processes occur as well (e.g. two climatic zones), then the link between sampling locations A and B (dashed line) should be removed and that between C and D added (bold line).

appropriate in all circumstances, either because the quantitative data cannot be accurately classified into discrete classes (e.g. mixed forest dominated by deciduous or by coniferous species), or their spatial location is uncertain or inaccurate (e.g. telemetry data), or the data measurement is approximated (e.g. vegetation percentage cover reported in classes). Fuzzy classification and fuzzy *k*-means (see Burrough & McDonnell 1998 for mathematical details) have been suggested to be more appropriate spatial clustering methods in such cases (Jacquez *et al.* 2000). These methods are based on fuzzy set theory (Zadeh 1965) where the membership function is not a discrete dichotomy (0 or 1, an integer value) but rather a real number ranging from 0 to 1, called a 'possibility'. The possibility of being a member of a cluster is based either on expert knowledge or on spatial location uncertainty. The membership function can take several different shapes (e.g. linear, sigmoidal, symmetric, asymmetric) defined over a range of values of a variable, called transition zones. The advantage of the fuzzy classification approach is that it may more adequately reflect ecological processes and species' responses to environmental conditions. The drawback is that it requires more knowledge about the processes and any user-defined decisions can introduce a lot of subjectivity resulting in

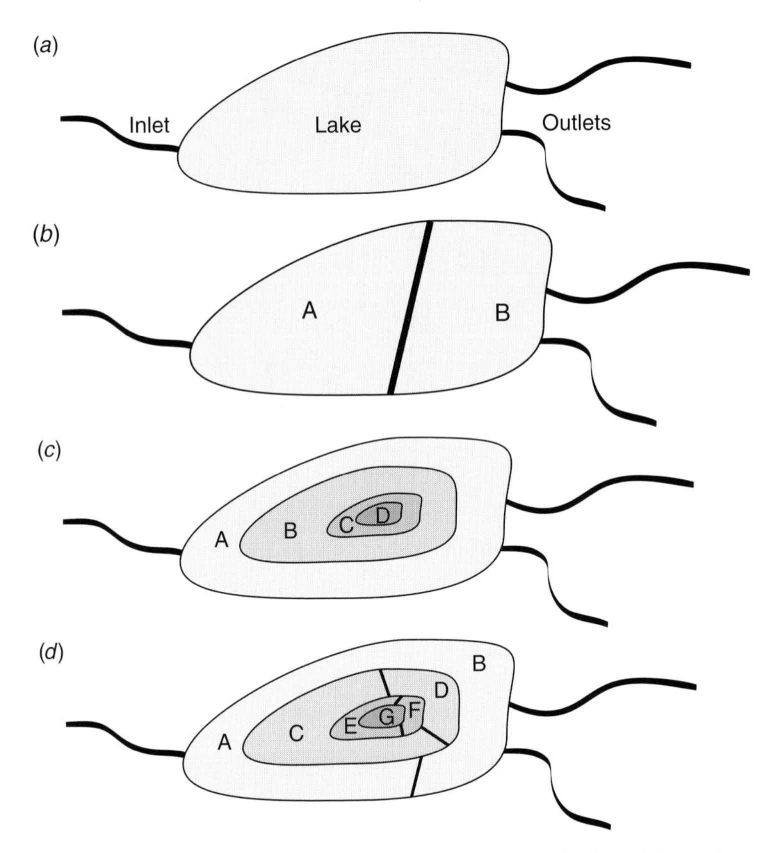


Figure 4.7 Spatial constraints in an aquatic system: (*a*) in shallow lakes where only local effects occur, spatial adjacency may correspond to the processes resulting in two spatial clusters A and B (*b*). (*c*) In deep lakes, bathymetry may play an important role and spatial constraints may not be appropriate. In fact, bathymetric isolines may reflect better the spatial constraints of the lakes forming four spatial clusters A, B, C and D. (*d*) Both spatial adjacency and bathymetry can be considered important, in which case seven spatial clusters will be formed (A, B, C, D, E, F and G).

non-optimal spatial clusters. To reduce the amount of subjectivity, the researcher can employ the fuzzy $k$-means approach (also referred to as fuzzy $c$-means, Burrough & McDonnell 1998). This method is also an iterative procedure that minimizes the within-cluster variability, but a fuzzy exponent is added to allow the overlap of clusters. When this fuzzy exponent is set to zero, it is equivalent to the $k$-means algorithm; when it is too high, all the clusters overlap. McBratney & de Gruijter (1992) suggested a value of 2 as a starting point for the fuzzy exponent. Using the same tree species data, the fuzzy $k$-means (based on five classes and a fuzzy exponent of 2) produces a different spatial partition from the equivalent spatial $k$-means

| 0.4 | 0.2 | 0.4 | 1-E | 1-E | 0.4 |
|---|---|---|---|---|---|
| 0.7 | 0.7 | 0.2 | 0.4 | 1-E | 1-E |
| 0.7 | 0.2 | 0.4 | 0.7 | 0.7 | 0.7 |
| 0.2 | 0.7 | 0.7 | 0.7 | 0.7 | 1-D |
| 0.0 | 0.7 | 0.7 | 1-B | 0.7 | 1-D |
| 0.7 | 1-B | 1-B | 1-B | 0.7 | 1-D |
| 1-B | 0.7 | 0.7 | 0.4 | 0.7 | 0.7 |
| 1-B | 0.7 | 0.7 | 0.4 | 1-C | 0.7 |
| 1-B | 0.2 | 0.7 | 0.7 | 0.7 | 1-C |
| 1-B | 0.7 | 0.7 | 0.7 | 0.7 | 0.7 |
| 0.7 | 0.7 | 0.7 | 1-A | 0.7 | 0.7 |
| 0.7 | 0.7 | 1-A | 0.7 | 1-A | 0.7 |
| 0.7 | 0.7 | 0.7 | 0.7 | 0.7 | 1-A |
| 0.7 | 0.7 | 0.7 | 0.7 | 0.7 | 0.7 |

Figure 4.8 Fuzzy $k$-means spatial clusters of the tree abundance data of 84 sampling locations using $k = 5$. The sampling locations parts of a spatial cluster have a membership possibility of 1 and are indicated by 1-A, 1-B, 1-C, 1-D and 1-E, where A, B, C, D and E signify the five spatial clusters. The other sampling locations have a membership possibility of belonging to a spatial cluster varying form 0 to 0.9999. For illustration purposes, the membership possibilities were classed into four categories: 0.7 (from 0.5 to 0.9), 0.4 (from 0.3 to 0.49), 0.2 (from 0.1 to 0.29) and 0.0 for 0. The five spatial clusters each contain very few sampling locations. In fact, most of the sampling locations almost belong to a spatial cluster (as indicate by 0.7) and illustrate the spatial extent (width, area) of the boundary zones between the spatial clusters.

clustering example (Figure 4.8). Here, the five crisp spatial clusters are smaller than with those obtained previously, but are surrounded by a gradient of fuzzy membership values. For illustration purposes, the possibility values were classified into five categories: 1 (1.0: 100% possibility to be part of the cluster), 2 (from 0.5 to 0.9), 3 (from 0.3 to 0.49), 4 (from 0.1 to 0.29) and 5 (0.0). Then, category 1 (100% possibility to be in a cluster) was divided into the five requested spatial clusters (A, B, C, D and E). The advantage of the fuzzy $k$-means method is that it allows determination of fuzzy boundary zones between spatial clusters.

Fuzzy set theory has been applied to detect fuzzy boundaries (among others, see Leung 1987; Edwards & Lowell 1996; Brown 1998; Burrough & McDonnell

1998). Jordan (2002) compared two fuzzy sets approaches, fuzzy classification and fuzzy boundaries, and found these methods to be complementary.

## 4.2 Boundary delineation

Several disciplines have developed analytic tools to detect boundaries: ecology, to detect ecotones and ecological boundaries, and remote sensing and medicine to detect an object's edge. Here we will focus on those most relevant to ecological studies (Figure 4.3).

### *4.2.1 Ecological boundaries*

In ecology, the development of boundary detection methods has a long tradition in studies associated with ecotone delineation. Ecotones are of interest in ecology because they are at the interface between two communities or ecosystems, where the exchange of nutrients and other forms of 'information' occurs. Ecotones have distinct structural and functional properties that differ from the adjacent systems (Holland *et al.* 1991; Hansen & di Castri 1992). The structural properties of ecotones are directly related to the type and the strength of the underlying processes that either generate or maintain them. Thus, ecotones, or ecological boundaries, represent linear responses to steep gradients in the environmental conditions or non-linear responses, such as thresholds, to environmental gradients (Table 4.1). Our ability to detect ecological boundaries depends therefore on the ecological process(es) under investigation (Gosz 1993; Dungan *et al.* 2002) as well as on the sampling design and analytic tools employed (Fortin & Drapeau 1995; Fortin 1997, 1999b; Fortin *et al.* 2000). Sampling designs for detecting boundaries should include sufficient sampling locations over a transect, or an area, so that not only the boundary itself but also the adjacent patches that it separates are covered.

### *4.2.2 Boundary properties*

Before examining the geometrical properties of boundaries, we need to define some terms. The term edge refers to a sharp demarcation. In image segmentation there are three major types of edges: the step edge, the roof edge and the spike edge. The step edge is the ideal case in which two well-defined and almost uniform patches of different types meet (e.g. forest and agriculture land), while the roof edge occurs when either or both patches are spatially autocorrelated. The spike edge, where an abrupt change in intensity occurs only locally, is rarer in ecology than in image processing. In ecology, the term edge is mainly used to refer to the step edge from

Table 4.1 *Processes and environmental factors creating and maintaining ecological boundaries*

| Environment and landscape structure changes | Processes and factors creating or maintaining boundaries | Type of boundary | Ability to detect edges and the underlying processes and factors |
|---|---|---|---|
| Sharp environmental changes | Geomorphology, topography, biogeochemistry, climate | Sharp, narrow, persistent | Possible to detect abrupt changes in diversity or abundance |
| Gradual environmental changes | Geography, climate, species' ranges (species physiological limits), species interactions | Blurred, wide, persistent or transient | Difficult to detect changes in biomass and abundances; possible to detect compositional changes |
| Spatial heterogeneity within large disturbances | Fire, storm, drought, species interactions, succession | Sharp to smooth, transient | Possible to difficult depending on the intensity of the disturbance |
| Spatial heterogeneity within small gaps | Treefall, species interactions, succession | Blurred, transient | Difficult to detect due to qualitative and quantitative noise |
| No environmental changes | Species interactions, dispersal ability | Sharp, persistent | Possible to difficult depending on species interactions |

man-made origins. There are several equivalent terms employed to refer to step edges, such as sharp, crisp or line boundaries (Figure 4.9*a*, *b*). The opposite of sharp step edges are gradual, intermediate, fuzzy boundaries or transition zones (Figure 4.9*c*, *d*). The term boundary includes both edge (line) and gradual (zone) demarcation. Herein, we will use the term boundary to refer to both sharp and gradual edges.

Related to the notion of boundary sharpness is the boundary's width, either narrow or wide. In an ecological context, it is quite probable that the width of a boundary varies asymmetrically along the length of the boundary as well as on each side of it (Figure 4.9*c*, *d*). Hence there are locations where ecological processes are more likely to be sharper than others. This can create only localized boundaries that do not enclose an area. Such boundaries are called 'difference' or 'open' boundaries

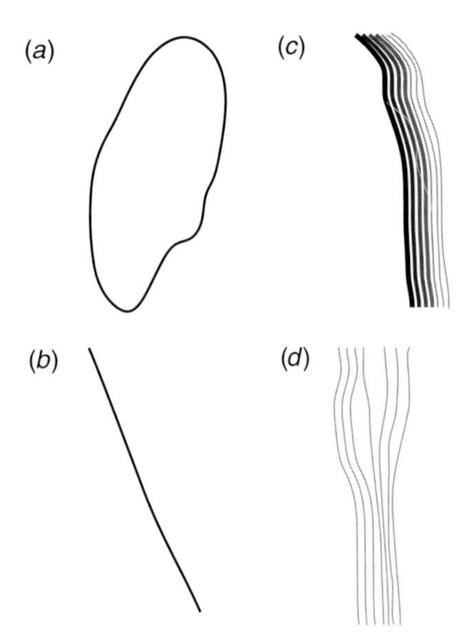


Figure 4.9 Boundary properties: sharp (crisp, step edge, line) boundaries (*a* and *b*); gradual (intermediate, fuzzy, transition) boundaries (*c* and *d*), open (difference) boundaries (*b*, *c* and *d*) and closed (area) boundary (*a*).

(Figure 4.9*b–d*), whereas those that surround and delimit an area completely (e.g. a patch) are called 'area' or 'closed' boundaries (Figure 4.9*a*).

Man-made boundaries are more likely to be straight. Boundaries originating from natural processes, on the other hand, are more likely to be sinuous – forming peninsulas, hence a series of convex and concave shapes. The degree of straightness/sinuosity of boundaries can be measured using the fractal dimension (Burrough 1981; Mandelbrot 1983). The accuracy of the fractal dimension, however, is directly affected by the spatial resolution of the sampling units.

Finally, all the structural and functional properties of boundaries are scale dependent (Gosz 1993; Fortin 1999b; Csillag *et al.* 2001; Handcock & Csillag 2002). Boundary studies need to acknowledge the scale of the ecological processes under study as well as the effects of the spatial resolution, of both the sampling unit and the extent of the study area, on the accuracy of detecting boundaries.

### *4.2.3 Boundary detection based on several variables*

Boundaries can be detected using either categorical information or continuous quantitative data. Here, with quantitative data, one operational definition of a boundary is the location in space where the change in intensity of a set of variables is the highest (Burrough 1986; Fortin 1994). With qualitative data, it is the location in space where species turnover is the highest.

#### *4.2.3.1 Multivariate methods*

One important property of ecotones is that this interface between two communities, or ecosystems, creates a unique assemblage (e.g. a new community or a new ecosystem) characterized by the change, usually an increase, in species diversity. Whittaker (1972) introduced three different types of diversity: alpha ($\alpha$) diversity refers to the diversity of species within a particular habitat or community; beta ($\beta$) diversity is a measure of the rate and extent of change in species along a gradient from one habitat to another; and gamma ($\gamma$) diversity is the species richness of a range of habitats in a geographical area that depends upon both $\alpha$ and the extent of $\beta$. Hence, the measure of beta diversity, also known as differentiation diversity, has been used to establish locations where the rate of species turnover is the highest, which are often associated with the presence of ecotones or ecological boundaries.

For quantitative data of species abundance, two statistics have been developed to measure beta diversity: the half-changes (Whittaker 1972) and the 'gleason' (Wilson & Mohler 1983). Half-change units are analogous to the notion of half-life for radioactive elements and were developed to evaluate the numbers of species in common at the two extremes of a gradient. The major inconvenience with the measure is that it provides no information about the rate of change along the gradient (Wilson & Mohler 1983). Therefore, these authors proposed a new measure of beta diversity, the gleason, which measures the rate of species turnover at any point along a gradient. This method assumes that the changes are continuous along an environmental gradient and re-scales each sampling location along the gradient space. The most widespread re-scaling method utilized to detect boundaries is the detrended correspondence analysis, DCA (see Legendre & Legendre 1998 for mathematical details, and Choesin & Boerner 2002 for an example). This re-scaling step, using an ordination technique, can be meaningless and inappropriate when disturbances induce discontinuities along a gradient, or when more than one gradient exists. Such disturbances are almost always found in second-growth forests, especially when area data are collected, since it is rare to find an area without natural boundaries or human disturbances.

#### *4.2.3.2 One-dimensional transect data*

The simplest and most effective way to detect ecological boundaries from quantitative data is to apply a moving split-window technique (Webster 1973; Johnston *et al.* 1992). This technique consists of computing the difference between two halves of the window. The window size can vary containing minimally only one sampling location per half. Various metrics can be used to measure the differences between the two adjacent window-halves (left and right halves) such as: discriminant

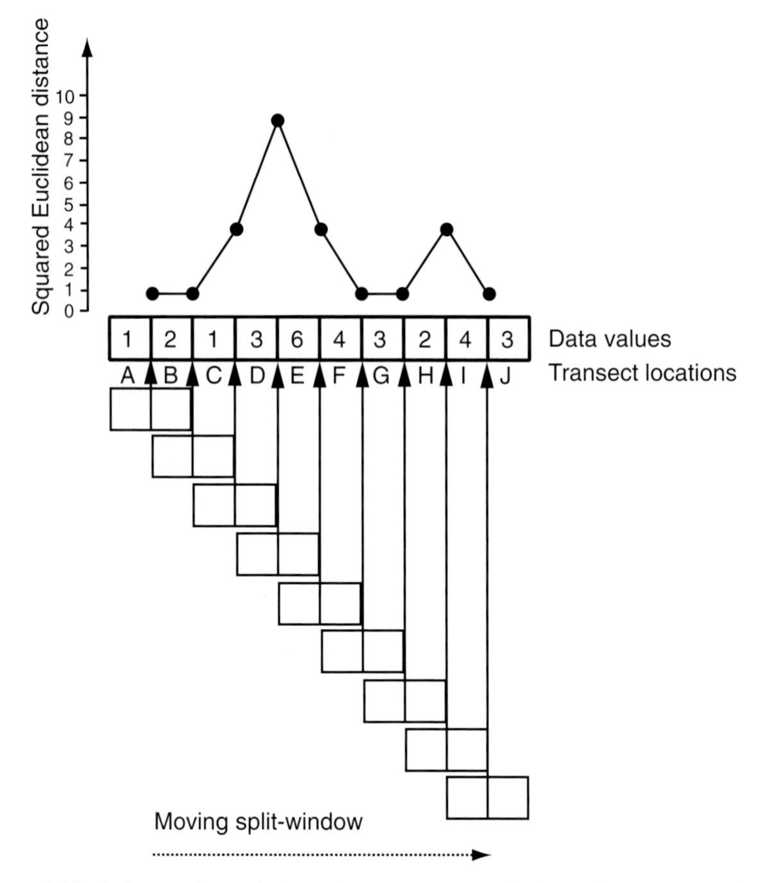


Figure 4.10 Split-moving window along a transect of 10 contiguous sampling locations. The split-moving window size is 2 sampling units (one sampling unit in each half). The square Euclidean distance computed for each pair of sampling units resulting in 9 values where the strongest peak (i.e. boundary) is between the sampling locations D and E and the second, weaker peak between sampling locations H and I.

functions, Mahalanobis distances and squared Euclidean distance (Ludwig & Cornelius 1987; Brunt & Conley 1990). The squared Euclidean distance, SED, is by far the most commonly used metric:

$$\mathrm{SED}(x_1, x_2) = \sum_{i=1}^{p} (z_{1i} - z_{2i})^2, \tag{4.2}$$

where $x_1$ and $x_2$ are the two sampling locations to compare and $z_{1i}$ and $z_{2i}$ the values of the $p$ variables at these two locations. The window is then slid along the entire transect, one sampling location at a time, so that all adjacent sampling locations can be compared (Figure 4.10). In the example illustrated in Figure 4.10, each half contains the data from one sampling location and resulting measures are located between the sampling locations for a total of $n - 1$ difference values. Sharp boundaries occur where high and narrow peaks identify the location of ecological boundaries, whereas gradual boundaries occur where peaks are low and wide. By computing differences based on adjacency, the moving split-window is in essence a local boundary detector. The drawback of all local boundary detectors is that they are sensitive to local noise in the data. To minimize the undesirable effects of local noise in the data, computation of the differences can be performed using windows of increasing sizes. The results can then be drawn on the same plot where peaks corresponding to ecologically meaningful boundaries will persist while peaks due to local noise will be smoothed out. Ecological boundaries will produce peaks at the same locations. The similarity to analysis with a Haar wavelet is obvious (Dale *et al.* 2002).

Comparable methods are available for presence : absence data collected at contiguous sampling locations. Basically, the observed number of species present at a given location is compared against a random distribution derived from a Monte Carlo procedure: Dale's method (1986, 1988) computes the amount of spatial overlap at each location, while the McCoy *et al.* (1986) technique is based on the probabilistic similarity between pairs of sampling locations.

One-dimensional transect data allow only detection of the sharpness and width of boundaries. To determine the other properties of boundaries, two-dimensional area data are needed. As sampling techniques and data acquisition by remotely sensed imaging gain popularity in ecology, it is more and more common to have such lattice data. The next section will present boundary detection methods for two-dimensional area data.

#### *4.2.3.3 Two-dimensional area data*

Two-dimensional data can be either a complete lattice or a sample of an area. We will therefore refer to lattice data when the spacing between any adjacent sampling units is constant, except for the outer limits. Each sampling unit has four connected, adjacent, sampling units directly north, south, east and west, forming a checkerboard-type pattern. The main advantage of using two-dimensional area data over one-dimensional transect data is that all the boundary features can be estimated (sharpness, width, shape, sinuosity, etc.).

As for the one-dimensional boundary detection methods, most boundary detectors for two-dimensional area data compute some difference among locally neighbouring sampling locations using either moving windows or kernel filters. The difference between this dichotomy of techniques at first seems subtle, but it is not. Window approaches compute a metric based on the values from adjacent sampling locations forming a square (e.g. a window size of $2 \times 2$ sampling locations) that quantifies the degree of difference among the four values (see the lattice-wombling section below). After having slid the window over the entire area (e.g. grid, lattice), the resulting number of different values is less than the original number of sampling locations (Figure 4.11*a*). Given that four sampling locations are needed to compute a difference, with lattice data having $n$ rows and $m$ columns there will be $n - 1 \times m - 1$ rates of change. The shape of the window can vary and the difference can also be computed based on three adjacent sampling locations forming a triangle using triangulation-wombling (described below).

Kernel filters operate differently. Indeed, kernels are usually squares of various sizes (e.g. $3 \times 3$, $5 \times 5$, $7 \times 7$, etc.), where each cell of the kernel contains a value that is multiplied with the correspondent location in the original lattice and where the resulting summation of all these multiplications is assigned to the centre cell. This type of procedure is called a convolution procedure in GIS and remote sensing. As for the window approaches, the kernel filter is then slid over the entire area but unlike the former, the kernel filters produce a new value for each cell of the lattice (e.g. the original number of sampling locations: $n_{\text{rows}} \times n_{\text{columns}}$; Figure 4.11*b*). There are several filters (mathematical operators, formulations) available to enhance the edges of an object (Pitas 2000). Here we will present only one of these filters (see Section 4.2.6).

*Lattice-wombling* With quantitative lattice data, the difference (hereafter called rate of change) in values among the four adjacent sampling locations forming a square (i.e. a $2 \times 2$ square window) can be estimated by computing the first partial derivative of a variable in the $x$ and $y$ spatial direction (Womble 1951):

$$m = \sqrt{\left[\frac{\partial f(x, y)}{\partial x}\right]^2 + \left[\frac{\partial f(x, y)}{\partial x}\right]^2}, \tag{4.3}$$

where $f(x, y)$ is a bilinear function, in the $x$ and $y$ spatial direction, of values $z_i$ at the four sampling locations ($i = 1, 2, 3$ and 4):

$$f(x, y) = z_1(1 - x)(1 - y) + z_2 x(1 - y) + z_3 xy + z_4(1 - x)y. \tag{4.4}$$

This formulation assumes that the distance between these four sampling locations is small. For convenience, the actual $x$ and $y$ coordinates are scaled to range from 0 to 1. The value of the rate of change, $m$, is computed for the centroid of a square window. Subsequently, the square window is slid over the entire lattice by shifting one sampling location at a time. In a multivariate context, the difference among the four adjacent sampling locations is the average, $\bar{m}$, of the absolute values of the first derivatives of each variable, $m$. When there is only one variable, the detected boundaries reflect high difference in the values of that variable; when

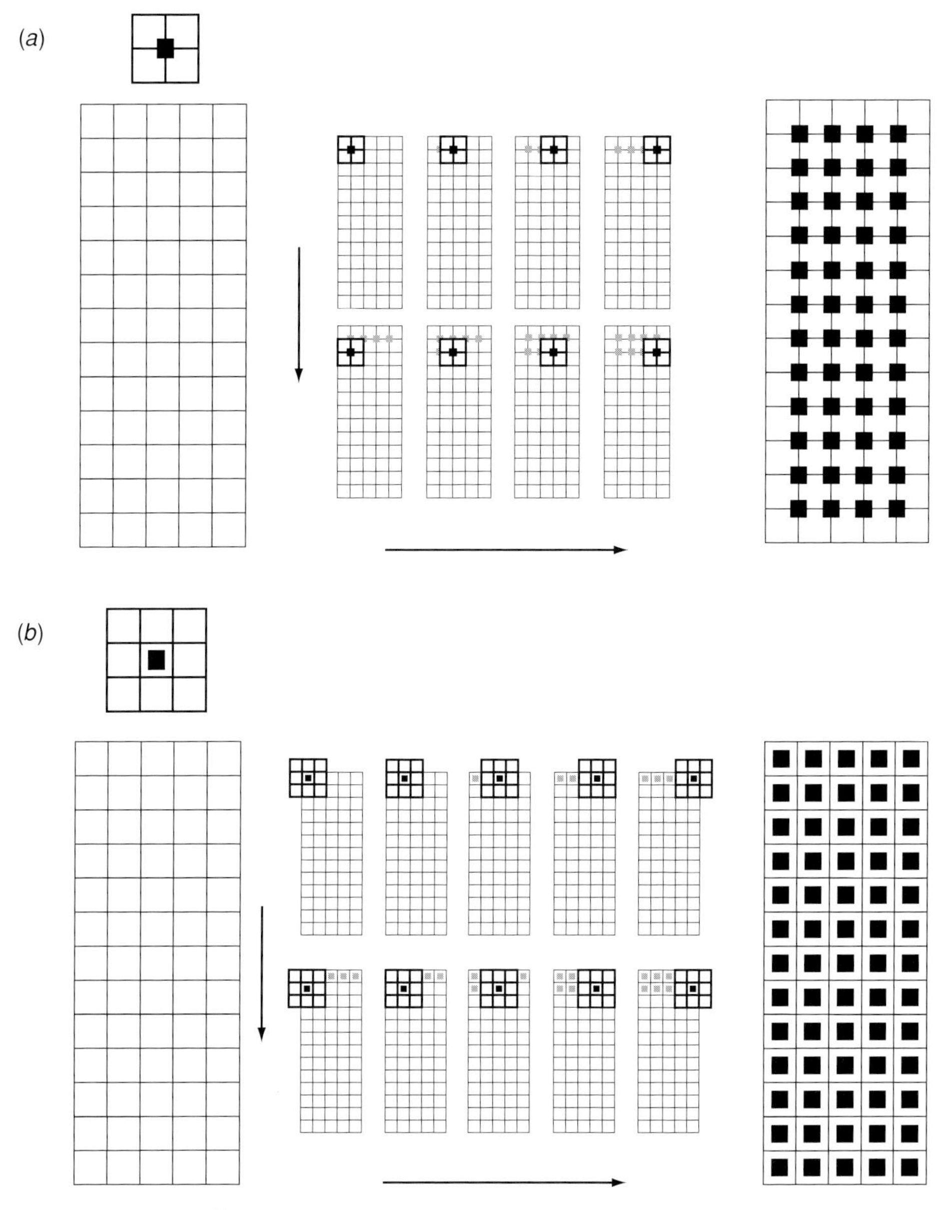


Figure 4.11 Difference measures computed by window (*a*) and kernel filters (*b*) to detect boundary. The window or the kernel, respectively, is slid one sampling unit at a time in the $x$ and then in the $y$ direction over the entire study area, and the difference at the centroid location of the window (the square in (*a*)) or at the central cell of the kernel (the square in (*b*)) computed. By doing so, there are fewer difference values computed with the window approach than in the original data ($n_{\text{rows}} - 1 \times n_{\text{columns}} - 1$) but not in the kernel approach where there are the same number of difference values as number of sampling locations.

several variables are available, both the amount of species turnover as well as their difference in values affect the location of the boundaries.

Given that the rate of change is computed according to spatial direction, it is also possible to compute the orientation, angle, of the change (Barbujani *et al.* 1989):

$$\theta = \tan^{-1}\left[\frac{\left(\frac{\partial f}{\partial x}\right)}{\left(\frac{\partial f}{\partial y}\right)}\right] + \Delta, \tag{4.5}$$

where

$$\Delta = \begin{cases} 0^\circ, & \text{if } \left(\frac{\partial f}{\partial x}\right) \geq 0, \\ 180^\circ, & \text{otherwise.} \end{cases} \tag{4.6}$$

The orientation of the gradient is established by first doubling the angles to avoid slopes of opposite direction cancelling each other, then averaging and halving the result. When the highest rate of change in one variable occurs in a north–south direction and that in another variable in a south–north direction, we do not want these two directions to cancel but to re-inforce this axis.

In essence, the magnitude of the rate of change $m$ is the slope of the plane that can be fitted to the values of the variables at the four adjacent sampling locations (Figure 4.12). Boundaries, defined as the spatial location where high rates of change occur, correspond to steep gradients among the four values of a variable. Weak differences among adjacent values will result in low (close to zero) values for rates of change. Such adjacent locations of low rates of change can be considered as a spatial homogeneous cluster: a patch. The major problem is to decide the threshold value of rate of change for boundary detection (Figure 4.12). When an arbitrary threshold is used, say the highest 10th percentile (Fortin 1994, 1997, 1999b), the rates of change are then called 'potential' or 'candidate' boundary elements. Subsequently boundary properties (length, width, shape, etc.) can be measured in terms of spatially connected candidate boundary elements using the boundary statistics (Section 4.2.4). The selection of the threshold depends on the context (strength of the boundaries and their number) and the number of sampling locations (Fortin 1999b). We will comment on this issue in Section 4.2.4.

Using 26 tree abundance data (Fortin 1992), lattice-wombling rates of change and their orientation were computed (Figure 4.13). The rate of change values were ranked in decreasing order and classified into 10 categories each being a 10th percentile: the highest percentile is indicated by 1, and the lowest by 10. Here, the number of candidate boundary elements at the 10% threshold is 7 out of 65 rates of

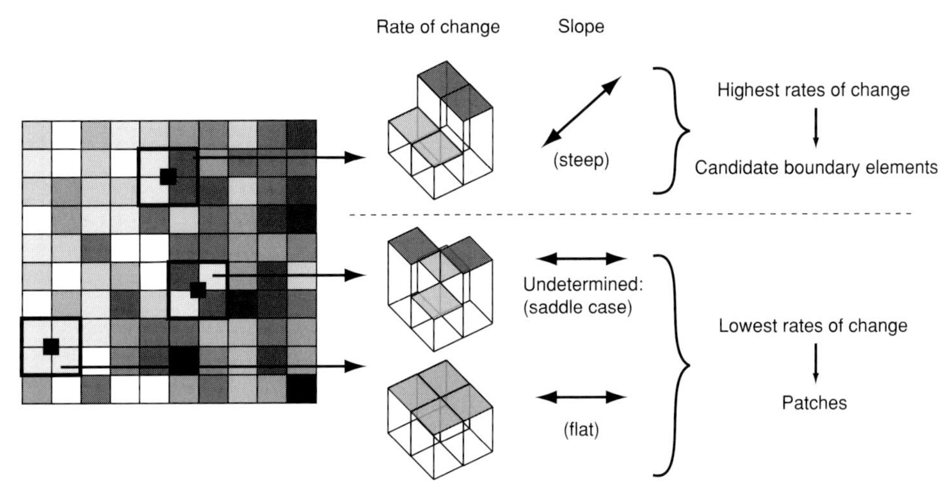


Figure 4.12 Lattice-wombling algorithm. Rates of change based on lattice-wombling are in essence the slope of the plane that fits the four values of the variable at the sampling locations. The orientation of the slope could also be useful in some studies. When the rates of change are ranked in decreasing order of magnitude (highest values of the slope), the candidate boundary elements are determined using an arbitrary threshold. When the rates of change are very low, close to zero, the sampling locations are most likely part of a patch.

change ($14 - 1 \times 6 - 1 = 65$ rates of change). There are cases, however, where at the cut-off rank between two percentiles, the rate of change value is exactly the same as the highest percentile. In such circumstances, in order to avoid bias in the selection of one rate of change location against another, it is recommended to continue to rank all rates of change that have the same values with the same percentile class and to include them all as candidate boundary elements. Here we selected the arbitrary threshold to be 10% (indicated by 1) to designate the candidate boundary elements. The candidate boundary elements that are adjacent to one another are linked in Figure 4.13*a*. It is found that there are candidate boundary elements in the lower part of the study area as well as in the middle. At a threshold of 10%, the candidate boundary elements are 'difference/open' boundaries that demarcate the strongest local boundaries. Looking at the subsequent highest rates of change (say 2, 3 and 4 ranked values in Figure 4.13*a*), these locations coincide with the demarcation among the spatial clusters based on the *k*-means algorithm with 20 clusters (Figure 4.5). Spatial clustering and boundary detection can be used as complementary methods to highlight different particularities of a study area: i.e. complete spatial partitioning (spatial clusters) and the spatial location and properties of the boundaries (width, shape, sinuosity) between the patches (Fortin & Drapeau 1995).

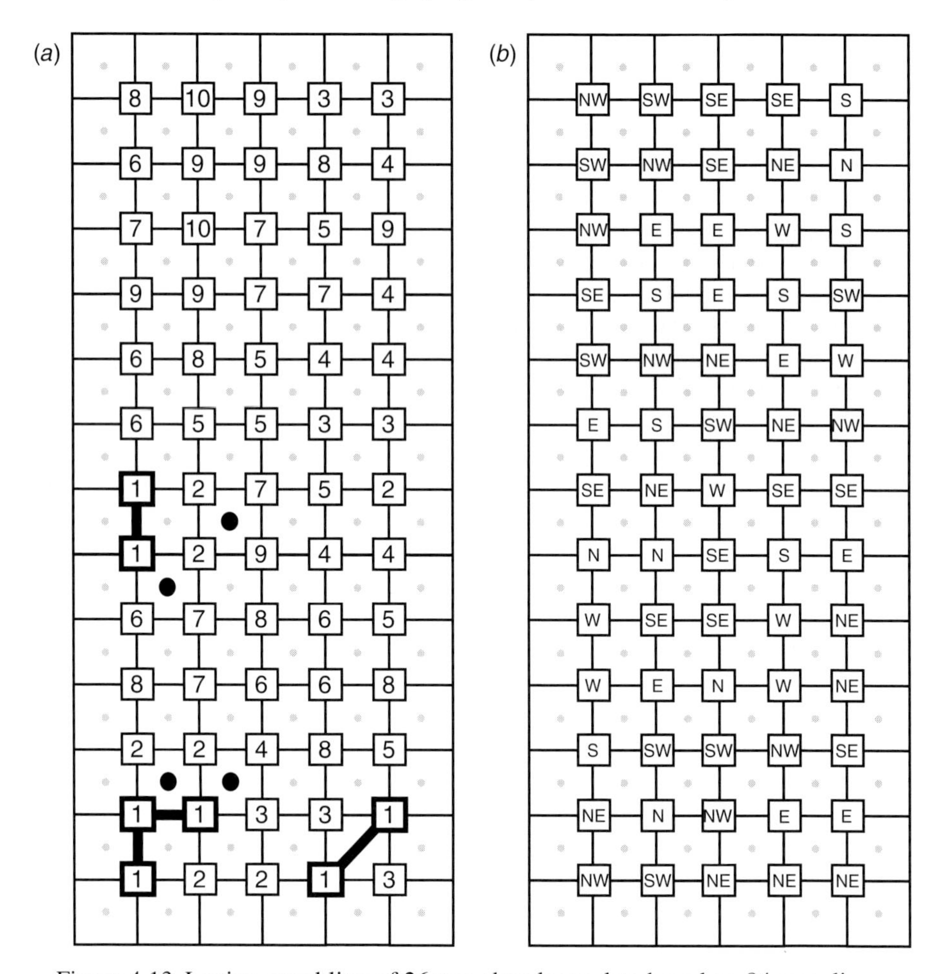


Figure 4.13 Lattice-wombling of 26 tree abundance data based on 84 sampling units (represented by squares with solid lines). (*a*) The 65 rates of change ($14_{\text{rows}} - 1 \times 6_{\text{columns}} - 1 = 13 \times 5$) are classified in 10 classes of 10th percentile each: the highest rate of change values are indicated by 1 and the lowest by 10. The candidate boundary elements are determined using the arbitrary threshold of the highest 10th percentile values (1 in bold) for a total of seven (rounded value). The bold lines link the candidate boundary elements that are spatially adjacent so that they can be connected into boundaries. There are three boundaries: two in the lowest part of the plot and one in the middle. The 10% second-highest derivatives (10% of 48 values = 5) are indicated by circles and they mark the end of the boundaries. (*b*) Orientation associated with each rate of change classified in eight directions (N, north; NE, northeast; E, east; SE, southeast; S, south; SW, southwest; W, west; and NW, northwest).

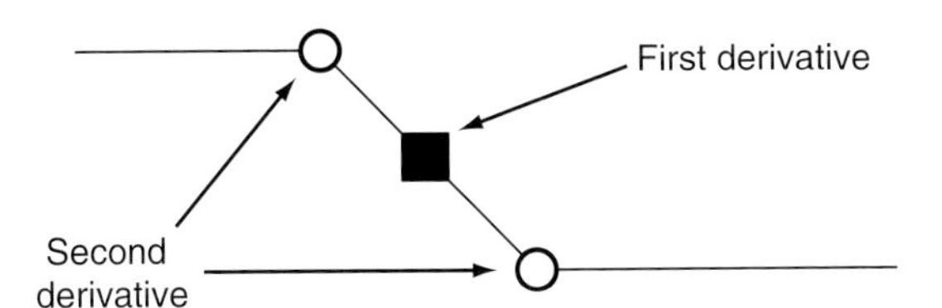


Figure 4.14 Lattice-wombling first derivatives identify boundary as the magnitude of the slope of a plane (square), whereas the second derivates identify the inflection point where the boundary ends (circle). The bold lines link the candidate boundary elements that are spatially adjacent so that they can be connected into boundaries. There are two boundaries: one in the lowest part of the plot and one in the middle.

As mentioned above, rates of change measure the magnitude of the slope of the gradient among adjacent sampling locations, demarcating boundary elements at the steepest part of the gradient (Figure 4.14). To obtain a better indication of a boundary's width and where it ends, second derivatives can be computed to identify the inflection point location corresponding to the limit of a boundary (Figure 4.14). Again, selecting only the highest 10% values of second derivatives (closed circles in Figure 4.13*a*), their location indicates where the boundaries end. The orientation associated with the rates of change is not very informative in the present case (Figure 4.12*b*). Indeed, this woodlot has several local gaps that make orientation of the rates of change uninteresting. In studies dealing with large-scale processes, as in Barbujani *et al.* (1989) where they investigated the migration path of human populations in Europe, the orientation of the rates of change may provide interesting insights.

The significance of each candidate boundary element cannot be tested using a complete spatial randomization procedure. The reason is that complete spatial randomness tests assume that each datum is independent. This is not the case in areas where patches and boundaries are suspected to occur because, within the patch and around the boundary zone, nearby sampling locations are more likely to have similar values (Fagan *et al.* 2003). When a complete randomization procedure is applied, the value of one patch can be placed next to that of another patch, creating a higher rate of change than in the observed data. Complete randomization is therefore much too conservative to test the significance of candidate boundary elements (Oden *et al.* 1993). A restricted randomization test, which considers the degree of spatial structure in the data, is therefore recommended (Fortin & Jacquez 2000). Also, when several variables are analysed, one could determine whether the candidate boundary elements are significant using a binomial test (Barbujani & Sokal 1991; Fortin 1994). This tests the significance of each candidate boundary element separately from all the other candidate boundary elements. For example, using the arbitrary threshold of the 10th percentile, each rate of change has a probability $p = 0.1$ of being classified as a candidate boundary element. If $a$ variables out of $b$ variables analysed are candidate boundary elements at a given locality, the probability that this location is overall significant for the rates of change of all the variables is given by a binomial test:

$$\Pr(a|b) = \binom{b}{a} 0.1^a 0.9^{b-a},$$

where $\binom{b}{a}$ is the number of possible ways to choose $a$ elements out of $b$. The location is said to be significant when the binomial probability of the actual count $a$, given the maximum possible number $b$, is less than or equal to the 5% level. To test whether connected candidate boundary elements form cohesive boundaries, boundary statistics should be used (see Section 4.2.4).

Finally, there are cases when the bilinear algorithm will not be able to provide accurate estimates of the gradient among the four adjacent sampling location values. For example when two diagonal values are high and two others are low, hence creating a saddle shape (see Figure 4.12), the gradient computed at the centroid point will misrepresent the data behaviour at those four sampling locations. Also, depending on the degree of spatial autocorrelation within patches, some boundaries may fall within patches, which makes the detection of cohesive ecological boundaries among patches more difficult (Csillag *et al.* 2001). One way to minimize within-patch boundaries and to maximize detection of between-patch boundaries is to carry out boundary detection at several spatial resolutions of the sampling units. Such scaling procedures allow identification of the degree of boundaries' persistence across scales (Fortin 1999b; Csillag *et al.* 2001; Handcock & Csillag 2002).

*Triangulation-wombling* In the field, ecological data are rarely completely surveyed on a lattice but rather data are sampled where the sampling locations are irregularly spaced. With such a data set, lattice-wombling cannot be carried out unless the data are initially interpolated onto a regular lattice. This is not recommended, however, because as presented in Chapter 3 most interpolation techniques smooth out the data, which in turn can diminish the strength of boundaries. Fortin (1994) proposed the use of a triangular window instead of a square one, where the three nearby sampling locations can be determined using a Delaunay triangulation algorithm that links sampling locations on the basis of triangles (see Chapter 2). By doing so, a plane can be fitted to the values of a variable observed at the vertices of the triangle. The magnitude of rate of change, $m$, based on the values of the three nearest sampling locations 1, 2 and 3 forming a triangle, is computed using the same equation as the one for lattice-wombling (Eqn (4.3)), but where $f(x, y)$ is:

$$f(x, y) = ax + by + c, \tag{4.7}$$

and

$$\begin{bmatrix} a \\ b \\ c \end{bmatrix} = \begin{bmatrix} x_1 & y_1 & 1 \\ x_2 & y_2 & 1 \\ x_3 & y_3 & 1 \end{bmatrix}^{-1} \begin{bmatrix} z_1 \\ z_2 \\ z_3 \end{bmatrix}. \tag{4.8}$$

The position of the centroid is at the location:

$$\left(\frac{x_1 + x_2 + x_3}{3}\right), \quad \left(\frac{y_1 + y_2 + y_3}{3}\right). \tag{4.9}$$

As for the lattice-wombling algorithm, the average rate of change can be computed as well as the orientation of the rates of change with Eqn (4.5). By using a triangle instead of a square window, the saddle problem that can occur with the square window is impossible. On the other hand, when the four nearest sampling locations are laid out as a perfect square, there are two possible combinations of triangles that can be selected, two triangle windows need to be selected (arbitrarily or not) for the analysis. When the four values are more or less similar, the selection of two triangles instead of two others will not have a big impact. However, if one or two values are very high and the others not, the rates of change will differ greatly and may affect the detection of boundaries.

Figure 4.15 illustrates the triangulation-wombling results based on a subset of 42 sampling locations from the 84 original ones. The triangulated systematic sampling design of the 42 sampling locations facilitates the visualization of the 64 triangle windows that were identified by the Delaunay connectivity algorithm (grey dashed lines in Figure 4.15). As in lattice-wombling, the number of calculated rates of change is smaller than the number of sampling locations, however, unlike lattice-wombling, there is no formula to evaluate this number because the number of triangles depends on the spatial arrangement of the sampling locations but it is usually around six (see Chapter 2). The candidate boundary elements, based on the highest 10th percentile (i.e. 10% of 64 triangles = 6), are all connected and located in the lower left part of the study area. Therefore, both lattice- and triangulation-wombling are congruent in their detection of boundaries (Fortin & Drapeau 1995).

*Categorical-wombling* It is common to have species presence : absence data over a two-dimensional area. In such cases, boundaries are located at the locations in space that have high-species turnover. These boundaries can be established using either spatial clustering as mentioned above, or by computing a match–mismatch measure between adjacent sampling locations (Oden *et al.* 1993). This last method is

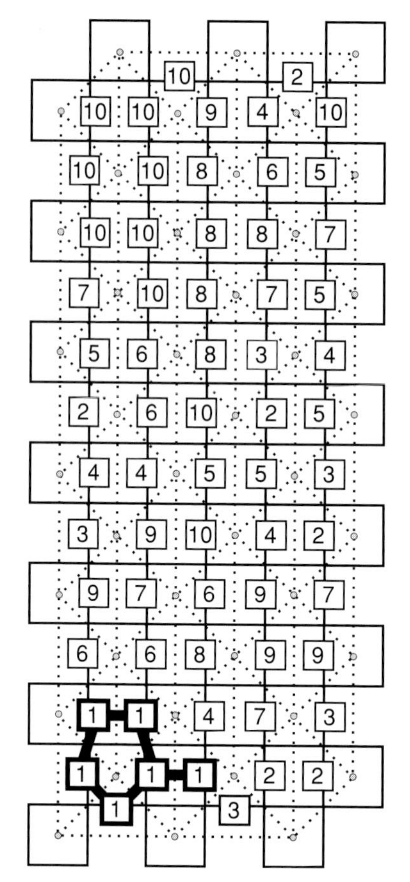


Figure 4.15 Triangulation-wombling of 26 tree abundance data based on 42 sampling units (represented by squares with solid lines). The 64 rates of change (one in each triangle that can be established using the Delaunay links – the dashed lines) are classified in 10 classes of 10th percentile each: the highest rate of change values are indicated by 1 and the lowest by 10. The candidate boundary elements are determined using the arbitrary threshold of highest 10th percentile values (1 in bold) for a total of six (rounded value). The bold lines link the candidate boundary elements that are spatially adjacent so that they can be connected into boundaries. There is only one boundary in the lowest part of the plot linking the six candidate boundary elements together. The locations of these boundaries coincide with those found with the lattice-wombling algorithm in Figure 4.19.

known as categorical-wombling (Oden *et al.* 1993, Fortin & Drapeau 1995), where mismatch values between adjacent sampling locations (i.e. not the same species in adjacent sampling locations or one species present in one sampling location and absent in another) are summed over all the categorical variables (here species). Adjacent sampling locations can be obtained using any connectivity algorithm (Chapter 2). The number of mismatches can be ranked as with the lattice- and triangulation-wombling algorithms and the highest values are represented at the midpoint between the linked sampling locations (Figure 4.16). Here there are 105 Delaunay links for a total of 11 candidate boundary elements, i.e. the highest 10th percentile. Given that the number of elements of candidate boundaries is higher than the two previous wombling methods (categorical-wombling = 11, lattice-wombling = 7 and triangulation-wombling = 6), boundaries are detected not only in the lower part of the study area but also in the middle.

When we want to find boundaries using only presence : absence data of one species (e.g. species geographical range limit), the categorical-wombling method is not appropriate and instead home-range delimitation methods based on kernel approaches should be used. We will not present these home-range methods here but refer the reader to Blundell *et al.* (2001) among others.

### 4.2.4 Boundary statistics

Boundary detection techniques can be as subjective as spatial clustering methods: in the former, the researcher decides at which threshold to consider rates of change as candidate boundary elements; whereas in the latter, the researcher determines both the degree of similarity and the number of clusters to use. To reduce the degree of subjectivity made in the choice of threshold and to test whether the candidate boundary elements at a given threshold form cohesive boundaries, Oden *et al.* (1993) developed subboundary statistics (hereafter only referred to as boundary statistics). These boundary statistics try to capture the desirable properties of cohesive boundaries in terms of connected candidate boundary elements (Fortin & Drapeau 1995; Bowersox & Brown 2001): i.e. a few boundaries (number of boundaries, Figure 4.17*a*) that are long (the length of a boundary) and that divide an area into patches (Figure 4.17*b*):

- The number of boundaries: the number of connected and isolated candidate boundary elements (four boundaries – A, B, C and D – in Figure 4.17*a*). When there are cohesive boundaries in the study area, the number of boundaries should be low. When there are weak difference boundaries, the number of boundaries is high.
- The number of singletons: the number of isolated, unconnected candidate boundary elements (one boundary, D, in Figure 4.17*a*). When there are cohesive boundaries in the study area, the number of singletons should be low. When there are weak difference boundaries, the number of singletons is high.
- The maximum length of the longest detected boundary in the study area: the number of connected candidate boundary elements (six candidate boundary elements in A, Figure 4.17*a*).
- The mean length of all the boundaries in the study area: ((6 + 3 + 2 + 1)/4 boundaries) = 3.

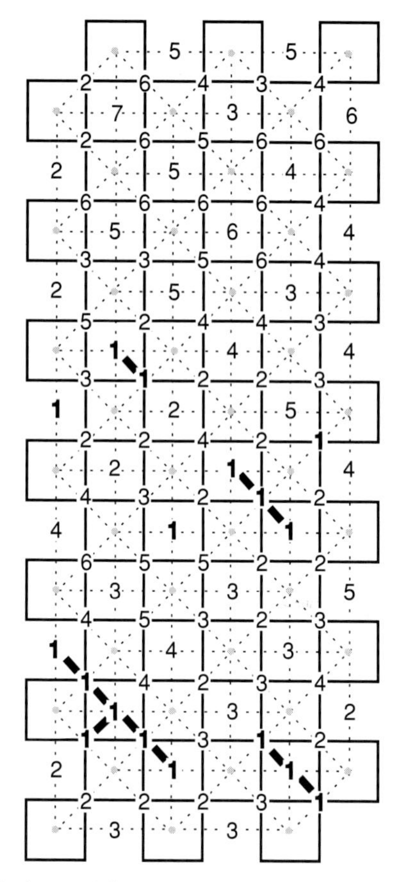

Figure 4.16 Categorical-wombling of 26 tree presence : absence data based on 42 units (represented by square with solid lines). The 105 rates of change (one for each Delaunay link – the dashed lines) are classified in 10 classes of 10th percentile each: the highest rate of change values are indicated by 1 and the lowest by 10. The candidate boundary elements are determined using the arbitrary threshold of the highest 10th percentile values (1 in bold) for a total of 11 (rounded value). However, given that several of the rate of change values are similar, there are 17 candidate boundary elements in total. The bold lines link the candidate boundary elements that are spatially adjacent so that they can be connected into seven boundaries: one in the lowest part of the plot linking six candidate boundary elements, three small boundaries (linking two or three candidate boundary elements) and three singletons (boundaries with only one candidate boundary element) either in the lower or middle part of the plot. The locations of these boundaries coincide with the ones found with those of the lattice-wombling algorithm in Figure 4.13 and of the triangulation-wombling algorithm in Figure 4.15.

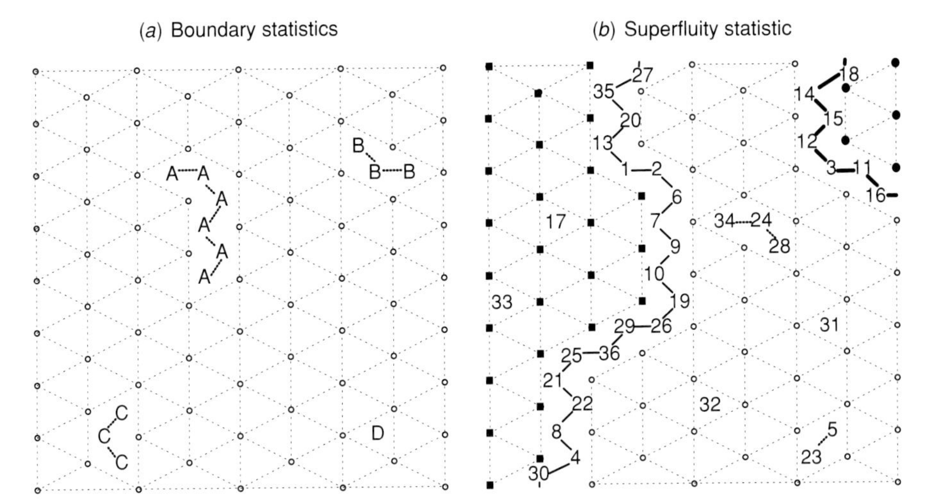


Figure 4.17 Boundary and superfluity statistics for 81 sampling locations (open circles). The Delaunay links are the dashed lines from which 128 triangles can be formed. Using the triangulation-wombling algorithm with a 10% threshold, there are 13 candidate boundaries elements and boundary statistics as follows: (*a*) There are four boundaries (A, B, C and D) where one is a singleton (boundary D). The maximum length and maximum diameter are both for boundary A with six candidate boundary elements in length and five in diameter. The mean length is 3.25 and the mean diameter is 2.5. (*b*) To compute the superfluity statistic, the rates of change need to be ranked in decreasing order (1 being the highest). The first boundary that partitions the study area into patches is formed of only seven rates of change (bold lines) but it was necessary to go down the list to the rank of 18th percentile. The remaining 11 rates of change are part of the next boundary that divides the big patch into two and one rate of change is part of a difference boundary in the lower right part of the plot. The superfluity statistic for the first partition is 0.63 and for the second it is 0.45. To partition the study area into three patches (open circles, closed circles and closed squares) it was necessary to go down the list of rates of change to the 36th rank. The rates of change not part of the two partitioning boundaries are small difference boundaries or singletons.

- The maximum diameter among the detected boundaries: the minimum distance, in terms of connected candidate boundary elements, between the two most extreme locations of the candidate boundary elements in a boundary (five candidate boundary elements in A, Figure 4.17*a*). These statistics will be the same as the maximum length when the boundaries are in a straight line; they will differ, however, when a boundary completely encloses a sampling unit: the length will be high but the diameter small. These two boundary statistics are therefore quite useful to discriminate among local and small discontinuities from more pronounced and cohesive boundaries.
- The mean diameters of all the boundaries in the study area: $((5 + 3 + 2 + 1)/4$ boundaries$) = 2.75$.
- The superfluity statistic: a measure of the efficiency of the boundaries to divide a study area into patches (i.e. the number of unnecessary rates of change, those whose removal

does not change the number of patches) divided by the number of necessary rates of change that separate a study area into patches (those whose removal decreases the number of patches). The value of superfluity is low when there are only a few major cohesive boundaries because the majority of the rates of change will be necessary to divide an area into two, or more, patches. The value of superfluity is high when there are several weak differences and singleton boundaries dispersed throughout an entire study area so that the majority of the rates of change do not contribute to the partition into patches. In Figure 4.17*b*, the first boundary to separate the area into two patches contains seven rates of change and 11 were unnecessary ones (in ranking in descending order the rates of change, it took up to the 18th percentile to divide the area into two patches): the superfluity statistic is 0.63. The second boundary dividing the area into three patches (open circles, closed circles and closed squares) is formed of 20 rates of change and nine were unnecessary, giving a superfluity statistic of 0.45.

Cohesive ecological boundaries will have significant boundary statistics that have extreme values: high ones for the number of boundaries, maximum length, maximum diameter, mean length and mean diameter and low ones for the number of singletons and the superfluity statistic. Significance tests assess whether the connected candidate boundary elements (i.e. boundaries) are likely to have occurred by chance or not. Oden *et al.* (1993) compared the two types of randomization procedures to assess the significance of the boundary statistics: a complete randomization test and a spatially restricted one considering the spatial structure of the data. They found that a null distribution of boundary statistics generated from a random spatial pattern led to a conservative test that falsely rejected the null hypothesis fewer times than when tested using a spatially autocorrelated null distribution.

### 4.2.5 Overlap statistics

Once cohesive boundaries have been delineated and their significance tested using boundary statistics, interesting ecological questions can be investigated using overlap statistics (Jacquez 1995; Fortin *et al.* 1996) that quantify the degree of spatial relationship between the locations of boundaries. Do boundaries directly overlap? Are boundaries spatially associated or do they repulse one another? In studying animal responses to forest boundaries, overlap statistics can be used to identify and to test which type of spatial relationships prevail. There are four overlap statistics: one measures the perfect spatial overlap between boundaries while the three others account for small spatial lags between the two boundaries due to sampling measurements errors:

- The direct overlap statistic, $O_s$, is the number of the candidate boundary elements that are at the same location. In Figure 4.18, $O_s = 7$.

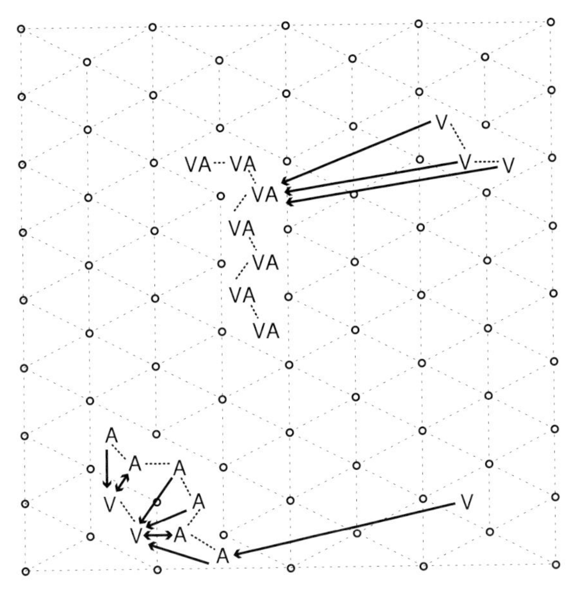


Figure 4.18 Overlap statistics between boundaries based on forest vegetation and animal abundance data for 81 sampling locations (open circles). The Delaunay links are the dashed lines from which 128 triangles can be formed. Using the triangulation-wombling algorithm with a 10% threshold, there are 13 candidate boundaries elements for a total of four vegetation boundaries (indicated by V) and a total of two animal boundaries (indicated by A). The lines with an arrow at one end indicate the minimum nearest distances between the two types of boundaries (from one type to the other), while the lines with double arrows indicate the cases where the minimum nearest distances are symmetric in both directions. The overlap statistics are: $O_s$ (direct overlap) is 7 candidate boundary elements (as indicated by the VA); the mean minimum nearest distance statistic, $O_V$, is 15.5 units; the mean minimum nearest distance statistic, $O_A$, is 6.5 units; and the overall mean minimum nearest distance statistic, $O_{VA}$, is 11.0 units. This example illustrates well that the animal boundaries are closer to the vegetation ones than the reverse; suggesting that animal boundaries are spatially associated with vegetation boundaries, but that vegetation boundaries are not spatially associated with the animal ones.

- The mean minimum nearest distance statistic, $O_1$, is an asymmetric measure of the distance from boundary 1 to boundary 2:

$$O_1 = \frac{\sum_{i=1}^{n_1} \min(d_{i.})}{n_1},$$

where $n_1$ is the number of candidate boundary elements in boundary 1 and $\min(d_{i.})$ is the minimum Euclidean distance between the $i$th candidate boundary element of boundary 1 to a candidate boundary element of boundary 2. In Figure 4.18, the minimum distance between vegetation boundaries and animal boundaries is 15.5 units.

- The mean minimum nearest distance statistic, $O_2$, is an asymmetric measure of the distance from boundary 2 to boundary 1:

$$O_2 = \frac{\sum_{j=1}^{n_2} \min(d_{.j})}{n_2},$$

where $n_2$ is the number of candidate boundary elements in boundary 2 and $\min(d_{.j})$ is the minimum Euclidean distance between the $j$th candidate boundary element of boundary 2 to a candidate boundary element of boundary 1. In Figure 4.18, the minimum distance between animal boundaries and vegetation boundaries is 6.5 units.

- The overall mean minimum nearest distance statistic, $O_{12}$, between boundaries 1 and 2:

$$O_{12} = \frac{\sum_{i=1}^{n_1} \min(d_{i.}) + \sum_{j=1}^{n_2} \min(d_{.j})}{n_1 + n_2}.$$

In Figure 4.18, the overall mean minimum distance between the two boundaries is 11.0 units.

The statistic $O_s$ allows us to test whether boundaries spatially coincide with one another, whereas the three other statistics, $O_1$, $O_2$ and $O_{12}$, can help discriminate between boundaries that are spatially associated (small significant values) or repulsing one another (large significant values). To test the significant spatial relationship between boundaries, it is recommended to randomize the rates of change, rather than the raw data, because the rates of change already include the inherent spatial structure of each variable and that is what is of interest. As an example, candidate boundary elements (20% threshold) based on tree species and shrub species are mapped in Figure 4.19 as well as the minimum distance between them: $O_s$ is 6 ($p = 0.1089$), $O_{\text{tree}}$ is 8.6 m ($p = 0.0990$), $O_{\text{shrub}}$ is 17.4 m ($p = 0.2178$) and $O_{\text{tree–shrub}}$ is 13.0 m ($p = 0.1188$). Although these overlap statistics are not significant, their trends indicate that the boundaries based on trees (8.6 metres) are closer to those based on the shrubs (17.4 metres). This indicates that tree-canopy opening due to gaps affect the spatial responses of both trees and shrubs, but that shrub changes (isolated singletons) do not affect trees.

The overlap statistics have been used to investigate the spatial relationship between forest edges and soil discontinuities (Fortin *et al.* 1996), as well as to test the relationship between bird and forest boundaries (Hall & Maruca 2001; St-Louis *et al.* 2004). These overlap statistics offer a new means of investigating forest edge effects on other wildlife and environmental variables.

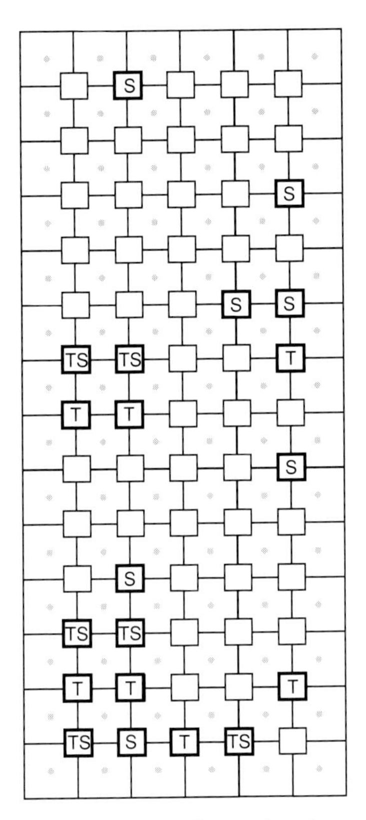


Figure 4.19 Overlap statistics between 26 tree abundance and 9 shrub abundance candidate boundary elements (using a 20% threshold; so a total of 13) based on the lattice-wombling algorithm and 84 sampling locations: $O_s$ is 6 ($p = 0.1089$), $O_{\text{tree}}$ is 8.6 m ($p = 0.0990$), $O_{\text{shrub}}$ is 17.4 m ($p = 0.2178$) and $O_{\text{tree–shrub}}$ is 13.0 m ($p = 0.1188$). These results are not significant but can be used, combined with map observations, to indicate that tree-canopy gaps affect the spatial responses of both trees and shrubs but changes in shrubs (isolated singletons) do not affect trees.

### 4.2.6 Boundary detection based on one variable

With only one quantitative variable (e.g. vegetation productivity based on the normalized difference vegetation index, NDVI), the lattice-wombling algorithm, being a local boundary detector, may not accurately detect a boundary, especially when there is local noise and spatial autocorrelation in the data. In such circumstances, hierarchical global edge detectors and kernel filters are more appropriate, and we will describe those next.

#### *4.2.6.1 Hierarchical global partitioning using wavelets*

A hierarchical global boundary detector, such as the wavelet transform analysis, can be used to identify boundaries from quantitative data along a transect (Redding *et al.*

2003) or an area (see Csillag & Kabos 2002 for mathematical details). In Chapter 2, wavelet variance was introduced as a method to characterize and determine the scales of spatial patterns. Wavelet analysis is also used to compress an image to use less storage (Daubechies 1993) by using only a few wavelet transformation coefficients that can model the structure of relatively homogeneous subregions of an image. This feature can consequently be used to detect boundaries. This is achieved by partitioning and characterizing an image into relatively homogeneous areas using as many waveforms as are needed to model local pattern. Relatively homogeneous areas require few coefficients of low values whereas contrasting locations, such as edges, require more coefficients with larger values. Wavelet analysis allows local multiscale analysis of the data by partitioning the data into relatively homogeneous spatial subareas. The determination of these spatially homogeneous areas is obtained using a hierarchical procedure based on 'quadtree' decomposition (Csillag & Kabos 1996). Quadtree is a recursive algorithm that partitions an area into four initial quadrants and continues to divide each quadrat into four smaller quadrants in a hierarchical way until relatively homogeneous subareas are obtained. Depending on the spatial structure of the data, only one hierarchical partition will be sufficient to create a homogeneous subarea where a few wavelet transformation coefficients adequately describe the structure; while for other subareas, it will be necessary to add more partitions. The resulting subareas are often referred to as the leaves of the quadtree partition. The smallest possible leaf is the sampling unit itself. At each partition, wavelet transform coefficients are added to describe the structure and to indicate the level (scale) of partitions needed. For each leaf there is an equation. To obtain a higher degree of fit with the spatial pattern of data, more waveforms (wavelet transform coefficients) need to be kept at each partition level. For example, using the lattice-wombling of rates of change based on the 26 tree abundance data, the number of homogeneous areas increases when more coefficients are kept in the equation (Figure 4.20*a*, *b*; Table 4.2). For image compression purposes, however, there is a trade-off between the amount of resolution retained and the storage of wavelet transformation coefficients at each scale and leaf. Usually more precision implies more leaves, thus more coefficients. Because the extent of the tree data is small (13 rows $\times$ 5 columns = 65 values), the partition cannot be performed for more than two hierarchical levels. Using black spruce percentage cover data over a larger region (425 rows $\times$ 350 columns = 148,750 values), the partition can be extended up to the four hierarchical levels (Figure 4.21).

#### *4.2.6.2 Edge enhancement with kernel filters*

The kernel filters are local edge detectors used to enhance the contrast of the adjacent pixel of an image in order to detect the edges of objects. There are several

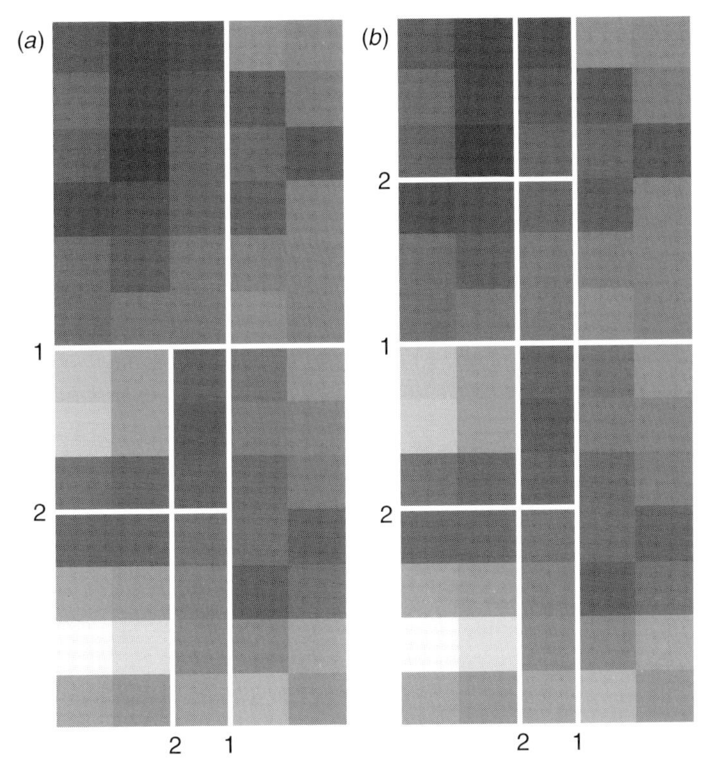


Figure 4.20 Spatial partition using the wavelet transforms at two hierarchical levels based on the rates of change of the lattice-wombling algorithm for the 26 tree abundance data (high values of rate of change are in white and low values in black): (*a*) seven subareas were found using only a few coefficients (10%) to describe the spatial structure in each; whereas in (*b*) 10 subareas were detected when using more coefficients (40%). Given the small sample size, only two partitions could be computed. Both in (*a*) and (*b*) the lower and middle of the plots show boundaries. When more coefficients are kept (*b*), other boundaries are identified in the upper left section of the plot.

algorithms, called operators, which are available in most GIS and remote sensing software packages (Pitas 2000). The first ones to be developed aimed to measure the gradient on adjacent pixels using first- and second-order derivatives, such as the Laplacian filter. The $3 \times 3$ discrete approximation kernel version of the second-derivative Laplacian is:

| | | |
|---|---|---|
| 1 | 1 | 1 |
| 1 | –8 | 1 |
| 1 | 1 | 1 |

,

Table 4.2 *Summary of the spatial analysis methods presented in Chapter 4*

| Spatial analysis method | Template |
|---|---|
| Spatial clustering | 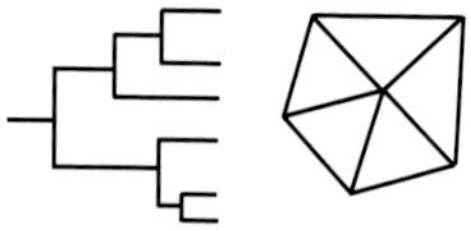 |
| Split-moving window | 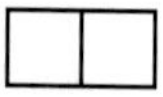 |
| Lattice-wombling | 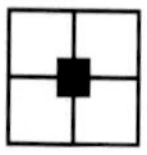 |
| Triangulation-wombling | 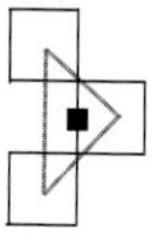 |
| Categorical-wombling | 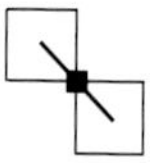 |
| Enhancement filter | 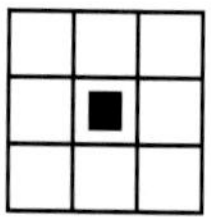 |
| Wavelets | 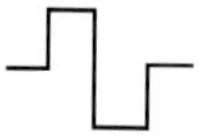 |
| Scale-space | 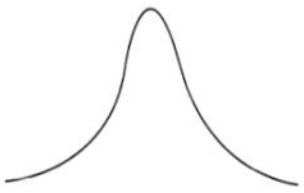 |

where the sum of the kernel equals zero. Note that this Laplacian filter is the same as the template of the 9TLQV as in Figure 2.32. With such a filter, the resulting values are all zero except at the locations where an edge begins and ends, hence, boundaries are easier to detect. The major problem with the Laplacian operator, however, is that it is sensitive to noise, making it necessary to smooth the data first. Smoothing the data can be achieved either by aggregating adjacent cells, obtaining fewer larger cells (see Fortin 1999b), or by using a Gaussian filter that preserves the same number of cells. The most efficient kernel filters both reduce the noise and detect edges, such as the Canny adaptive filter (Canny 1986) or the scale-space techniques using the Laplacian or Gaussian algorithm (Marr & Hildreth 1980; Lindeberg 1994; Faghih & Smith 2002). The scale-space techniques perform a series of smoothing

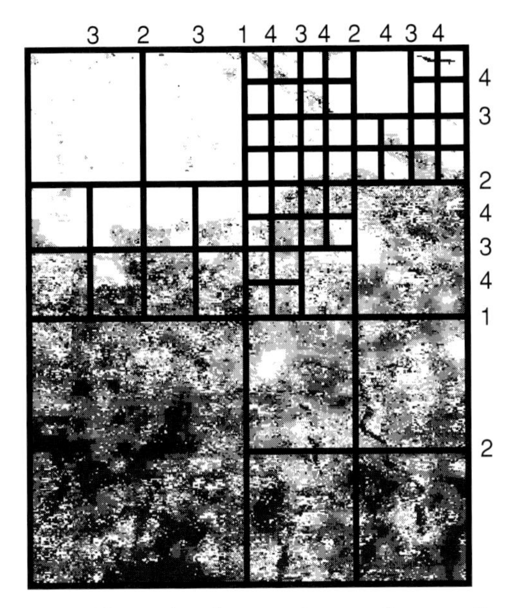


Figure 4.21 Spatial partition using the wavelet transforms at four hierarchical levels based on percentage coverage of black spruce in Quebec boreal forest (high values in white and low values in black). Most of the spatial partitions are in the upper right quadrant where the juxtaposition of high (white) and low (black) values necessitates more partition, i.e. coefficients, to describe the data. Indeed, trend patterns such as those seen in the upper left quadrat are easier to characterize than several small patches of low percentage coverage in forests of high percentage coverage.

using a Gaussian kernel of increasing size, allowing detection of the persistence of boundaries across scales. Figure 4.22 combines the smoothed data based on a Gaussian filter using a scaling factor of 40 cells and the delineated edges based on the Laplacian filter. Unlike the wavelet analysis that finds partitions based on the entire region (Figure 4.21), the scale-space approach identifies many more local boundaries because all the edges are mapped. As mentioned, the Laplacian algorithm localizes edges where the sum of the kernel is not zero. Thus, the boundaries can be either strong or weak, but will be treated similarly. Note that the use of overly large kernel size can distort the spatial partitioning by smoothing the data in an isotropic fashion. The scale-space approach has been used in forestry to identify individual trees in a forest from high spatial resolution aerial imagery data (Brandtberg 1999).

Several other kernels are available, such as non-linear ones based on polynomials or global thresholding kernels. The reader interested in kernel filters should read the computer vision and image recognition literature applied to remotely sensed data and medical imagery.

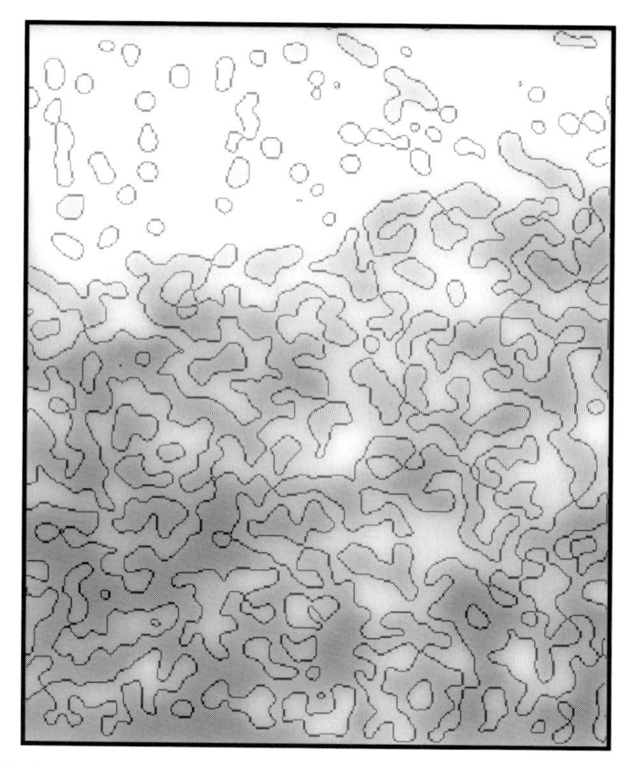

Figure 4.22 Spatial partition using the scale-space technique of the percentage coverage of black spruce in Quebec boreal forest as in Figure 4.21. First the data are smoothed using a Gaussian filter with a scaling factor of 40 cells (high values in white and low values in black) and then edges are delineated using a Laplacian filter (solid lines). All the edges are mapped without knowing which ones indicate the sharpest difference among cells.

## 4.3 Concluding remarks

The boundary detection methods presented in this chapter are only the tip of the iceberg. Here we focus only on the methods that have been developed, or more commonly used, by ecologists (Table 4.2). There are, however, numerous other methods mostly developed for image recognition that enhance edges by smoothing and thresholding noise in an image. These sophisticated kernels are useful when only one variable is available. Caution is advised, however, because by seeking to smooth out noise using larger kernel sizes the spatial pattern can be deformed by imposing an isotropic shape.

In ecological contexts where several variables are used to detect ecotones or cohesive ecological boundaries, several conceptual and methodological aspects need to be addressed. From an ecological point of view, all variables and species may not all have the same weight in the detection of boundaries. Hence, rare or omnipresent species may be not as important as indicator species or species responding to specific environmental conditions. From a methodology perspective, novel ways to measure the spatial overlap between boundaries are needed so that we can compute the distance between a line boundary (vector mode) and a difference boundary (raster mode).

Finally, the most challenging issue to be addressed is the need to integrate both ecological concepts and statistical theory. What are the most appropriate ways to generate restricted randomization procedures that can test if boundaries are cohesive ones? Indeed, as presented above, complete spatial randomness is not an appropriate procedure to test the significance of rates of change, boundary statistics and overlap statistics. Restricted randomization that captures the spatial structure of the data is recommended. With any ecological data, restricted randomization procedures need to reflect our ecological understanding of the processes. In the particular case of ecological boundaries, several processes are acting: the processes that created the patches and those that generated the boundaries. Usually, they are all different, or at least the processes that generated the patches on each side of the boundary that separates them are different. Consequently, a given study area where patches and boundaries occur is the perfect example of a non-stationary situation where global spatial statistics and global randomization cannot be performed. Hence we are in a 'chicken and egg' situation where the proper way to restrict the randomization by area is first to identify these subareas having the same stationarity: but this is exactly what we are seeking by doing spatial partitioning. Issues related to randomization procedures with ecological data are discussed in more depth in Chapter 7.

# 5

# Dealing with spatial autocorrelation

## Introduction

The familiar procedures of parametric statistics are based on the assumption of independence of the individual observations in the data under scrutiny, but in ecological data the assumption of independence is often violated and we need to understand the effects of a lack of independence. A lack of independence can arise because, in the natural world, things (samples, observations, etc.) that are closer together sometimes have a tendency to be more similar than those that are further apart, a phenomenon known to geographers as 'Tobler's Law' (Tobler 1970, and see Chapters 1 and 3). We refer to this lack of independence as 'spatial dependence' in the data (see Chapter 1), whatever the cause. One source of this phenomenon is autocorrelation in the data due to causal interactions within the measured variable itself; for example, in studying species distribution and abundance, the abundance of a single species may be spatially autocorrelated because of constraints on the organisms' mobility and dispersal. This kind of autocorrelation is sometimes called 'true autocorrelation' (see Chapter 1), but it might be more accurate to refer to it as 'inherent autocorrelation' ('autogenic' might be even more accurate, perhaps, but more unwieldy). The descriptor is to distinguish this phenomenon from 'induced spatial dependence' (see Chapter 1), where the observed variable (e.g. species abundance) has a functional dependence on an underlying variable (e.g. soil moisture or nutrient content), which is itself autocorrelated (cf. Legendre *et al.* 2002). It may not be easy to distinguish the two in ecological studies, and it is possible that both may occur in a single example. In fact, for many types of ecological study, this may be a common situation with both the biological response variable and an underlying environmental factor having some inherent spatial autocorrelation. In this chapter, we will concentrate on inherent autocorrelation and induced spatial dependence, describing their characteristics and effects in some detail.

Autocorrelation is a 'fact of life' for ecologists, because the data we analyse almost always come from particular physical locations, and natural systems almost always have autocorrelation in the form of patchiness or gradients, both of which can occur over a wide range of spatial and temporal scales. As Legendre (1993) suggests, ecologists must learn to deal with autocorrelation, whether it is viewed as a nuisance that causes trouble for statistical testing, or as part of a new understanding of the importance of spatial structure to ecological processes. As with most ecological phenomena, autocorrelation is scale or distance dependent; the values of a particular variable may be more similar than expected at short distances (positive autocorrelation) but less similar than expected at greater distances (negative autocorrelation). A useful first step, therefore, is to consider some of the ways in which spatial autocorrelation can vary with distance.

This chapter is a bit different from those preceding, because of the importance of models (and their relation to underlying processes) to the discussion. In organizing the material for this chapter, it became clear that models are needed to understand the underlying concepts of the subject, and not just for technical details. We will begin by discussing simple models of data in one dimension (spatial or temporal), and later we will describe more general versions of these structures. Historically, this approach was developed first for time series and was applied only later to spatial series. The first models apply to observations at a series of $n$ locations, $x_1, x_2, \ldots, x_i, \ldots, x_n$, with the following structures:

Model 1a, complete independence:

$$x_i = \varepsilon_i, \quad \varepsilon_i \approx N\big(0, \sigma_\varepsilon^2\big). \tag{5.1a}$$

Here, $\varepsilon_i$ is an independent 'error' term following some statistical distribution, such as a normal distribution with a mean of zero and variance $\sigma_\varepsilon^2$. Figure 5.1*a* illustrates the fact that the expected spatial autocorrelation between observations is 0. The term $\varepsilon_i$ is usually assumed to follow a normal distribution, but that is not necessary for independence, and other distributions might be considered. An alternative model of independence can include functional dependence, but retaining spatial independence:

Model 1b, spatial independence:

$$\begin{cases} x_i = \beta z_i + \varepsilon_i, \\ z_i = \xi_i. \end{cases} \tag{5.1b}$$

Both $\varepsilon_i$ and $\xi_i$ are independent 'error' terms following some statistical distribution, and $\beta$ is the linear regression parameter. The expected correlation between

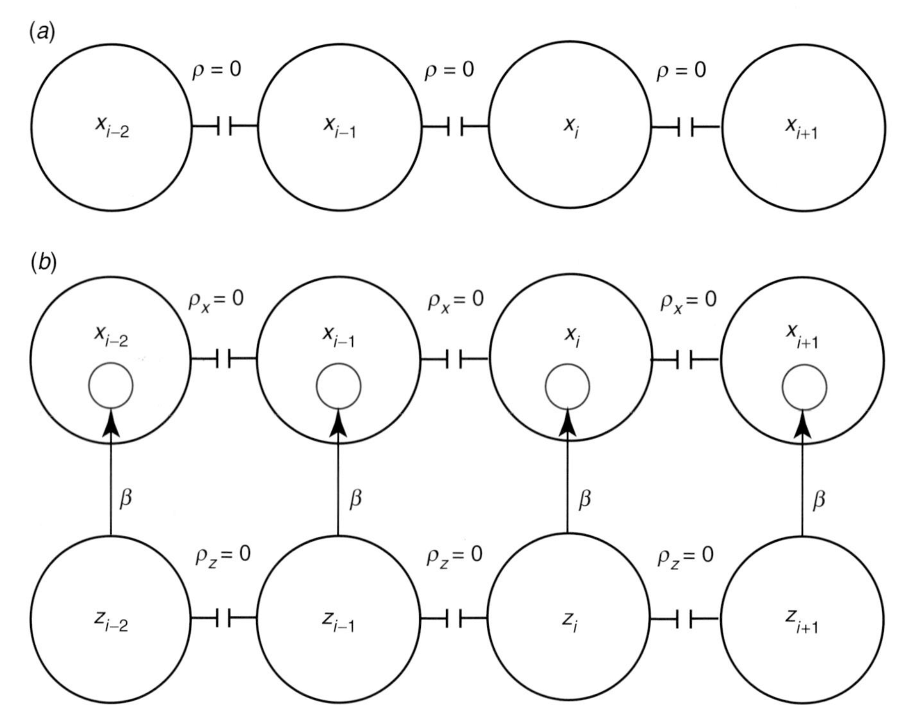


Figure 5.1 (*a*) Model 1a: the series of values is independent of each other and so the expected correlation between adjacent values is 0 (see Eqn (5.1a)). (*b*) Model 1b: the series of $x$ values is dependent on the $z$-series, which are independent of each other. The $x$ values remain spatially independent and so the expected correlation between adjacent values is 0 (see Eqn (5.1b)).

adjacent values of $x$ and between adjacent values of $z$ is zero, as illustrated in Figure 5.1*b*.

Model 2, inherent autoregressive:

$$x_i = \rho x_{i-1} + \varepsilon_i, \quad -1 \leq \rho \leq +1. \tag{5.2}$$

Here, $\varepsilon_i$ is an error term as above, and $\rho$ is the autocorrelation parameter that determines the strength of the autocorrelation. Figure 5.2 illustrates the model; the expected correlation between adjacent values is $\rho$.

Model 3, induced autoregressive:

$$\begin{cases} x_i = \beta z_i + \varepsilon_i, \\ z_i = \rho z_{i-1} + \xi_i, \end{cases} \text{where} \quad \xi_i \approx N\left(0, \sigma_\xi^2\right). \tag{5.3}$$

In this model, both $\varepsilon_i$ and $\xi_i$ are normally distributed error terms, $\beta$ is the usual regression parameter and $\rho$ is again the autocorrelation parameter. The correlation

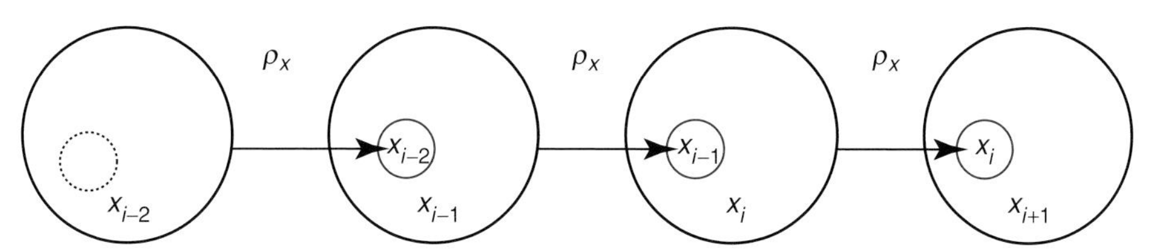


Figure 5.2 Model 2: the values in the series are not independent, with each being directly dependent on its predecessor and thus indirectly dependent on all preceding values. The expected correlation between adjacent values cannot be 0, but $\rho$ (see Eqn (5.2)).

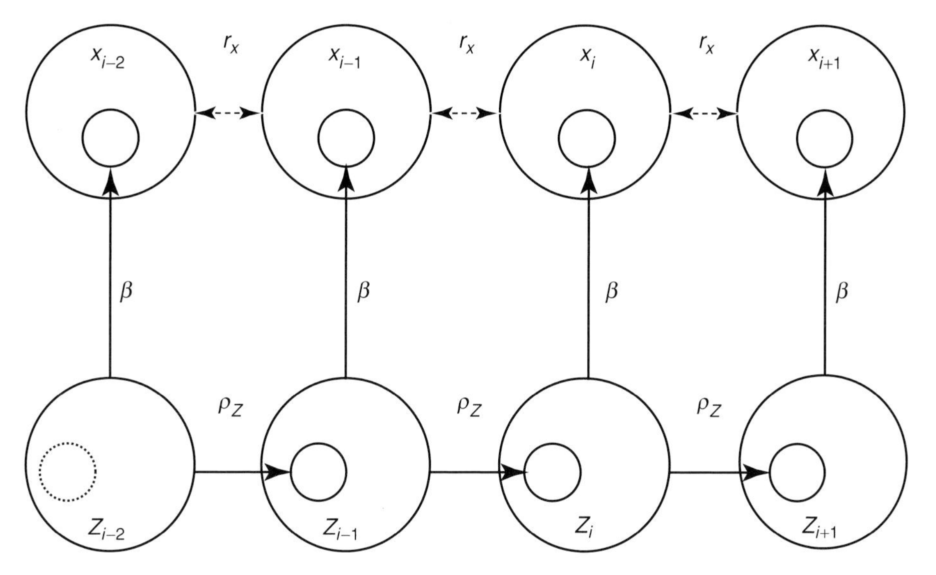


Figure 5.3 Model 3: the values in the $x$-series are not independent, but their observed autocorrelation, $r_x$, is induced by their linear dependence on the $z$-series, which is governed by a Model 2 process (see Eqn (5.3.)).

between adjacent values of $x$ is not expected to be 0, but it is a function of the values of $\beta$ and $\rho$, as illustrated in Figure 5.3. Again, the second error term, $\xi_i$, need not follow a normal distribution, but many discussions will assume that it does.

Model 4, doubly autoregressive:

$$\begin{cases} x_i = \beta z_i + \rho_x x_{i-1} + \varepsilon_i, \\ z_i = \rho_z z_{i-1} + \xi_i. \end{cases} \tag{5.4}$$

The symbols are as in previous models, but there are now two autocorrelation parameters, one for $x$ and one for $z$. This more complicated model is referred to

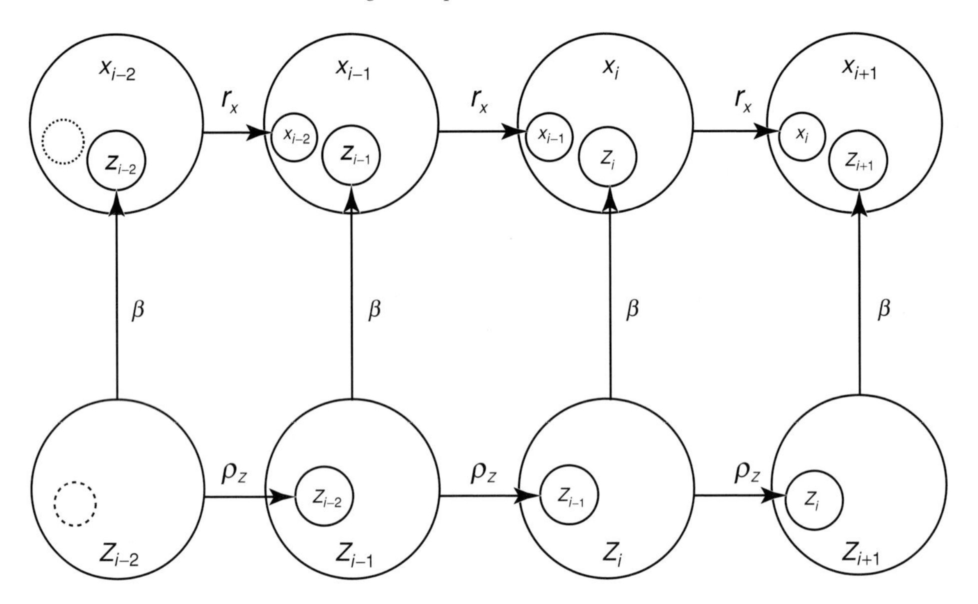


Figure 5.4 Model 4: Correlation of $x$ values comes from two sources, inherent in $x$ itself and induced by dependence on $z$, giving a doubly autoregressive model.

as doubly autoregressive because it includes both inherent and induced forms of spatial dependence; it is illustrated in Figure 5.4.

In Model 1 (both 1a and 1b), the values of $x$ are independent of each other and Model 2 gives rise to what is called a first-order autoregressive structure, which is probably the most frequently studied form of autocorrelation. In that case, the correlation of any two variates, $x_i$ and $x_j$, depends only on the separation between them (here distance can be measured by the number of intervening steps):

$$\mathrm{Cor}(x_i, x_j) = \rho^{|i-j|}. \tag{5.5}$$

For example, it is easy to show, by substituting for $x_{i-1}$ in (5.2), with the second version of that equation, that:

$$x_i = \rho(\rho x_{i-2} + \varepsilon_{i-1}) + \varepsilon_i = \rho^2 x_{i-2} + \rho\varepsilon_{i-1} + \varepsilon_i, \tag{5.6}$$

so that $\mathrm{cor}(x_i, x_{i-2}) = \rho^2$. The first-order autoregressive structure discussed here is the simplest of a more general $k$th-order autoregressive model (Cressie 1991):

$$x_i = \sum_{j=1}^{k} \rho_k x_{i-k} + \varepsilon_i. \tag{5.7}$$

In Model 3, the dependence observed in $x$ is induced by its linear dependence on $z$ and $z$'s inherent autoregressive structure. With more effort, we can show that

Table 5.1 *Matrix of covariance among positions for a first-order autoregressive inherent correlation structure*

| | | | | | | |
|---|---|---|---|---|---|---|
| 1 | $\rho$ | $\rho^2$ | $\rho^3$ | $\dots$ | $\rho^{n-1}$ | |
| $\rho$ | 1 | $\rho$ | $\rho^2$ | $\dots$ | $\rho^{n-2}$ | |
| $\rho^2$ | $\rho$ | 1 | $\rho$ | | $\rho^{n-3}$ | $\sigma_\varepsilon^2/(1-\rho^2)$ |
| $\rho^3$ | $\rho^2$ | $\rho$ | 1 | | $\rho^{n-4}$ | |
| $\vdots$ | | | | $\ddots$ | $\vdots$ | |
| $\rho^{n-1}$ | $\rho^{n-2}$ | $\rho^{n-3}$ | $\rho^{n-4}$ | $\dots$ | 1 | |

Table 5.2 *Matrix of covariance among positions for a first-order autoregressive induced correlation structure. The constant is* $\kappa = \beta^2\rho/(\beta^2\sigma_\xi^2 + (1-\rho^2)\sigma_\varepsilon^2)$ *from Eqn (5.3).*

| | | | | | | |
|---|---|---|---|---|---|---|
| 1 | $\kappa\rho$ | $\kappa\rho^2$ | $\kappa\rho^3$ | $\dots$ | $\kappa\rho^{n-1}$ | |
| $\kappa\rho$ | 1 | $\kappa\rho$ | $\kappa\rho^2$ | $\dots$ | $\kappa\rho^{n-2}$ | |
| $\kappa\rho^2$ | $\kappa\rho$ | 1 | $\kappa\rho$ | | $\kappa\rho^{n-3}$ | ${\sigma_\xi}^2$ |
| $\kappa\rho^3$ | $\kappa\rho^2$ | $\kappa\rho$ | 1 | | $\kappa\rho^{n-4}$ | |
| $\vdots$ | | | | $\ddots$ | $\vdots$ | |
| $\kappa\rho^{n-1}$ | $\kappa\rho^{n-2}$ | $\kappa\rho^{n-3}$ | $\kappa\rho^{n-4}$ | $\dots$ | 1 | |

the correlation produced by this model is of a similar form:

$$\mathrm{Cor}(x_i, x_j) \propto \rho^{|i-j|}, \quad \text{for } i \neq j. \tag{5.8}$$

The $n \times n$ variance–covariance matrix for the $n$ $x$s is expected to be as shown in Table 5.1 for Model 2 and as in Table 5.2 for Model 3.

The value of the variance is different for the two models, being proportional to $\sigma_\varepsilon$ for Model 2 and to $\sigma_\xi$ for Model 3, but the basic structure is the same. If both these variances are 1, for example, for $\rho = 0.4$ and $\beta = 0.6$, the expected correlation of adjacent observations for Model 3 is 0.12, considerably less than the 0.48 in Model 2. The covariance structure for Model 4 (doubly autoregressive) is, not unexpectedly, considerably more complicated, but under the same parameter values ($\beta = 0.6$; $\rho_x = \rho_z = 0.4$), the expected correlation of adjacent values is approximately 0.5.

The models discussed so far seem to have been formulated for temporal rather than spatial series. This impression results from the fact that the models make explicit the dependence of $x_i$ on $x_{i-1}$ (Model 2) and thus on all preceding values of $x$. This may seem inappropriate for spatial data, even in only one dimension, where we would expect the dependence to be equal in both directions. Remember, however, that in this apparently directional model, the correlation between any two variates depends only on the distance between them, not on the direction (Eqn (5.5)). Consider the following thought experiment: we provide a data series generated by Model 2, and a second series which is the first series reversed; what criteria could be used to distinguish the original? In fact, Cressie (1991) commented that while the autocorrelation structure given by Eqn (5.5) can be generated by Model 2, it can also be associated with spatial data not generated in that way (Cressie 1991, p. 14). For example, if the amount of pollen produced by a single isolated tree declines with distance, $d$, such that the amount is proportional to $(1/2)^d$, measurements every metre might resemble the output of Model 2, but Model 2 might not be a good description of how the observed pattern arises. Therefore, we can continue to use these models to provide insight into the characteristics of spatial dependence, for which we may not know the underlying process that generates the correlation structure we detect.

If we only have a single spatial data set, such as one series of $n$ observations of a variable recorded for a single time, it is like having only a single realization of an underlying model or process and we cannot calculate the covariance of two individual values, such as $\text{cov}(x_3, x_5)$, for example. We can, however, calculate an observed covariance for all values separated by distance 1, by distance 2 and so on, as estimates of the underlying covariances, and that is what a covariogram or correlogram does (see Chapter 3). The details of the calculation are as follows. The estimated covariance for lag $d$ is:

$$C_x(d) = \frac{\sum_{i=1}^{n-d} x_i x_{i+d} - \sum_{i=1}^{n-d} x_i \sum_{j=d+1}^{n} x_j / n}{n-d-1} = \frac{\sum_{i=1}^{n-d} (x_i - \bar{x}_{(1,n-d)})(x_{i+d} - \bar{x}_{(d+1,n)})}{n-d-1}, \tag{5.9}$$

where the sample means used are for subsets of the data based on the first and the last $n - d$ values in the series, which we will refer to as the 'regional' means. For example, in Figure 5.5, for $n = 12$ and a lag of 3, the sample means of $x_4$ to $x_{12}$ and of $x_1$ to $x_9$ would be used. It is tempting to calculate the correlation for lag $d$ using the regional sample variances (for example, of $x_4$ to $x_{12}$ and of $x_1$ to $x_9$ again in Figure 5.5), as we would for any usual calculation of a correlation coefficient:

$$r_x^*(d) = \frac{C_x(d)}{\sqrt{s^2_{x(1,n-d)} s^2_{x(1+d,n)}}}, \tag{5.10}$$

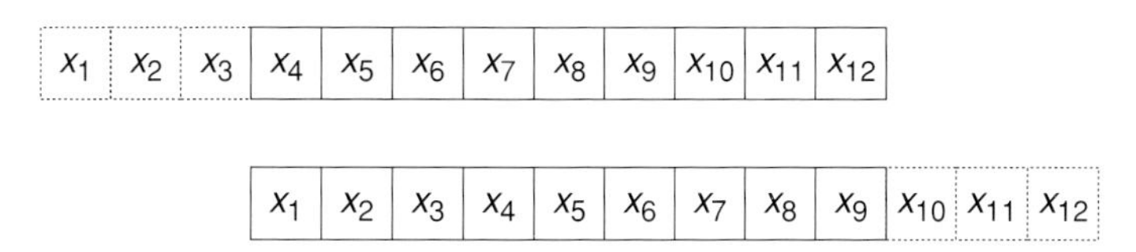


Figure 5.5 The calculation of autocovariance or autocorrelation for $n = 12$ and lag $d = 3$: the 'regional' sample means are calculated from $x_1$ to $x_9$ and from $x_4$ to $x_{12}$. Using the regional values of sample variance results in poor estimates of the autocovariance or autocorrelation.

but this leads to poor estimates (Legendre & Legendre 1998, Jenkins & Watts 1968). A better estimate is:

$$r_x(d) = \frac{C_x(d)}{s^2_{x(1,n)}} = \frac{C_x(d)}{C_x(0)}. \tag{5.11}$$

This is the formulation for calculating autocorrelation that we will use in investigating some of the characteristics of a range of autocorrelation models, including those already described.

The discussion so far has focused on data in one spatial dimension, but many of the comments apply equally well to data in two or more dimensions. For example, in two dimensions, we usually estimate the covariance using distance classes, as described above, but in some cases, where anisotropy is a concern (see Chapters 1 and 3), direction classes may be used as well. In two or more dimensions, however, although the concepts related to spatial autocorrelation are the same as those in one dimension, the technical aspects of setting up models of spatial autocorrelation in additional dimensions tend to be more complicated. For example, Whittle (1954) shows that in one dimension bilateral dependence (where $x_i$ depends on both $x_{i-1}$ and $x_{i+1}$) can be reduced to unilateral dependence, which is more tractable, but in two dimensions, a similar reduction is not easily achieved.

To investigate further the characteristics of models of autocorrelation in one dimension, we need to generate not one, but a large number of realizations of these models on a computer, using $n = 500$ and $\rho = 0.4$. For each realization, we calculate the sample mean, $\bar{x}$, and sample variance, $s^2$, and then the $t$ statistic to test the null hypothesis that the true mean of $x$ is zero: $H_0 \equiv (\mu_x = 0)$:

$$t = \frac{\bar{x} - 0}{s/n}. \tag{5.12}$$

For 10,000 realizations of each model, we count the number of times the test statistic is less than the 0.05, 0.5, 2.5 and 5% critical values of $t_{n-1}$ and the number of times it is greater than the 95, 97.5, 99.5 and 99.95% critical values. For Models 3

Table 5.3 *Results of 10,000 simulations for Models 1 to 4 ($\rho = 0.4$; $\beta = 0.6$)*

| Critical value (%)[a] | 0.05 | 0.5 | 2.5 | 5 | 95 | 97.5 | 99.5 | 99.95 |
|---|---|---|---|---|---|---|---|---|
| Expected count | 5 | 50 | 250 | 500 | 500 | 250 | 50 | 5 |
| Model 1a | 4 | 43 | 227 | 472 | 531 | 262 | 66 | 3 |
| Model 1b | 6 | 54 | 272 | 628 | 442 | 198 | 42 | 6 |
| Model 2 | 116 | 419 | 913 | 1,317 | 1,575 | 1,086 | 490 | 126 |
| Model 3 ($x$) | 32 | 181 | 497 | 857 | 617 | 369 | 124 | 24 |
| Model 4 ($x$) | 309 | 648 | 1,252 | 1,600 | 1,635 | 1,210 | 624 | 237 |
| Model 3 or 4 ($x'$)(residuals) | 0 | 0 | 0 | 0 | 0 | 0 | 0 | 0 |

[a] The values tabulated are the number of trials in which the test statistic was more extreme than the critical value associated with the probability given by the column heading. The null hypothesis (of mean 0) is true and, therefore, these rates represent Type I error.

and 4, which include induced autocorrelation, we can examine the results both for the original variable $x$ and for its residual after the linear dependence on $z$ is removed: $x' = x - (a + bz)$, where $a$ and $b$ are estimated from the data using the standard linear regression techniques. Table 5.3 shows some typical results.

For the independence model (Model 1a or 1b), the rates are close to the expected counts (as they should be), but for Model 2 and Model 3 ($x$), the rates are much higher. This comparison illustrates the effect of positive spatial autocorrelation on standard statistical tests: they become too liberal, producing more apparently significant results than the data actually justify (Cliff & Ord 1981, among many). This effect is one of the main topics of this chapter and the subject of much of the discussion that follows, but we should interject a comment about the last line of Table 5.3. Given that in Model 3 ($x$), the autocorrelation appears in $x$ because of its linear dependence on $z$, which has inherent autocorrelation, a reasonable prediction would be that removing that dependence would just remove the induced autocorrelation, and its effect from $x$, so that the last line should resemble the first. The fact that it does not may be a bit surprising, and we will return to the topic of removing linear (and other) dependence later in this chapter. The concept is closely related to the process used in time series analysis called 'pre-whitening' in which trends in the data are removed, supposedly leaving only pure error or 'white noise'.

The main message of Table 5.3, putting aside the last line of interesting zeros, is that positive spatial autocorrelation, whether inherent or induced, produces 'too many' significant results, and a lot too many. For a two-sided test with $\alpha = 0.05$, as is often used in ecological studies, i.e. using the 2.5 and 97.5% critical values, Model 2 gives almost 2,000 apparently significant results, four times as many as the nominal rate of 500 in 10,000 trials. For Model 3 ($x$), with the same value of $\rho$, it is at least twice the nominal rate, and for Model 4, it is about eight times the nominal rate. Clearly, the magnitude of the effect could lead to serious errors in decision making based on the test results. An intuitive understanding of this effect can be based on the fact that because the $n$ observations are not completely independent of each other, we do not get a full $n$ units of information, but something less. The 'effective sample size' is the equivalent number of independent observations that would provide the same amount of information as $n$ non-independent observations. Here the effective sample size, $n'$, is less than $n$ and so when we use $s/n$ in the divisor of the $t$ statistic, rather than $s/n'$, we are dividing by a number that is larger than it deserves to be, thus underestimating the variance of the mean. This then produces a test statistic too great in magnitude. In going through the literature that deals with this topic, we can find many articles that illustrate the problem, but few that identify 'appropriate remedial action' (cf. Haining 1991). There are, however, several approaches that might be taken to deal with this problem, ranging from some fairly unsophisticated 'quick fixes', to sophisticated solutions that depend on the suitability of a particular underlying model, to a few robust techniques of more general applicability.

We will not keep the reader who is intrigued by the line of zeros at the bottom of the Table 5.3 in suspense (although they are pertinent to a later discussion). The zeros arise because by using $x' = x - (a + bz)$, with $a$ and $b$ estimated from the data, the mean of $x'$ is forced to be zero, causing the test statistic also to be zero in every case, and thus never outside the critical values. The topic of the effect of removing the dependence of one variable on another will return in another context. We will now proceed with our discussion of the main problem, the fact that spatial autocorrelation changes the rates at which statistical tests detect significant results.

## 5.1 Solutions

### *5.1.1 Quick fixes*

The simplest approach might be to acknowledge the existence of spatial autocorrelation and to adjust the Type I error rate, $\alpha$, to a more conservative value: e.g. $\alpha' = \alpha/5$. For example, in Dale & Zbigniewicz (1997), $t$-tests were carried out on plant density data from transects of 1,001 contiguous $10 \times 10$ cm quadrats. They used a 1% significance level, rather than the usual 5% because the critical value for 1% for large $\nu$ is 2.57 which is close to the 5% critical value for $\nu = 5$. Therefore, even if the autocorrelation in the data reduced the effective sample size by an order of magnitude or more, from 1,001 to as low as 5, an $\alpha$ value of less than 0.05 was assured. This may not be the best approach, because we do not know by how much the nominal error rate needs to be adjusted to give a true error rate of the desired value. Depending on the true autocorrelation structure underlying the data, there is a real danger of using a test that is much too conservative, as we will discuss further below.

One feature of the commonly invoked first-order autoregressive structure (Model 2) is that autocorrelation declines exponentially with distance. This suggests that for distances greater than some particular value, the autocorrelation is effectively zero and observations further apart can be treated as independent. If this is true, and there is a great abundance of data, it should be possible to use only a widely spaced subset of the data for analysis to ensure independence. For example, Ostendorf & Reynolds (1998), in an analysis of landscape patterns in two dimensions, determined that autocorrelation did not extend beyond 20 pixels in their data, and therefore used what they considered to be a non-autocorrelated subset of pixels 20 units apart, 1/400 of the pixels available. This approach has two major draw-backs: first, it seems very wasteful of data (Legendre & Legendre 1998, p. 14) and second, the concept of a 'distance to independence' may be mistaken for real spatial data, where non-zero autocorrelation may have an effect even if it is not significantly different from zero. We will elaborate in the next section.

### 5.1.2 Adjusting the effective sample size

As we described above, autocorrelation in data modifies the effective sample size to be something other than the number of samples, $n' \neq n$. Positive autocorrelation reduces the effective sample size and therefore the estimated degree of spatial autocorrelation can be used to determine how much smaller the effective sample size is than the number of observations (Clifford *et al.* 1989; Cressie 1991; Dutilleul 1993b). Let us begin by considering tests concerning the mean, as described above. In the absence of spatial autocorrelation, the variance of the mean is estimated as the sample variance divided by the sample size:

$$\mathrm{Var}(\bar{x}) = s^2/n. \tag{5.13}$$

In the presence of spatial autocorrelation, $n$ in (5.13) is replaced by $n'$, the effective sample size, which is what we wish to determine. In general, as Cressie (1991) explains, the variance of the mean of the observations, $x_1, x_2, \ldots, x_n$, can be adjusted to correct for autocorrelation using the covariances of the $x$s, '$\mathrm{Cov}(x_i, x_j)$' (Cressie 1991, Eqn (1.3.4)):

$$\mathrm{Var}(\bar{x}) = n^{-2} \sum_{i=1}^{n} \sum_{j=1}^{n} \mathrm{Cov}(x_i, x_j). \tag{5.14}$$

Therefore, we can get an estimate of the effective sample size by equating the right-hand sides of (5.13) and (5.14) and transposing to isolate $n'$:

$$n' = \frac{n^2 s^2}{\sum_{i=1}^{n}\sum_{j=1}^{n} \mathrm{Cov}(x_i, x_j)} = \frac{n^2}{\sum_{i=1}^{n}\sum_{j=1}^{n} \mathrm{Cor}(x_i, x_j)}. \tag{5.15}$$

For example, for a first-order autoregression correlation structure (Table 5.1) with parameter $\rho$, i.e. $\mathrm{Cor}(x_i, x_j) = \rho^{|i-j|}$, the effective sample size is:

$$n' = \left[1 + 2\frac{\rho}{1-\rho}(1 - 1/n) - 2\frac{\rho^2}{(1-\rho)^2}\frac{1-\rho^{n-1}}{n}\right]^{-1} \times n. \tag{5.16}$$

For large $n$, this becomes:

$$n' \cong n\frac{1-\rho}{1+\rho} = n\Theta, \tag{5.17}$$

where $\Theta$ is the approximate correction factor $\Theta = (1-\rho)/(1+\rho)$. For example, if $n = 1{,}000$ and $\rho = 0.4$, then $n' = 429$. Numerical simulations by computer, such as those described above, using artificial data with this autoregressive structure confirm the correctness of using $n' = n\Theta$ for one- and two-sample $t$-tests and for ANOVA $F$-tests for comparisons among means. Computer simulations also show that the same correction will apply to paired sample $t$-tests (Dale & Fortin 2002). Of course, for this to be useful, the autoregressive model has to be a good description of the autocorrelation structure, or the approach has to be robust to departures from that model. We will discuss the question of robustness further, because it is extremely important if the method of adjusting the effective sample size is to be used in real data analysis.

Our next step is to investigate further the equation for the effective sample size based on the correlation matrix, **R**, and its elements $r_{ij}$ estimated from the correlation calculated for each lag distance, $r(d)$. Because we cannot estimate the correlation of individual pairs of variates, our best estimate is to calculate the correlation for each lag, $d$, for which there are $(n-d)$ pairs. In the correlation matrix of individual pairs of variates (see Table 5.2 as an example) there are $n$ 1s on the main diagonal, and then two of each of the other entries, on either side of the diagonal. Therefore the estimate based on these correlations is:

$$n'(\mathbf{R}) = \frac{n^2}{\sum_{i=1}^{n}\sum_{j=1}^{n} r_{ij}} = \frac{n^2}{n + 2\sum_{d=1}^{n-1}(n-d)r(d)}. \tag{5.18}$$

Table 5.4 *First-order serial autocorrelation structure*

$$\begin{vmatrix} 1 & \rho & 0 & 0 & \cdots & 0 \\ \rho & 1 & \rho & 0 & \cdots & 0 \\ 0 & \rho & 1 & \rho & & 0 \\ 0 & 0 & \rho & 1 & & 0 \\ \vdots & & & & \ddots & \vdots \\ 0 & 0 & 0 & 0 & \cdots & 1 \end{vmatrix}$$

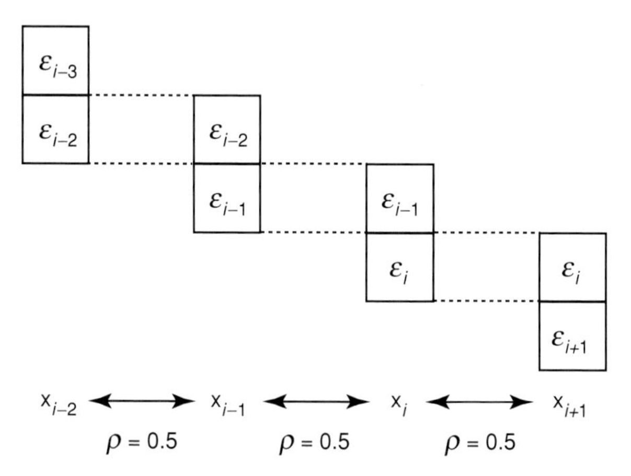


Figure 5.6 Model 5: $x_i = \varepsilon_i + \varepsilon_{i-1}$, a first-order moving average model. Half the information in $x_i$ is contained in $x_{i-1}$ and the other half in $x_{i+1}$ (see Eqn (5.19)). Unlike Model 2, there is no dependence beyond lag 1. The correlations are: $\mathrm{Cor}(x_i, x_{i-1}) = 0.5$ and $\mathrm{Cor}(x_i, x_j \mid j \neq i; j \neq i \pm 1) = 0$.

To evaluate the effective sample size, $n'(\mathbf{R})$, we will examine other autocorrelation structures. For example, a first-order serial autocorrelation structure has a correlation of $\rho$ between adjacent observations, but it is 0 for all other pairs (Table 5.4). The effective sample size is $n' = n^2/(n + 2(n - 1)\rho)$. Model 5 is a first-order moving average model:

$$x_i = \varepsilon_i + \varepsilon_{i-1}, \text{ (see Figure 5.6)}, \tag{5.19}$$

it produces $\rho = 0.5$, and for large $n$, $n' = n/2$. Figure 5.6 gives an intuitive illustration of why this is so. If half the information in $x_i$ is contained in $x_{i-1}$ and the other half is contained in $x_{i+1}$, then only every second one of the $x$s are needed to recover all the information in the series, and $n' = n/2$.

Model 5 is the simplest member of a class of autocorrelation models called 'moving average' models, which, with $k$ being the model's order, have the general

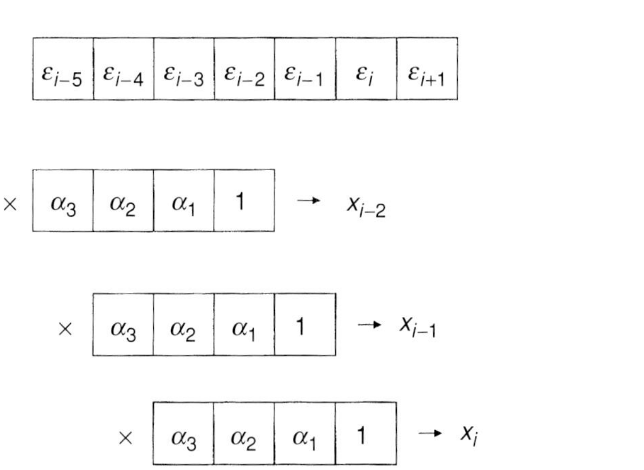


Figure 5.7 Illustration of a general moving average (MA) model, here of order 4. A template with weights moves along the data series and calculates a weighted average at each position to create the value of $x$.

form (Chatfield 1975):

$$x_i = \varepsilon_i + \sum_{j=1}^{k} \alpha_j \varepsilon_{i-j}. \tag{5.20}$$

Figure 5.7 illustrates the appropriateness of the name, with a moving window creating a possibly weighted average of the $\varepsilon$s. The question of directionality is not as puzzling for these models as for the autoregressive. For example, for Model 5, we could re-label all the $x_i$s as $x_{i-1}$, producing forward-looking rather than backward-looking dependence, without changing anything else in the characteristics of the data produced.

Extending autocorrelation beyond the first neighbours using the second model of this moving average series, the correlation between adjacent observations is $\rho_1$, $\rho_2$ for pairs at one remove, and 0 for all other pairs. The effective sample size is:

$$n' = n^2/[n + 2(n-1)\rho_1 + 2(n-2)\rho_2]. \tag{5.21}$$

The values $\rho_1 = 0.67$, $\rho_2 = 0.33$ can be generated by the model (Model 6):

$$x_i = \varepsilon_i + \varepsilon_{i-1} + \varepsilon_{i-2} \text{ (see Figure5.8).} \tag{5.22}$$

For these values and large $n$, $n' = n/3$.

Table 5.5, similar to one presented in Dale & Fortin (2002), shows the results for a range of autocorrelation structures, generated in the same fashion as Models 5 and 6. With the exception of the last two lines, computer experiments show that the effective sample sizes in the last column are ‘correct’ in that, with 10,000 realizations as in Table 5.3, the rates at which the true null hypothesis is rejected are close to the nominal values, when the derived effective sample size is used. The

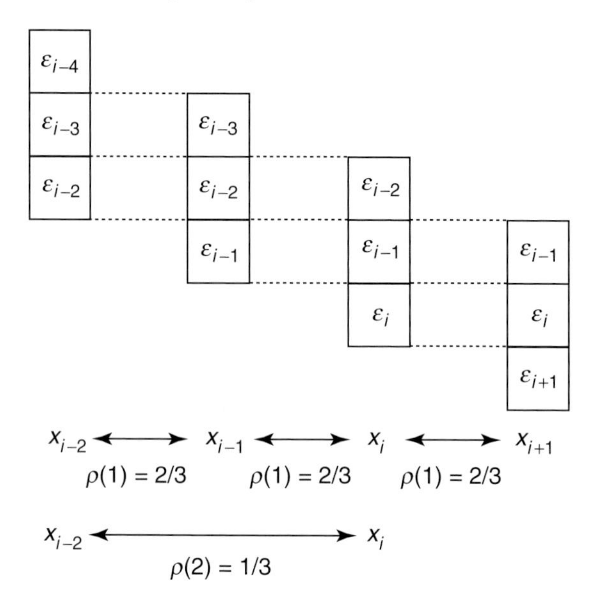


Figure 5.8 A particular second-order moving average model: $x_i = \varepsilon_i + \varepsilon_{i-1} + \varepsilon_{i-2}$, $\mathrm{Cor}(x_i, x_{i-1}) = 2/3$ and $\mathrm{Cor}(x_i, x_{i-2}) = 1/3$.

Table 5.5 *Effective sample sizes based on autocorrelation values, $n'(\rho)$, for artificial data from 'moving average' (MA) models of autocorrelation*

| $\rho(d)$, $d = 0, 1, 2, \ldots$ | $\Sigma\Sigma\rho_{ij}/n$[a] | $n'(\rho)$ |
|---|---|---|
| 1, 0.5, **0**[b] | 2 | 250 |
| 1, 0.67, 0.33, **0** | 3 | 167 |
| 1, 0.8, 0.6, 0.4, 0.2, **0** | 5 | 100 |
| 1, 0.25, 0, −0.25, **0** | 1 | 500 |
| 1, 0.5, 0, −0.17, −0.33, −0.17, **0** | 2/3 | 750 |
| 1, 0.4, −0.2, −0.4, −0.2, **0** | 1/5 | 2500 |
| 1, −0.83, 0.67, −0.5, 0.33, −0.17, **0** | ≅0 | ? |
| 1, −0.75, 0.5, −0.25, **0** | ≅0 | ? |

[a] The actual sample size is $n = 500$.
[b] **0** means that all correlations at this and larger lags are zero.

most interesting autocorrelation structures are those which have some negative autocorrelation added. (For example, the model $x_i = \varepsilon_i + \varepsilon_{i-1} - \varepsilon_{i-2} - \varepsilon_{i-3} - \varepsilon_{i-4}$, cited in the third to last line of Table 5.5, produces apparently cyclic behaviour in the data, but it is aperiodic because the autocorrelation is close to zero beyond lag 4, see Figure 5.9.) Using the general formula for effective sample size and depending on the parameter values chosen, this can actually increase the effective sample size to an extent that it is greater than $n$. This means that strong positive autocorrelation

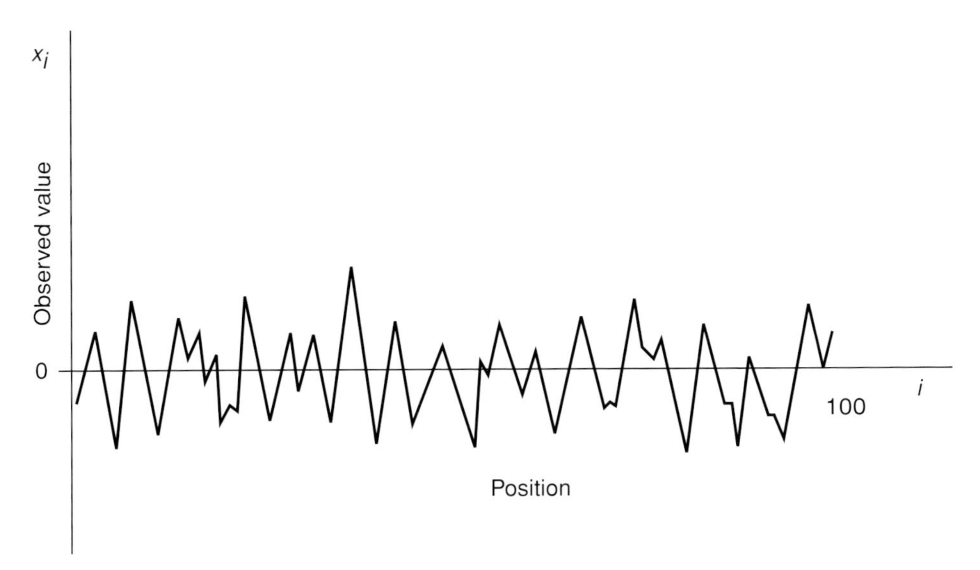


Figure 5.9 Cyclic behaviour of the variable $x$ induced by the moving average model $x_i = \varepsilon_i + \varepsilon_{i-1} - \varepsilon_{i-2} - \varepsilon_{i-3} - \varepsilon_{i-4}$. The behaviour of $x$ is, however, aperiodic, with autocorrelation expected to be 0 at lags of 5 and beyond.

at small scales does not necessarily compromise statistical tests if there is cyclic behaviour that produces negative autocorrelation at larger scales. In fact, the test statistic may require inflation rather than deflation to achieve significant results at the correct nominal rates. This fact has implications for testing ecological data in which repeating structure (spatial pattern) is very common. One implication is that Dale & Zbigniewicz (1991), who attempted to correct for autocorrelation effects by using $\alpha = 0.01$ rather than the usual $\alpha = 0.05$, may have greatly overcorrected based on the short-range positive autocorrelation but leaving out of consideration the longer range negative autocorrelation.

An important point is that autocorrelation at all lags must be included in the calculation of effective sample size, even if the individual value does not seem to be itself statistically significant. It is certainly tempting, and may seem logical to suggest using only significant values, but that can lead to errors. As an illustration, consider the first-order autoregressive model (Model 2) with $\rho = 0.3$. For $n = 400$, the effective sample size is $400 \times 0.7/1.3 = 215.4$. What happens when we omit the values of autocorrelation that are not significant? Ignoring for the moment that tests for different distances are not independent, using the inverse of the $z$ transform of the correlation coefficient with $n - 3$ degrees of freedom (Sokal & Rohlf 1995), the critical value to determine significance is

$$\tanh \frac{1.96}{\sqrt{397}} = 0.098. \tag{5.23}$$

The sequence of autocorrelation coefficients for increasing lag is 1, 0.3, 0.09, 0.0027,... Of these, only the first two are significant, so that the sum of the correlations is:

$$n \times 1 + 2(n-1) \times 0.3 = 639.4,$$

giving an effective sample size of:

$$n' = \frac{n^2}{\sum_{i=1}^{n}\sum_{j=1}^{n}\rho_{ij}} = 400^2/639.4 = 250.2. \qquad (5.24)$$

Therefore, the effective sample size calculated from significant values is 16% greater than that calculated using all lags. A more extreme example would be a situation with autocorrelation of 0.3 at lag 1, 0.1 at lag 2 and −0.004 for all lags thereafter. If only 'significant' values are used, the effective sample size seems to be 222.5, but when all the small negative values are included, the effective sample size is 1,840! These examples illustrate that omitting the 'non-significant' values can strongly affect the effective sample size calculated. Applying a Bonferroni-type correction (see Chapter 3) to the critical value used, in acknowledgement of the lack of independence of the tests of different lags, does not improve the situation, because it makes the criterion for significance more stringent, thus tending to detect and thus to omit even more 'non-significant' values. One implication of this discussion is that a large number of small negative values for autocorrelation at greater distances may be able to counteract the effects of larger positive values at short distances. This situation may arise frequently in ecological data, if the system being studied exhibits patchiness, which can give rise to cycles of positive and negative autocorrelation.

Based on the 'correctness' of the effective sample sizes for a range of models given in Table 5.5, it is tempting to suggest that the solution is to calculate the autocorrelation matrix, **R**, from the data and then to use its values to find the correct effective sample size. Alas, the computer runs that gave rise to Table 5.5 demonstrated the real problem with trying to adjust tests of data with spatial autocorrelation using estimates from the data themselves. The real problem is that the realizations of a very simple structure, such as that generated by $x_i = \varepsilon_i + \varepsilon_{i-1}$, can have very different estimates of $n'$. For example, in a set of 1,000 runs with $n = 500$ of that simple model (the first line of Table 5.5), while the average effective sample size calculated from the data was 302 (which looks fine), the range for individual realizations was from 48 to 492. Clearly, this approach cannot be used for even artificial data with a simple underlying structure and how much more dangerous might it be for real data with an unknown and possibly complex structure.

Another feature became apparent from the computer runs, which is that while $\Sigma\Sigma\rho_{ij}$ cannot be less than 0, from a simple algebraic argument, in rare cases, estimates from data can be. The argument that the overall sum cannot be less than zero is:

$$\sum_{i=1}^{n}\sum_{j=1}^{n}\rho_{ij}\sigma^2 = \sum_{i=1}^{n}\text{Var}(x_i) + 2\sum_{i=1}^{n-1}\sum_{j=i+1}^{n}\text{Cov}(x_i, x_j) = \text{Var}\left(\sum_{i=1}^{n}x_i\right), \tag{5.25}$$

which, being a variance, cannot be negative. For the models in the last two rows of Table 5.5, the expected sum of the autocorrelations is very close to zero (for example, see Figure 5.7, which gives rise to the last row of the table).

In the absence of justification for proceeding from the data to an estimate of effective sample size, it may seem that we should abandon this approach, but that is not exactly true, as we will show. Before we return to a discussion of general solutions (if any) to the problem, however, we will describe other kinds of models that produce spatial autocorrelation, as important background information, and then we will present some particular examples of corrections available in the literature.

### *5.1.3 Other kinds of models*

The discussion so far has dealt implicitly with continuous variables, often with a normally distributed error term, for which the calculation of correlation was a logical approach. We need also to consider the concept of autocorrelation as it applies to discrete variables, the most simple being data that consist of sequences of 0s and 1s. The introduction of autocorrelation in such data can be achieved by having the value at a particular location being dependent on the values at preceding locations; for example, the probability of a 1 could decrease with the length of the preceding run of 1s. These structures are called Markov models and are described in greater detail below.

The discussion so far has also implicitly and explicitly used directionality in the description of models: for example with $x_i$ as a function of $x_{i-1}$. This direction of apparent dependence is logical in time series, but does not seem to have the same intuitive appeal for spatial data, particularly when we consider two dimensions rather than just one. The unidirectionality of the one-dimensional models is apparent in the first-order autoregressive model (Model 2):

$$x_i = \rho x_{i-1} + \varepsilon_i. \tag{5.26}$$

The question is whether, in spatial models, both the 'forward' and 'backward' neighbours should be considered; for example:

$$x_i = \rho x_{i-1} + \varepsilon_i + \varphi x_{i+1}. \tag{5.27}$$

Clearly, to implement this structure in a practical way requires the simultaneous solution of a set of $n$ equations for the $n$ values of the $x$s. This is not as direct as starting with a random value for $x_0$, and then generating $x_1$ to $x_n$ each from the preceding value using Eqn (5.26).

In this one-dimensional example, each location has two neighbours, but in two dimensions, usually more neighbours are considered, whether in a regular lattice or not, and their effects on the location's value may have different weights depending on the neighbours' positions and distances. The weights are often given in a 'proximity matrix', **W**, with $w_{ij}$ being greater than zero if the value at location $i$ is not independent of the value at location $j$. For example, in a regular square lattice, all 'queen's move' (or 'king's move' if it is only one step) neighbours might receive equal weighting (say $1/8$) in a proximity matrix, with all others being 0. There is a large number of different ways in which autocorrelation can be introduced in two-dimensional data, and we will describe only two, the simplest versions of two approaches: simultaneous autoregressive models (SAR) and conditional autoregressive models (CAR). These approaches are most easily explained using matrices (indicated by bold font), and our explanation of SAR borrows heavily from Bailey & Gatrell (1995) who describe this model with great clarity.

Simultaneous autoregressive models are based on the concept illustrated in Eqn (5.27), in which the equation defining $x_i$ contains $x_{i-1}$ and $x_{i+1}$, each of which has their own defining equations containing other $x$s. Therefore there is a system of *simultaneous* equations to be solved. We begin with the model in which the measured variable $x$, given as a vector, **x**, is linearly dependent on some independent underlying variables, $z_1, z_2, z_3, \ldots$, given as a matrix **Z**:

$$\mathbf{x} = \mathbf{Z}\beta + \mathbf{u}, \tag{5.28}$$

where **u** is a vector of possibly non-independent errors with a mean of zero and a variance–covariance matrix **C**. Spatial autocorrelation is introduced into the model by having the errors given in **u** autocorrelated:

$$\mathbf{u} = \rho\mathbf{W}u + \varepsilon, \tag{5.29}$$

where $\varepsilon$ is a vector of independent error terms: $\varepsilon_i \approx N(0, \sigma_\varepsilon^2)$. The matrix **W** is the neighbour weights (standardized to row totals of 1), described above. In this case, **W** is not necessarily symmetric, making it possible to include the effects of water currents, prevailing winds or other factors that might impose directionality on the autocorrelation effects. The model is now:

$$\mathbf{x} = \mathbf{Z}\beta + \rho\mathbf{W}(\mathbf{x} - \mathbf{Z}\beta) + \varepsilon \tag{5.30}$$

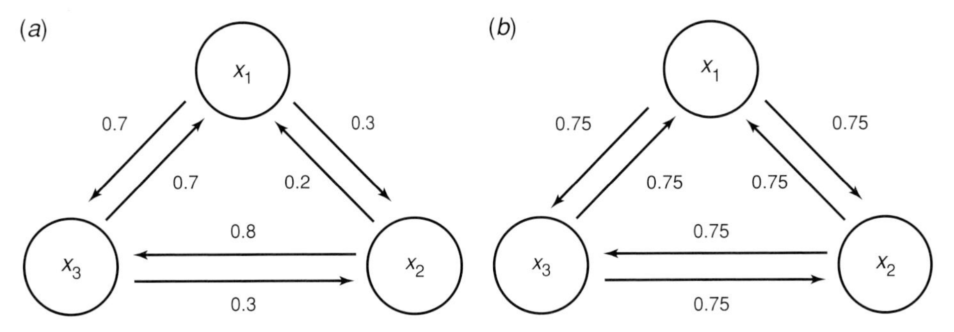


Figure 5.10 (*a*) Illustration of the asymmetric weight matrix given in Eqn (5.32). (*b*) Illustration of the symmetric weight matrix given in Eqn (5.37).

and the variance–covariance matrix (associated with **u**) is:

$$\mathbf{C} = \sigma^2[(\mathbf{I} - \rho\mathbf{W})^{\mathrm{T}}(\mathbf{I} - \rho\mathbf{W})]^{-1} \tag{5.31}$$

(see Bailey & Gatrell 1995 for the derivation). Bailey & Gatrell gave a simple example for $n = 3$ and $\rho = 0.5$ (see Figure 5.10*a*): if

$$\mathbf{W} = \begin{bmatrix} 0.0 & 0.3 & 0.7 \\ 0.2 & 0.0 & 0.8 \\ 0.7 & 0.3 & 0.0 \end{bmatrix}, \tag{5.32}$$

then

$$\mathbf{C} = \sigma^2 \begin{bmatrix} 1.81 & 0.97 & 1.31 \\ 0.97 & 1.69 & 1.15 \\ 1.31 & 1.15 & 1.90 \end{bmatrix}. \tag{5.33}$$

In this example, the variances (the elements on the main diagonal) are not all the same, but they are all greater then 1.0 because the variances are 're-inforced' by the correlation with the other values. The variance of the third unit is the greatest because it has the largest values of the neighbour weights.

Although the SAR model, just described, is used extensively in quantitative geography, for some technical reasons, many statisticians emphasize the use of the CAR model instead. The conditional autoregressive model is not based on the linear dependence of the value at a particular location on the values of its neighbours, but the probability that it takes a particular value is *conditional* upon the neighbour values:

$$P(x_i = x) = P(x_i = x|\{x_j; w_{ij} > 0\}). \tag{5.34}$$

It is not that much different from the SAR model, but it requires that the weight matrix, $\mathbf{V}$, be symmetric. The model is:

$$\mathbf{x} = \mathbf{Z}\beta + \mathbf{u}, \tag{5.35}$$

where $\mathbf{u}$ is a vector of errors with a mean of zero and a variance–covariance matrix that includes the autocorrelation parameter $\varphi$:

$$\mathbf{C} = \sigma^2(\mathbf{I} - \varphi\mathbf{V})^{-1}. \tag{5.36}$$

A simple CAR example for $n = 3$ is illustrated in Figure 5.10*b* (the values are not required to be equal):

$$\mathbf{V} = \begin{bmatrix} 0.0 & 0.75 & 0.75 \\ 0.75 & 0.0 & 0.75 \\ 0.75 & 0.75 & 0.0 \end{bmatrix}, \tag{5.37}$$

with $\varphi = 1/3$, this gives

$$\mathbf{C} = \sigma^2 \begin{bmatrix} 1.2 & 0.4 & 0.4 \\ 0.4 & 1.2 & 0.4 \\ 0.4 & 0.4 & 1.2 \end{bmatrix}. \tag{5.38}$$

Again, the values on the main diagonal are greater than 1.0, but here they all have the same magnitude.

A slightly more realistic looking example is:

$$\mathbf{V} = \begin{bmatrix} 0 & 0.8 & 0 & 0 & 0 & 0 & 0 \\ 0.8 & 0 & 0.8 & 0 & 0 & 0 & 0 \\ 0 & 0.8 & 0 & 0.8 & 0 & 0 & 0 \\ 0 & 0 & 0.8 & 0 & 0.8 & 0 & 0 \\ 0 & 0 & 0 & 0.8 & 0 & 0.8 & 0 \\ 0 & 0 & 0 & 0 & 0.8 & 0 & 0.8 \\ 0 & 0 & 0 & 0 & 0 & 0.8 & 0 \end{bmatrix} \tag{5.39}$$

then, with $\varphi = 0.625$,

$$\mathbf{C} = \frac{\sigma^2}{4} \begin{bmatrix} 7 & 6 & 5 & 4 & 3 & 2 & 1 \\ 6 & 12 & 10 & 8 & 6 & 4 & 2 \\ 5 & 10 & 15 & 12 & 9 & 6 & 3 \\ 4 & 8 & 12 & 16 & 12 & 8 & 4 \\ 3 & 6 & 9 & 12 & 15 & 10 & 5 \\ 2 & 4 & 6 & 8 & 10 & 12 & 6 \\ 1 & 2 & 3 & 4 & 5 & 6 & 7 \end{bmatrix}. \tag{5.40}$$

The two kinds of models are closely related and any SAR process is a CAR process with

$$\mathbf{V} = \mathbf{W} + \mathbf{W}^{\mathrm{T}} - \mathbf{W}^{\mathrm{T}}\mathbf{W}, \tag{5.41}$$

but the converse is not true (see Ripley 1981).

In the spatial analysis literature, the use of moving average models is surprisingly rare, although Haining (1978) advocated their advantages. For lattice or grid data, it seems natural to consider a moving average based on rook's move neighbours:

$$x_{ij} = \varepsilon_{ij} + (\rho_b \varepsilon_{i-1,j} + \rho_f \varepsilon_{i+1,j} + \rho_u \varepsilon_{i,j-1} + \rho_d \varepsilon_{i,j+1})/4$$

or, more simply:

$$x_{ij} = \varepsilon_{ij} + \rho(\varepsilon_{i-1,j} + \varepsilon_{i+1,j} + \varepsilon_{i,j-1} + \varepsilon_{i,j+1})/4. \tag{5.42}$$

The MA models have the advantage that autocorrelation can be made to decline sharply with distance and become more-or-less zero, whereas in autoregressive models it tends to persist over greater distances. In general form, the MA model is:

$$\mathbf{x} = \mathbf{Z}\beta + \rho\mathbf{W}\varepsilon + \varepsilon, \tag{5.43}$$

and the variance–covariance matrix is:

$$\mathbf{C} = \sigma^2[(\mathbf{I} - \rho\mathbf{W})^{\mathrm{T}} + (\mathbf{I} - \rho\mathbf{W})]. \tag{5.44}$$

This looks similar to the equation for the SAR model, but there is no inverse in the formula. For the same symmetric proximity matrix, **W**, the variance–covariance matrices that arise for the three different models will, in general, be different. For more on these models (SAR, CAR and MA), their properties and the estimation of their parameters, see Ripley (1981), Upton & Fingleton (1985), Griffith (1988) and Cressie (1991), Bailey & Gatrell (1995) among many. We will not go into the technical details here, with so many good references available. We will point out that these models can be viewed as an aid to understanding; the fact that we can estimate the parameters of a model and get a good agreement with the data does not mean that we know the underlying process. For real data, we probably do not really know even the proximity matrix, **W**. The other useful characteristic of such models, however, is that we can use them to generate artificial data of known structure, with which to compare what we have observed in the data we are trying to analyse. The use and comparison of these types of models seems to be the only way to approach the study of this phenomenon.

We now turn to some particular examples of solutions to the general problem of the effect of spatial autocorrelation on statistical testing.

### 5.1.4 Particular examples

*Tests for proportions*

Tests for proportions are carried out using contingency tables of counts and goodness-of-fit statistics such as Pearson's $X^2$ or the log-likelihood ratio, $G$. For example, the positive or negative association of two species can be tested based on the counts of their presence : absence in sample units. These counts are summarized in a $2 \times 2$ contingency table, and then the goodness-of-fit statistic is calculated and compared to the $\chi^2$ distribution with one degree of freedom. The question here, however, is how to account for the lack of independence among sampling units due to their spatial locations. Returning to the one-dimensional situation, if the sample units are contiguous (such as quadrats in a string or transect), the data will have spatial autocorrelation and their dependence might be well described by a Markov model. For example, suppose that each of two species is recorded in a string of contiguous sampling units as being in one of two possible states: 0 for absence and 1 for presence. A reversible Markov model of the sequences of presence and absence for species $s$ would be based on the underlying transition probabilities, $\tau_s(i, j)$, the probability that it makes the transition from state $i$ to state $j$ between one quadrat and its neighbour. The overall probability of state $i$ for species $s$ is $\pi_s(i)$. These two probabilities can be estimated from the data as $T_s(i, j)$, the frequency of transition from state $i$ to state $j$, and $p_s(i)$ as the overall probability of state $i$ for species $s$. The approach to accounting for the spatial dependence in the data is to use these estimated probabilities to determine how much the test statistic calculated from the $2 \times 2$ contingency table, derived from such data, should be deflated to give the correct rejection rate; i.e. to determine a value, $\Phi$, by which to decrease the test statistic.

For $2 \times 2$ contingency tables derived from data in which the serial correlation is due to a reversible Markov process, Tavaré (1983) provided a deflation factor for the test statistic, based on the non-unit eigenvalues, $\lambda$, of the transition probability matrices which can be estimated from the frequencies as:

$$\lambda_s = \sum_{i=1}^{n} \frac{T_s(i, j) - p_s(i)}{2 - 2p_s(i)}, \quad \text{with } s = 1 \text{ or } 2. \tag{5.45}$$

The deflation value is then:

$$\Phi = \frac{1 + \lambda_1\lambda_2}{1 - \lambda_1\lambda_2} \tag{5.46}$$

(Tavaré 1983; Tavaré & Altham 1983; see Upton & Fingleton 1989, p. 92).

The test statistic is calculated in the usual way but it is divided by the deflation factor before being compared to the reference distribution, $\chi_1^2$ in the case of a $2 \times 2$

contingency table. In the particular case of the goodness-of-fit statistic, deflating the statistic by $\Phi$ and reducing the effective sample size to $n' = n\Phi^{-1}$ are mathematically equivalent.

As a simple example, suppose that both species A and B have a transition probability from presence to absence or absence to presence of 0.1 and an overall probability of either state of 0.5. Under those circumstances,

$$\lambda_1 = \lambda_2 = \frac{0.9 - 0.5}{2 - 2 \times 0.5} + \frac{0.9 - 0.5}{2 - 2 \times 0.5} = 0.8,$$

and so

$$\Phi = \frac{1 + 0.8 \times 0.8}{1 - 0.8 \times 0.8} = 4.56. \tag{5.47}$$

This deflation factor is clearly large, but it results from the very high probability of adjacent quadrats having the two species in the same states, leading to a large amount of spatial autocorrelation in the data. If only one of the variables is spatially autocorrelated, deflation is not necessary (cf. Tavaré 1983).

Multiway $2^k$ tables arise in situations such as the testing of multispecies association, based on the presences and absences of $k$ species in sampling units (Dale *et al.* 1991). Again, if the data are from contiguous quadrats in transects, the spatial dependence may be well described by Markov models. Porteus (1987), following the Tavaré approach, provided the corresponding formulation for multiway tables, again based on an underlying Markov process, but the formula becomes quite complicated. For example, for a $2 \times 2 \times 2$ contingency table, the deflation factor is:

$$\Phi_3(\lambda) = \frac{\left(\frac{1 + \lambda_1\lambda_2\lambda_3}{1 - \lambda_1\lambda_2\lambda_3} + \frac{1 + \lambda_1\lambda_2}{1 - \lambda_1\lambda_2} + \frac{1 + \lambda_1\lambda_3}{1 - \lambda_1\lambda_3} + \frac{1 + \lambda_2\lambda_3}{1 - \lambda_2\lambda_3}\right)}{4}. \tag{5.48}$$

With three variables, of which only two have spatial autocorrelation, deflation is still necessary, but it reduces to Eqn (5.46). Cerioli (1997) provided a deflation factor for testing $2 \times 2$ contingency tables based on the correlation structure calculated from the data, rather than on the properties of an underlying Markov model. It looks very similar to the general approach of Clifford *et al.* (1989) and Dutilleul (1993b) and seems to be robust.

Following the evaluation of an entire contingency table using a goodness-of-fit test, standardized residuals are often calculated to determine which cells of the table contributed most strongly to a significant result. Where $o$ is the observed frequency and $e$ is the expected, the Freeman–Tukey standardized residual is calculated as: $z = \sqrt{o} + \sqrt{o + 1} - \sqrt{4e + 1}$. For a $2 \times 2$ table, the standardized residual can be compared to $\sqrt{\chi_1^2/4}$ to determine which values make important contributions to the overall significance (Sokal & Rohlf 1981). Dale *et al.* (1991) suggested that whatever the inflation factor, $\Phi$, is for the overall statistical test, the standardized residuals can be corrected by dividing by $\sqrt{\Phi}$. The argument for this procedure is that the sum of the squares of the residuals is approximately the test statistic and if it is to be deflated by $\Phi$, the residuals should be deflated by its square-root.

The main problem for this approach in an ecological context is that its applicability depends on how well a Markov model describes the characteristics of the data. Often, that will not be a good description of the data, and the test is not particularly robust to departures from the underlying assumption. To investigate this point, we need to understand the implications of a first-order Markov model for the pattern we see in the presence : absence data of a single species. If the probability of transition between different states is $\tau_\Delta = \tau(0, 1) = \tau(1, 0)$, the probability of an unbroken run of 0s or 1s of length $k$ is $(1 - \tau_\Delta)^{k-1}\tau_\Delta$. Therefore, the expected length of unbroken runs of 0s or of 1s is $1/\tau_\Delta$, but the distribution of lengths follows a geometric distribution with many short runs and very few very long ones. For example, with $\tau_\Delta = 0.1$, the expected length of a run is 10, but 25% of runs are expected to be of length less than 5, and only 8% will be longer than 40. Figure 5.11 contrasts the irregular appearance of field data from transects through sedge meadows on Ellesmere Island (see Dale *et al.* 1993; Young *et al.* 1999) with data generated by a Markov model. An assessment of the characteristics of the runs of 1s and 0s would be a good first step in the analysis of this kind of data (see Sokal & Rohlf 1995 for a 'runs' test of randomness).

To investigate the robustness of the proposed correction to account for spatial dependence, we can examine other models of presence : absence data, for which the mean lengths of runs and the overall frequencies are the same, but for which the run lengths are more restricted, for example, following a uniform distribution, rather than a geometric distribution.

As shown in Table 5.6, only when the data mimic the distribution that results from the Markov model fairly closely, as in uniform 1 to 19, is the deflation of the test statistic or adjustment of the effective sample size able to achieve the correct significance rates.

The Markov models described here are 'first order' in that the presence or absence at location $i$ depends only on the presence or absence at location $i - 1$, and not on locations further away. Higher-order Markov models do that and the interested reader should see Dale *et al.* (1993) for a discussion of the appropriateness of higher-order Markov models for data like those shown in Figure 5.11. For the purposes of this discussion on correcting for spatial autocorrelation in goodness-of-fit tests, the conclusion would be that the Tavaré approach should only be used when there is some confidence that the particular kind of Markov model is a good description of the characteristics of the data.

Table 5.6 *Robustness of Markov correction for 2 × 2 table test:* $\tau(0, 1) = \tau(1, 0) = 0.1$ *and* $\pi_1 = \pi_2 = 0.5$

| Critical value (%)[a] | 90 | 95 | 99 | 99.9 |
|---|---|---|---|---|
| Expected count[b] | 1000 | 500 | 100 | 10 |
| Distribution of lengths | | | | |
| Geometric (Markov model) | 1011 | 515 | 104 | 14 |
| Uniform 7 to 13 | 3895 | 3070 | 1756 | 813 |
| Uniform 5 to 15 | 1891 | 1191 | 397 | 83 |
| Uniform 1 to 19 | 1066 | 537 | 116 | 12 |
| Only 1, 2, 18 or 19 | 2149 | 1427 | 582 | 171 |
| Only 3, 4, 5 or 28 | 1993 | 1293 | 465 | 117 |

[a] Values from an average of 10 simulations of 10,000 trials each.
[b] Counts of the number of trials exceeding the column's critical value when the null hypothesis is true. $\Phi = 4.556$.

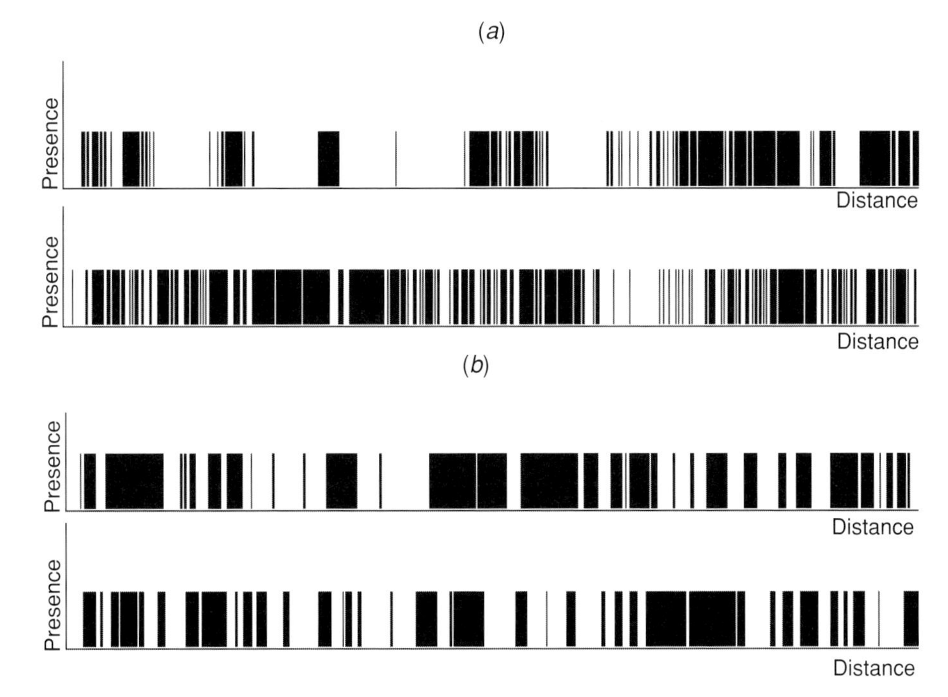


Figure 5.11 Field data from transects in a wet sedge meadow on Ellesmere Island (presence : absence of *Carex aquatilis* and *Eriophorum scheuchzeri*) in comparison with two realizations of a Markov model with similar overall density. There is a clear difference between the two pairs, in that the field data series have much more fine detail.

*Correlation and linear regression*

The correlation coefficient is a measure of the strength of the linear relationship between two variables. For the correlation coefficient between two independent variables each with an autoregressive correlation structure with parameters $\rho_1$ and $\rho_2$, Bartlett (1935) showed that its variance is approximately:

$$s_r^2 = \frac{1 + \rho_1\rho_2}{n(1 - \rho_1\rho_2)}. \tag{5.49}$$

This result suggests an effective sample size adjustment of:

$$n' = n\Psi = n\frac{1 - \rho_1\rho_2}{1 + \rho_1\rho_2}. \tag{5.50}$$

The structural similarity to the Tavaré and Cressie correction factors is striking. Tavaré & Altham (1983) suggest that under the null hypothesis, $r^2/s_r^2$ is asymptotically $\chi_1^2$. Computer experiments using artificial data with autoregressive structure confirm their suggestion. If there is autocorrelation only in one of the series, for example if $\rho_2 = 0$, then no correction is required (cf. Bivand 1980).

Clifford *et al.* (1989) suggested a method for using the *t*-test to assess the significance of the correlation coefficient in the presence of spatial autocorrelation. Dutilleul (1993b) refined this method and provided a generalized and exact form of their approximate method. Dutilleul's modification corrects problems that can occur with small sample sizes. This approach calculates the variance of the sample covariance from the (auto)covariance matrices and also provides an adjustment for the number of degrees of freedom: $n' = 1 + s_r^{-2}$. (It is not clear how closely related this method is to the formulae of Cressie and Bartlett, given above.)

The procedure is to calculate the covariance (not correlation!) matrices for $x$ and $y$ as estimates based on the distance classes $d = 0, 1, 2, \ldots$; call the matrices $\mathbf{S}_x$ and $\mathbf{S}_y$. Let $\mathbf{B}$ be the matrix with $b_{ii} = 1/n - 1/n^2$ on the main diagonal and $b_{ij} = -1/n^2$ elsewhere. Then:

$$n' = 1 + \frac{\mathrm{tr}(\mathbf{BS}_x)\mathrm{tr}(\mathbf{BS}_y)}{\mathrm{tr}(\mathbf{BS}_x\mathbf{BS}_y)}, \tag{5.51}$$

where 'tr' refers to the trace of the matrix, which is the sum of the elements on the major diagonal. The same correction can also be formulated using matrices of Moran's autocorrelation coefficient rather than the variance–covariance matrices (Legendre pers. comm.). With both $x$ and $y$ modelled as first-order simultaneous autoregressive processes on a lattice, Dutilleul (1993b) provided some examples of the effect of this correction. For a $10 \times 10$ lattice ($n = 100$) and $\rho_x = \rho_y = 0.1$, the effective sample size is 80; if the autocorrelation parameters are of opposite signs, $\rho_x = -\rho_y = 0.1$, the effective sample size is 119, illustrating again the 'positive' effects of negative spatial autocorrelation. Investigations by Legendre *et al.* (2002) confirm the effectiveness and robustness of Dutilleul's correction.

### 5.1.5 Restricted randomization and bootstrap

The general method of randomization is an approach to testing hypotheses by using the data themselves to generate a reference distribution (Manly 1997; Legendre & Legendre 1998). A test statistic calculated from the original data is compared with the distribution of the same statistic calculated after the data have been permutated or randomized in some way. In our applications, restricted randomizations are those in which the structure of the data (spatial, temporal, genetic and so on) is retained as much as possible, rather than having it erased as it would be by complete randomization (Figure 5.12; cf. Fortin & Jacquez 2000). These may therefore be useful in testing autocorrelated data. Figure 5.12 illustrates the difference between complete and restricted randomization. There are 400 cells in a $20 \times 20$ grid and each cell is classified into four density classes, which we will reduce to two, high and low, for analysis. There are 180 black (high density) cells in the original data (Figure 5.12*a*) and using a join count approach based on 'rook's move' neighbours (Chapter 3, cf. Pielou 1977), the observed number of black–black joins ($J_{BB} = 220$) is considerably greater than the number expected:

$$E(J_{BB}) = 760 \times \frac{180 \times 179}{400 \times 399} = 153.4. \tag{5.52}$$

This reflects the high degree of autocorrelation in the data, with obvious patches of low and high density. When the data are completely randomized by exchanging randomly chosen pairs of grid cells (Figure 5.12*b*), the spatial structure is destroyed, and the number of black–black joins falls to close to the expected value ($J_{BB} = 154$). In Figure 5.12*c*, the data have been randomized with a toroidal shift of 10 on the $x$ axis and 1 on the $y$ axis, which has preserved much of the spatial structure ($J_{BB} = 212$). This is one of the most commonly used restricted randomization techniques.

The applicability of any randomization procedure will depend on the nature of the data and the purpose of testing. For example, if we have data from a transect of contiguous quadrats, in which tree-canopy density and understorey cover are recorded, we could test the significance of their correlation, given their individual

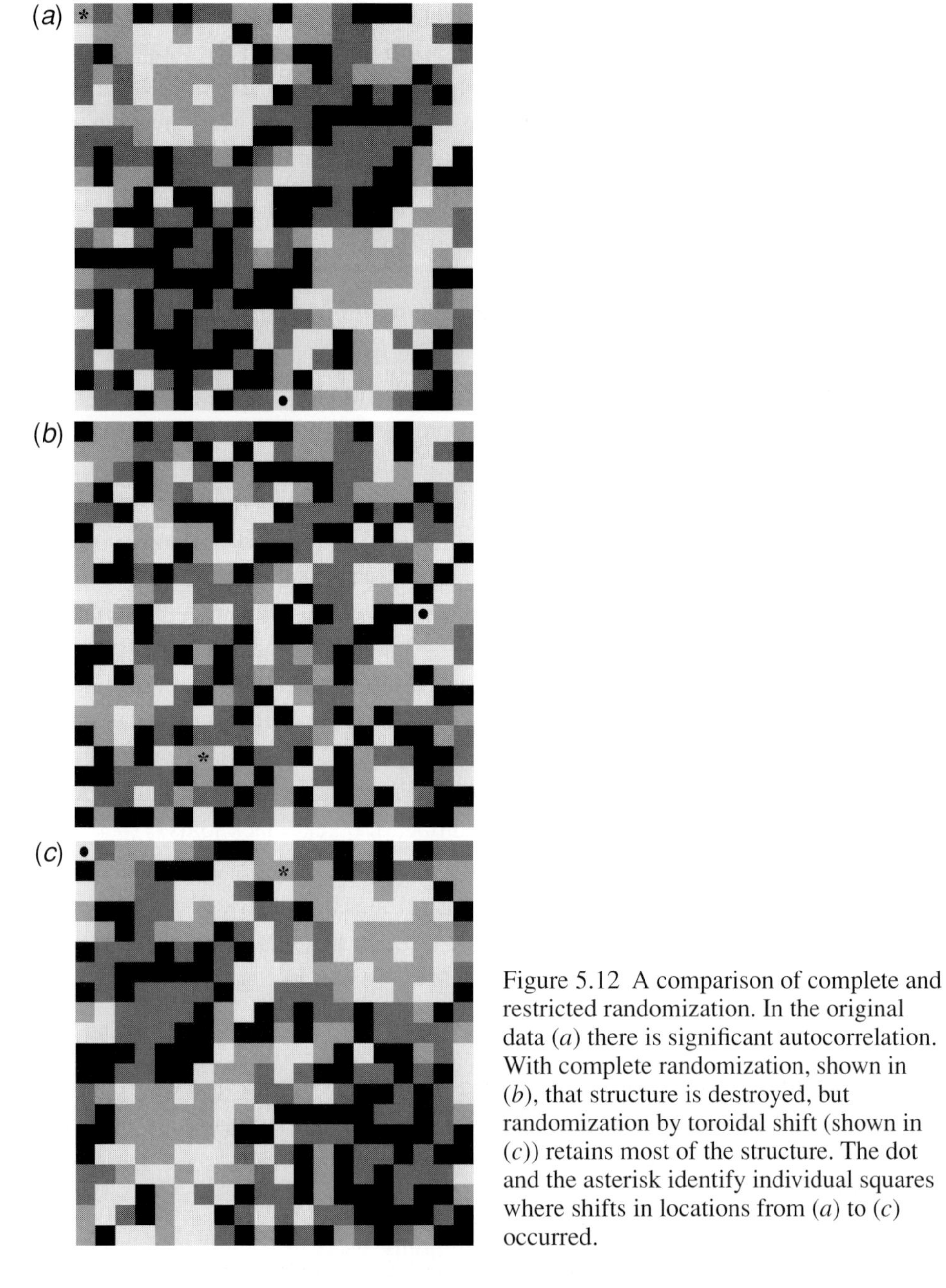


Figure 5.12 A comparison of complete and restricted randomization. In the original data (*a*) there is significant autocorrelation. With complete randomization, shown in (*b*), that structure is destroyed, but randomization by toroidal shift (shown in (*c*)) retains most of the structure. The dot and the asterisk identify individual squares where shifts in locations from (*a*) to (*c*) occurred.

spatial structures, by shifting the relative positions of the two data sets and recalculating the correlation for all possible relative positions (see Figure 5.13). This is sometimes referred to as 'caterpillar' randomization (after the tractor tread, not the insect larva) and it is the equivalent of 'toroidal shift' randomization (illustrated

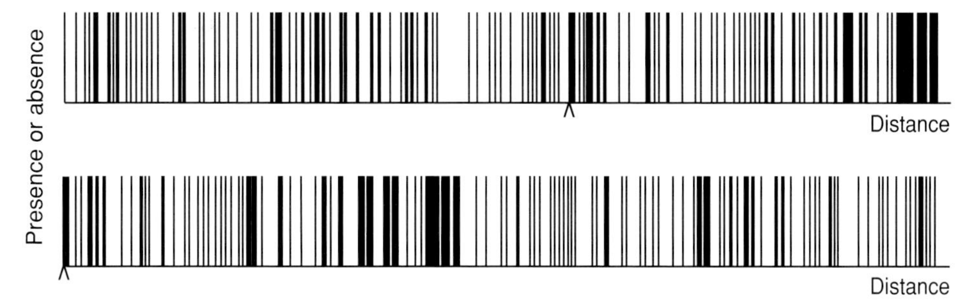


Figure 5.13 The 'caterpillar' randomization of one-dimensional data: the original data series (upper) is broken at a randomly chosen point (indicated by the arrowhead) and then recombined (lower). If there are two data sets to be related, either set can be shifted in this way to provide new relative positions.

above), but in only one dimension (cf. Upton & Fingleton 1985). If the original observed value is greater than 95% of the values from the shifted data sets, we would conclude that the observed value is significant. Longer transects with more quadrats will have more possible 'shifted' relative positions of the two data sets and greater sensitivity. Very short transects may not have sufficient numbers of relative positions to allow this kind of test. We can use restricted randomization for testing the correlation between variables in the same set of samples, but not for comparing the mean densities for two different transects. That is because there is no original natural pairing of the data in the two sets which can then be shifted in a restricted randomization test.

Legendre *et al.* (1990) described a contiguity-constrained permutation technique, which is a variant of spatially restricted randomization, for testing the significance of differences among regions in an ANOVA framework. Their numerical simulations show that, with their permutation technique, ANOVA is not very sensitive to spatial autocorrelation and provides a test with a correct Type I error.

Another application of randomization procedures is to derive variance estimators using jack-knifing estimation, as described by Lele (1991) and more recently by Heagerty & Lumley (2000). Jack-knifing is one of the randomization or permutation techniques introduced above. In particular, it derives the reference distribution by re-sampling the data, leaving out one observation in each iteration. This jack-knife approach enables confidence intervals to be derived for parameters of interest, for example, those of a regression model, which can then be tested for significant difference from 0. Cohn (1999) recommends a bootstrap procedure for comparisons of multivariate structures in the presence of serial correlation. Bootstrapping is another randomization technique, which uses re-samples of the data, allowing each datum to be used more than once. Bjørnstad & Falck (2001) proposed a bootstrap algorithm to create a confidence envelope for a non-parametric estimate of a spatial covariance function. There may be appropriate bootstrap approaches for other situations, but they are yet to be investigated.

### 5.1.6 Model and Monte Carlo

Another approach solving the autocorrelation problem is to develop a relatively simple model of the autocorrelation structure and then to use a Monte Carlo simulation to generate artificial data sets to compare with the observed data. This approach begins by finding a parametric model of the spatial dependence in the data by standard model selection procedures. The model can then be used in a Monte Carlo approach to find good confidence intervals for the test statistics (Manly 1997). This last approach is the one we advocate in the absence of a robust 'analytic' solution for a particular set of circumstances. It may seem somewhat indirect, but it does allow for tests of significance. It is also the procedure recommended by Mizon (1995) in the context of economic analysis. He suggested that we start with a very general model, which well describes the data no matter how many terms are necessary, and then test for valid reductions of it; i.e. we can determine which explanatory variables can be omitted. Overspecified models, those with more variables than they really need, do not lead to invalid inferences; they are merely inefficient (Mizon 1995). We could start, therefore, with a very general autoregressive model of the data, considering observations as far as 20 steps apart, such as:

$$x_i = \sum_{j=1}^{20} \beta_j x_{i-j} + \varepsilon_i, \tag{5.53}$$

where $\varepsilon_i$ is $N(0, \sigma^2)$.

By eliminating many of the variables, we might end up with a model in which only two or three of the $\beta$s (Fisher 1932) were significantly non-zero. (In this instance, it is appropriate to omit the non-significant terms.) As an illustration of this idea, we generated 10 realizations of the model $x_i = 0.4x_{i-1} - 0.2x_{i-4} + \varepsilon_i$, with $n = 100$. We then fit the model given in Eqn (5.53) and determined the best-fitting submodel using a maximum likelihood backward selection procedure. Table 5.7 gives examples of the results.

This table shows a range of possible outcomes for $n = 100$; many are 'close' to the original underlying model, particularly at lag 1, but the effective sample size ranges from one-third to double the 'true' value. Clearly, however, the best-fit model is not always the model that generated the data. Using a larger sample size, $n = 500$,

Table 5.7 *Coefficients of models fit to realizations of* $x_i = 0.4x_{i-1} - 0.2x_{i-4} + \varepsilon_i$

| $\beta_1$ | $\beta_2$ | $\beta_3$ | $\beta_4$ | $\beta_5$ | $\beta_6$ | $\beta_7$ | $n'$(approx.)[a] |
|---|---|---|---|---|---|---|---|
| 0.47 | −0.04 | −0.04 | −0.18 | 0.33 | •[b] | • | 22 |
| 0.56 | −0.22 | • | • | • | • | • | 44 |
| 0.30 | 0.19 | −0.04 | −0.21 | • | • | • | 59 |
| 0.30 | • | • | • | • | −0.26 | 0.08 | 77 |
| 0.21 | 0.05 | −0.08 | −0.32 | • | • | • | 128 |

[a] The effective sample size, based on the underlying model, is $n' = 65$ (approximately).
[b] Symbol '•' indicates a non-significant term.

the fit of the models is more similar to the original, showing the advantage of larger sample sizes, but they still exhibit considerable variability. Given our current state of knowledge, this does seem to be the best approach, but that raises the question of whether there are better methods yet to be discovered.

## 5.2 More on induced autocorrelation and the relationships between variables

Throughout this chapter, we have used the terms autocorrelation and autoregression, without drawing a clear distinction between correlation and regression, as is normally done in statistics texts. In general, correlation refers to the positive or negative relationship between two quantitative variables, both possibly measured with error, where it is not known that one has a direct causal effect on the other (Sokal & Rohlf 1995). For example, in the model of induced autocorrelation (Model 3), adjacent $x$s have non-zero correlation although they have no direct causal effect on each other. Correlation is a measure of the covariance of the two variables, relative to their variances:

$$r_{xy} = \frac{\mathrm{Cov}(x, y)}{\sqrt{\mathrm{Var}(x)\mathrm{Var}(y)}}. \tag{5.54}$$

Given three quantitative variables, $x$, $y$ and $z$, the partial correlation of $x$ and $y$ with $z$ held constant is:

$$r_{xy \cdot z} = \frac{r_{xy} - r_{xz}r_{yx}}{\left(1 - r_{xz}^2\right)\left(1 - r_{yz}^2\right)}. \tag{5.55}$$

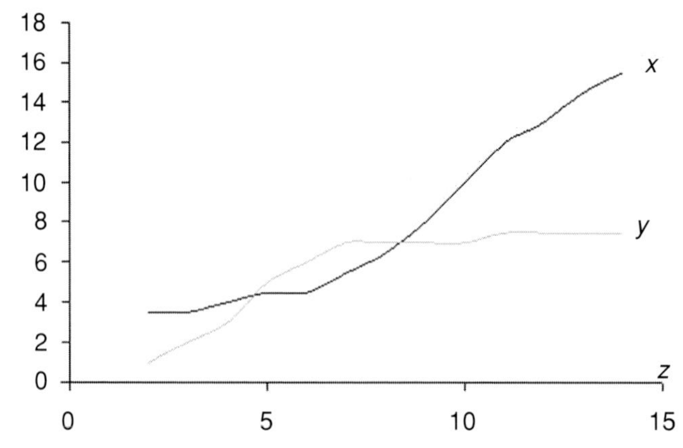


Figure 5.14 Both $x$ and $y$ are correlated with $z$ ($r_{xz} = 0.96$; $r_{yx} = 0.90$) as illustrated in part and positively correlated with each other ($r_{xy} = 0.76$); but they are negatively correlated with each other when their dependence on $z$ is controlled for ($r_{xy \cdot z} = -0.91$).

Figure 5.14 illustrates an artificial example, in which both $x$ and $y$ are positively correlated with $z$ (and apparently with each other), but they are negatively correlated with each other when the relationships with $z$ are removed: $r_{xy} = 0.76$ but, using the formula given above, $r_{xy \cdot z} = -0.91$.

In linear regression, by contrast, we are evaluating the strength of the linear dependence of a dependent variable on an independent variable, using a model as the underlying hypothesis, for example:

$$x_i = \alpha + \beta z_i + \varepsilon_i. \tag{5.56}$$

As we saw earlier in the chapter, when the linear dependence of $x$ on $z$ is controlled, what is left is the residual:

$$x_i' = x_i - (\hat{\alpha} + \hat{\beta} z_i). \tag{5.57}$$

In the artificial example of Figure 5.14, when the linear dependence of $x$ and $y$ on $z$ is removed, negative correlation of the residuals is obvious (Figure 5.15) and the correlation coefficient of $x'$ and $y'$ is $-0.91$. It is tempting to speculate that if autocorrelation in $x$ and $y$ could be attributed to their dependence on $z$, its effects could be similarly controlled by accounting for their dependence on $z$.

With real data, we observe spatial dependence in a variable of interest, $x$, but we may not always know its origins. It may be inherent, or it may be induced, or it may be both. We may not be able to distinguish among the possibilities. For Model 2, where autocorrelation is induced by an underlying autoregressive structure, the

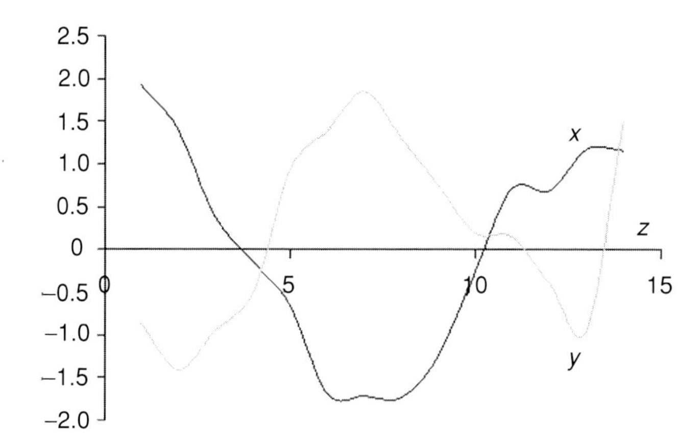


Figure 5.15 The residuals of $x$ and $y$ when the linear dependence on $z$ is taken into account.

correlation at lag $j$ is:

$$\mathrm{Cor}(x_i, x_{i-j}) = \rho^j, \tag{5.58}$$

and for Model 3 it is:

$$\mathrm{Cor}(x_i, x_{i-j}) = \frac{\beta^2 \rho^j \sigma_\xi^2}{(1-\rho^2)\sigma_\varepsilon^2 + \beta^2 \sigma_\xi^2}. \tag{5.59}$$

In both cases the autocorrelation declines exponentially with increasing distance, but that may not be true of all models of spatial autocorrelation. For the doubly autoregressive model (Model 4, refer to Figure 5.4), the correlation is much more complex, but it also declines exponentially with distance. Therefore, we may not even distinguish the possibilities of Models 2 and 3 from the more complicated Model 4, because they all have an exponential decay of autocorrelation with distance. In addition, given a single data set, it may not be possible to identify positively as being derived from an autoregressive or from a moving average model and, for field data, that distinction may not hold.

In describing the Dutilleul method for correctly testing the correlation coefficient of two variables, $x$ and $y$, in the presence of autocorrelation, we have begun to look at the evaluation of the relationships between variables under those conditions. In this section, we will investigate this topic further, by looking at a few models in which the relationship between $x$ and $y$ arises from their dependence on a third variable, $z$. There is obviously a number of ways in which this can happen.

Table 5.8 *Results of correlation tests of the residuals of $x$ and $y$, $x'(z)$ and $y'(z)$, with different combinations of autocorrelation. The values of $\beta$ were all positive. When $r > 0$, the rate of rejection of the null hypothesis is no longer relevant because it is no longer true, hence 'n.a.' in the table.*

| | Autocorrelation in $z$ | | |
|---|---|---|---|
| | 0 | MA | AR |
| Autocorrelation in $x$ and $y$ | | | |
| 0 | $r_{x'y'} = 0$<br>rates nominal | $r_{x'y'} = 0$<br>rates nominal | $r_{x'y'} = 0$<br>rates nominal |
| MA | $r_{x'y'} = 0$<br>rates inflated | $r_{x'y'} = 0$<br>rates inflated | $r_{x'y'} = 0$<br>rates inflated |
| AR | $r_{x'y'} > 0$<br>rate n.a. | $r_{x'y'} > 0$<br>rate n.a. | $r_{x'y'} > 0$<br>rate n.a. |

Let us begin with the simple model:

$$\text{Model 7:} \quad \begin{cases} x_i = \beta_x z_i + \varepsilon_i, \\ y_i = \beta_y z_i + \eta_i, \\ z_i = \xi_i. \end{cases} \tag{5.60}$$

We can then add autocorrelation to $x$ and $y$, or to $z$, in turn, either in the moving average (MA) or autoregressive (AR) form. We can then examine tests of correlation between the residuals of $x$ and $y$, after their linear dependence on $z$ is controlled, for non-zero correlation or for inflation of the rates of apparent significance in all nine combinations. The results are given in Table 5.8.

This table provides several clear messages:

(1) Controlling for the dependence on $z$ does not remove the effects of spatial autocorrelation in $x$ and $y$, in that the rates are inflated (MA) or the correlation is greater then zero (AR).
(2) The autocorrelation in $z$ is not an important factor in this context.
(3) The moving average and autoregressive models produce **qualitatively different** behaviour. (Models that combine AR and MA terms will also have a non-zero correlation.)
(4) Interpretation of a significant correlation will be especially difficult if it not obvious whether the AR or MA model is the better description of the data.

This discussion is especially important for the ecological context, in which we can reasonably expect some form of inherent autocorrelation in most of the biological variables we measure *and* some form of induced autocorrelation in those variables due to autocorrelation inherent in the underlying abiotic factors in the environment. Usually, we will not be able to determine the relative strength of these two sources.

When we then examine the relationship between two ecological variables, it is therefore probable that both will exhibit double autocorrelation.

## 5.3 Models and reality

The material of this chapter, more than in any other, relies heavily on the use of mathematical models. The reason for this feature is that it is the only way to gain insight into the effects of different forms of spatial autocorrelation on the testing procedures we use. As stated above, and well demonstrated, single realizations of even simple models can appear very different, especially for small values of $n$, and real data must have similar problems, even if the underlying processes are actually stationary (and worse if they are not!). Much of the treatment of spatial autocorrelation in the statistical literature is predicated on the simplest AR model, which produces an exponential decline in autocorrelation as a function of distance (Figure 5.16). In the geostatistical approach, it produces the 'typical' variogram depicted in Figure 3.10. Before proceeding to a discussion of how our current understanding of the effects of spatial autocorrelation can inform considerations for sampling or for experimental design, we should survey the information available to determine the common structures of spatial autocorrelation in ecologically interesting data.

In an unpublished survey, M. R. T. Dale & Y. Liang examined a large number of data sets and their published variograms in the literature. There were 320 available from various sources, and they were arbitrarily divided into categories such as 'trees' or 'soil/peat'. Most authors fit at least one of three geostatistical models (spherical, exponential and Gaussian) to their data. These three models, which all rise to an asymptote, may not be good descriptions, because repeating pattern will give rise to fluctuations in the variogram. We examined the goodness of the model fit, by examining the residuals for departures from randomness and for independence. We used a runs test on the signs of the residuals to look for series of values above or below the fitted line. We used a second runs test to look for series of increases or decreases in the residuals themselves. Fisher's (1932) sum of logs approach provided a meta-analysis of the results (see Sokal & Rohlf 1995, Chapter 18). For all of the groupings of data sets, the results of the tests were apparently significant for either the runs in the sign of the residuals or in the sign of the difference between adjacent residuals (or both). These results are only suggestive, not conclusive, because of the kinds of lack of independence in the calculations as discussed above in the context of PQV (pages 86 and 87). The suggestion is that the models were often not good descriptions and that there is a tendency for ecological variograms not to converge to an asymptote, but to rise and fall nonrandomly as a function of distance, as would result from patchiness in the variable of interest

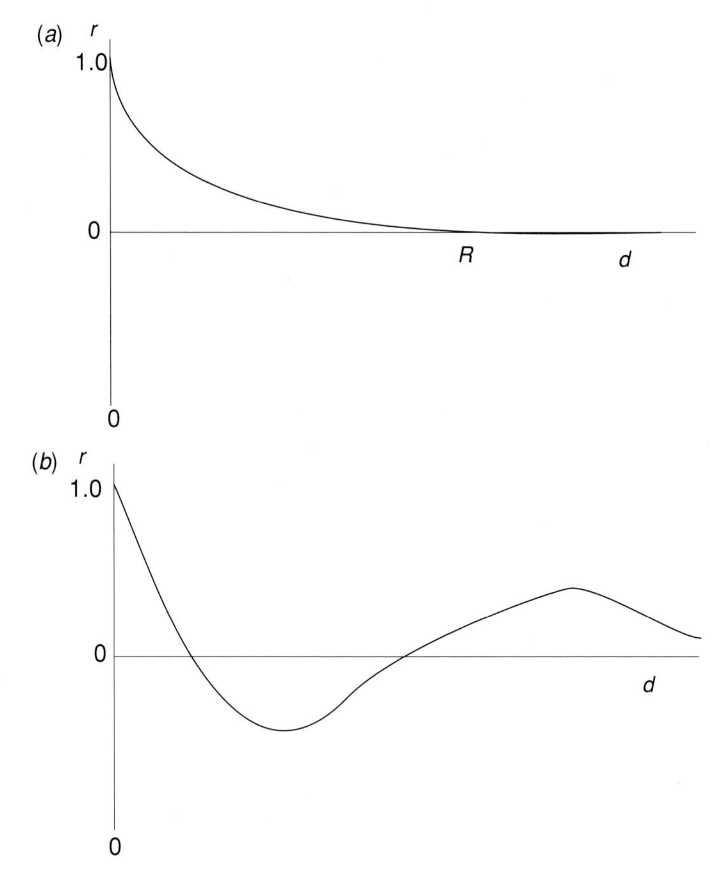


Figure 5.16 Two models of how autocorrelation may vary with distance. (*a*) Autocorrelation falls to zero at some distance, *R*, known in geostatistics as the range. This is equivalent to the 'exponential' model in geostatistical analysis. (*b*) Autocorrelation falls below zero and then fluctuates above and below zero.

("The natural world is a patchy place"! Dale 1999). If confirmed, this result may relieve some concerns about the effects of spatial autocorrelation on statistical tests, because the presence of negative autocorrelation may mitigate, at least somewhat, the reduction in effective sample size caused by positive autocorrelation. It also suggests, however, that simple corrections based on the first-order autoregressive model may often be incorrect or misleading.

## 5.4 Considerations for sampling and experimental design

### *5.4.1 Sampling*

All the preceding discussion has undoubtedly convinced the reader that spatial autocorrelation is an important characteristic in ecological data and one that must be considered in analysis and interpretation. It is probably also apparent to the reader that, if the underlying autocorrelation structure were known in advance, the design used for sampling or the design of an experiment could be adjusted to minimize the effect of spatial autocorrelation on the outcome of the study, before any analysis was conducted. For example, if the autocorrelation declines rapidly with distance and becomes effectively zero beyond some distance, $R$, (as in Figure 5.16*a*) samples or experimental units with spacing $R$ or greater may be treated as independent. This is the concept of 'distance to independence', which (alas) probably almost never applies in ecological studies. Ecological variables are typically patchy, often at more than one scale, so that their autocorrelation cycles between positive and negative with increasing distance (as in Figure 5.16*b*). Under those circumstances and with knowledge of the locations of regions of low values and regions of high values, sampling could then be stratified or the experimental units could be placed in positions of known characteristics. On the other hand, if autocorrelation declines only very slowly with distance, the distance between samples or experimental units may not be that important. In the intermediate case, where autocorrelation declines appreciably at the scale of the extent of the study, experimenters may wish to use a design that provides a balanced set of distances between units assigned the same treatment (van Es & van Es 1993; see also Dutilleul 1993a and Legendre *et al.* 2004).

Whatever the behaviour of the autocorrelation, the important first step is to find out its characteristics before designing the sampling scheme or the experiment, and to do so, a **pilot study** is required. In general, a pilot study will involve taking a number of samples according to some scheme that will allow an evaluation of autocorrelation, including its behaviour as a function of distance and whether it is anisotropic. We will consider these to be point samples of some kind, whether measurements of altitude, moisture or pH, estimates of population density or the presence of a particular substrate. There is a number of different ways in which these point samples can be arranged, and they all have advantages and disadvantages. Random placement has the advantage of simplicity, but the disadvantage of a lack of control over the spacing of the points, the coverage of the study plot and the range of lags available in each direction. A regular grid of equally spaced points has an inherent direction and scale of its own, which may interact with the actual pattern in the variable under study. The 'wagonwheel' design of radiating lines of sample points will produce a trend in sampling intensity and in circumferential distances from the centre of the wheel to the edge. A design based on the Fibonacci spiral avoids trend and directionality problems, but the full 'rose' of the spiral would require very high intensity of sampling and has a trade-off between the number of samples in the study area and the distances it can evaluate. This full spiral is illustrated in Figure 5.17 and its construction is described in detail below. The sampling designs that seem to combine most of the advantages with few disadvantages are

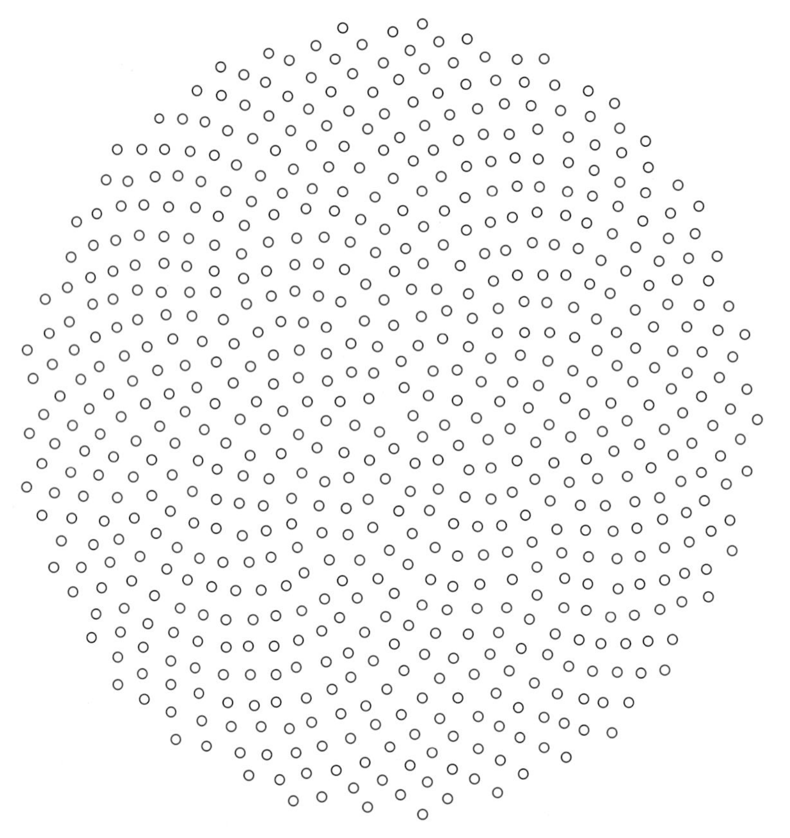

Figure 5.17 Fibonacci spiral defined by angle $\theta$ and radius $r$, with scaling parameter and 'golden mean' parameter: $\theta_i = 2\pi/\tau, r = \kappa\sqrt{i}$ and $\tau = (1 + \sqrt{5})/2$.

partial Fibonacci designs where not all points of the full 'rose' are used, but only a systematic or randomly chosen subset of it. We will now describe the details of this approach and give some examples.

The familiar Fibonacci sequence is formed by starting with two 1s and then each subsequent term is the sum of its two predecessors, so that the sequence is 1, 1, 2, 3, 5, 8, 13, 21, 34, 55, 89, . . . If Figure 5.17 reminds the reader of a sunflower head or similar botanical object, that is not surprising. The arrangement of plant parts are often well described by elements of the Fibonacci sequence (see Jean 1994), with spirals of 13, 21, 34 or 55 parts being common. The ratio of succeeding terms in the sequence approach the value $\tau = (1 + \sqrt{(5)}/2$; $\tau$, referred to as the 'golden mean' is one of those 'magic' irrational numbers that arise frequently in mathematics, like the more familiar number $\pi$. This number $\tau$ can be used to define an equiangular spiral with successive radii and successive segments following the Fibonacci sequence by calculating the angle, $\theta$, and radius, $r$, of each sample point as:

$$\theta_i = 2\pi/\tau,$$

and (5.61)

$$r_i = \kappa\sqrt{i},$$

where $\kappa$ is a scaling constant.

Figure 5.17 shows the full array of possible sample points arising from this structure, but the advantage of this is that subsets of these points may be used, and the subset can be chosen in advance for its properties. There is an almost infinite number of sampling designs that can be derived from the full array, either with a random component, for example with the probability that any point used is 0.035, or using a deterministic rule, such as using every 55th point in the full spiral. Figure 5.18 shows three examples. In the deterministic examples, three- and five-fold symmetry is evident in some, another reminder of the association between the Fibonacci spiral and the structure of flowers. Depending on circumstances, and the desirability of a particular distribution of distances between points and orientation of interpoint lags, a deterministic design may be randomly 'thinned' to reduce the sampling intensity, while retaining the useful characteristics. Whichever is chosen, these designs have the advantage of no directional bias, a good range of inter-sample distances and a lack of trend in sampling intensity.

We shall end this section of Chapter 5, with a summary of the findings of two recent studies on the effects of spatial structure on the design and analysis of field surveys and on the design of field experiments, by Legendre *et al.* (2002, 2004). Those studies were developed as part of the efforts of a working group at the National Center for Ecological Analysis and Synthesis in Santa Barbara, USA, of which the authors of this book were members. (In fact, the concept of this book was first discussed while there.) Both studies used simulations of an environment with one of several different kinds of spatial structure (gradient, waves, etc.) in the underlying environmental variable as well as autocorrelation in it and in the variable of interest to address questions about the effect of these structures on design and analysis.

In the study on surveys, the simulation structure can be described using $E_{ij}$ as the environmental variable, $\rho_{E,ij}(R)$ as its autocorrelation component with range $R$, and $\varepsilon_{ij}$ as a standard normal error term:

$$E_{ij} = S_{ij} + \rho_{E,ij}(R_E) + \varepsilon_{ij},$$

and (5.62)

$$V_{ij} = \beta E_{ij} + \rho_{V,ij}(R_V) + \eta_{ij}.$$

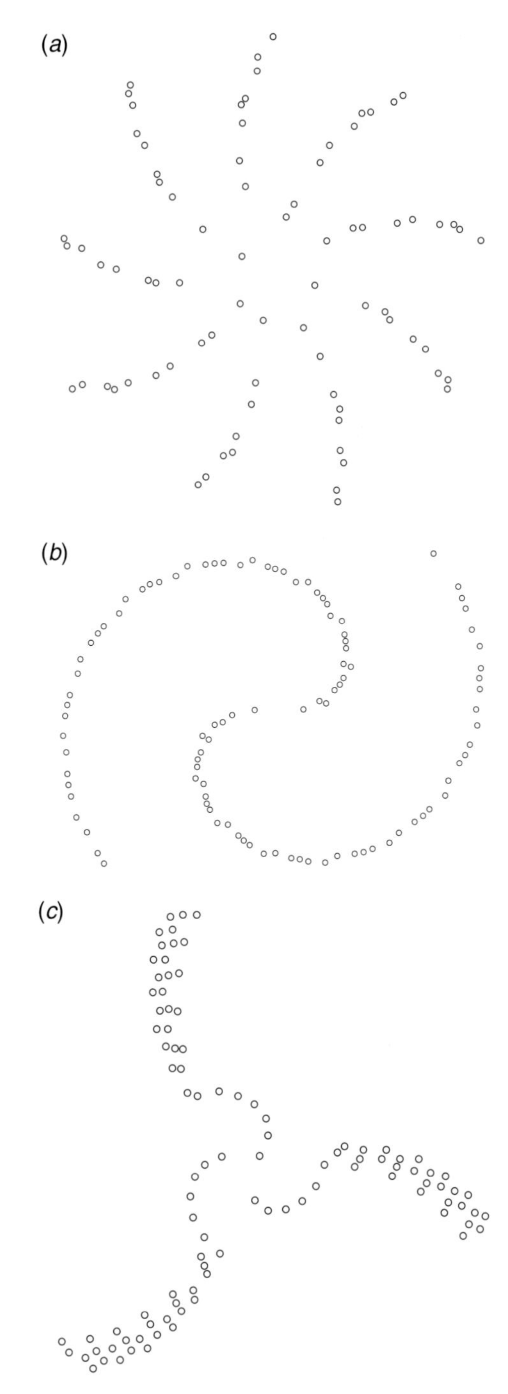


Figure 5.18 Three subsets of the Fibonacci spiral that could be used for sampling as part of a pilot study to determine the characteristics of spatial autocorrelation prior to a main study or experiment.

$V_{ij}$ is the variable of interest, $\rho_{V,ij}(R)$ is its autocorrelation component with range $R$, $\beta$ is a measure of its linear dependence on the environmental variable and $\eta_{ij}$ is a standard normal error term. Among the questions asked were these:

(1) Can the effect of spatial autocorrelation on tests be reduced by the survey design?
(2) Which designs provide the greatest power given a particular combination of spatial structure and autocorrelation range?

The study found that spatial autocorrelation in both the environmental and response variable affects the standard tests, spatial autocorrelation in only one does not. A broad-scale spatial structure in the underlying environmental variable, however, combined with spatial autocorrelation in the response variable inflated Type I error, just as spatial autocorrelation in both variables would. The major piece of advice to be derived from this study was to use a pilot study to identify the underlying structure. If there is a gradient, its effect can be accounted for by using it as a covariate. If there are different zones (really representing non-stationarity), a covariable that distinguishes the zones should be used. Another important finding of this paper was that Dutilleul's method correcting the $t$-test of the correlation coefficient is highly recommended as being robust to various spatial structures tested.

### 5.4.2 Experimental design

The study of experiments had a list of questions parallel to those for the survey study, and the approach used was similar. The only modification is that the variable of interest now includes a treatment effect for experimental units, $\tau$, which can be 0, (low), medium or high:

$$V_{ij} = \beta E_{ij} + \rho_{V,ij}(R_V) + \eta_{ij} + \tau_{ij}. \tag{5.63}$$

There are several general lessons to be learned from this study:

(1) If either spatial autocorrelation or repetitive structures like waves are present in the underlying environmental variable, randomly positioned experimental units should not be used. The use of blocks is recommended.
(2) For a set number of experimental units in the presence of spatial autocorrelation, using more, smaller blocks spread throughout the study provides greater statistical power.
(3) Short-range spatial autocorrelation (related to the size of the experimental units and the blocks of them) affects ANOVA tests more strongly than long-range spatial autocorrelation. For example, where the blocks were 3, 6 or more units in linear extent, autocorrelation with a range of 4 units (as opposed to 16 or 40 units) caused the greatest decrease in statistical power.

The approach used in these two studies is clearly a useful one and there is probably a lot more that we can find out about mitigating the effects of spatial autocorrelation on statistical tests by its further application. The 'bottom line', however, is that it is ESSENTIAL to have a good assessment of the nature and range of spatial autocorrelation before designing or applying a survey or experiment.

## 5.5 Concluding remarks

There are several major themes that run through the various sections of this somewhat complex chapter. The first is that spatial autocorrelation is an important characteristic of ecological data; it affects the outcomes of statistical tests and other statistical procedures and it cannot be ignored. Positive and negative autocorrelation affect the results in opposite ways, and autocorrelation at all distances within a study needs to be considered. Even apparently non-significant amounts of autocorrelation can have a serious cumulative effect. In addition, the autocorrelation in ecological variables may often have several sources, which may not be distinguishable. Therefore, this is a somewhat complex topic with a dearth of easy answers.

The second theme is that while the various models of autocorrelation can provide insight into its effects, they may not be reliable as good descriptions of ecologically important structures. In fact, because the natural world is often patchy, autocorrelation may fluctuate between positive and negative, with increasing distance, which poses its own kind of challenge for correction and interpretation. In addition, we might question whether the similarity of adjacent samples (contiguous quadrats in a transect, for example) really represents redundancy in the data, the way it would in a simple model of autocorrelation. Having two samples in the same patch tells us something more than sampling the same individual twice.

In considering solutions to the effects of this phenomenon on statistical tests, we can offer some observations. Thinning the data is not a good idea because it is wasteful of information and because it is based on the concept of distance to independence, which may well be wrong. Adjusting the effective sample size is a possible solution **if** the chosen model is a good description of the data's structure; we cannot usually calculate the effective sample size from the data. Randomization methods may work, but they must be applied carefully and with an awareness of the possible problems. Complete randomization of the data cannot and does not control the effects of spatial autocorrelation on statistical tests of significance. The 'model and Monte Carlo' method has several features that recommend it, but it is not perfect. As we showed with artificial data, the best-fitting model may not be the original model that generated the data; for field data, we will not know the relationship between the model fit and the underlying structure. Larger sample sizes may improve the accuracy of the modelling exercise, but they increase the risk of encountering non-stationarity if the extent of the sampling is also increased.

Studying the relationship between autocorrelated variables requires careful consideration of the structure of the variation, particularly in ecology, where autocorrelation may be both inherent and induced in a single variable, and where it may not be well described by one of the standard models. Of particular importance is the fact that removing the variables' dependence on an underlying factor does not avoid the problems associated with analysis in the presence of spatial autocorrelation.

In considering the design of sampling or experimentation, the important first step is to determine the characteristics of the spatial autocorrelation before doing anything else. Knowing that structure will enable the researcher to develop a design that avoids, or at least reduces, the effects of autocorrelation on subsequent analysis. Pilot studies are a necessity and may be combined with other prior knowledge to produce effective designs.

In 1993, Legendre posed the question 'Spatial autocorrelation: trouble or a new paradigm?' as the title of a paper that provided a wide-ranging discussion of this topic. The answer, 10 years later, is that it is both. It is troublesome, not in the sense of being just a nuisance, but because it is not an easy phenomenon to deal with, and it is certainly part of the current approaches to ecology, which include spatial structure (they must) or provide spatially explicit results. We must remember, however, that it is this sort of lack of independence through space and time that makes any prediction possible. We would be in trouble, indeed, if ecological phenomena were spatially and temporally independent (if that were possible). Clearly, the characteristics, effects and corrections for spatial autocorrelation, and other sources of spatial dependence, require and are worth a good deal more effort and thought, before we can suggest that we understand them truly and thoroughly.

# 6

# Spatio-temporal analysis

## Introduction

This chapter expands the discussion of the analysis of spatial structure to include the dimension of time, and the spatial dynamics of ecological processes and the resulting patterns. The intimate relationship between spatial structure and temporal change in ecological systems was eloquently described by Watt (1947) in his famous discourse on pattern and process in plant communities. His theme was that a plant community could be viewed as a working mechanism with dynamic behaviour of development, degradation and regeneration. In many plant communities, the various phases of the dynamic process coexist, and have an identifiable spatial relationship to each other (Figure 6.1). In communities of animals, the relationship between spatial locations and dynamic processes are even more obvious, as animals move through the spatial structure of their habitat to find resources or mates and to avoid predation. At the level of populations, we need to recognize that a population of a given density is not homogeneously distributed, and that the dynamics of different subpopulations' densities may be very different depending on location. At the level of the individual organism and its immediate environment, we need to realize that an individual is usually affected by very local, rather than global, conditions, and that these may change significantly over relatively small distances and over relatively small time periods. In considering almost any system, our concepts of spatial structure and its importance will include implicitly, if not explicitly, a temporal component.

The expansion to include time is not the same as the elaboration from one spatial dimension to two, or from two to three; time is not 'just another dimension' because of the direction of causality. Even when cyclic phenomena are under study, in which the building phase and the degradation phase may seem just to be mirror images of each other (Figure 6.1), we will often find that different processes are responsible for the two temporal 'directions', even if the resulting sequence of patterns appears similar, but reversed in time.

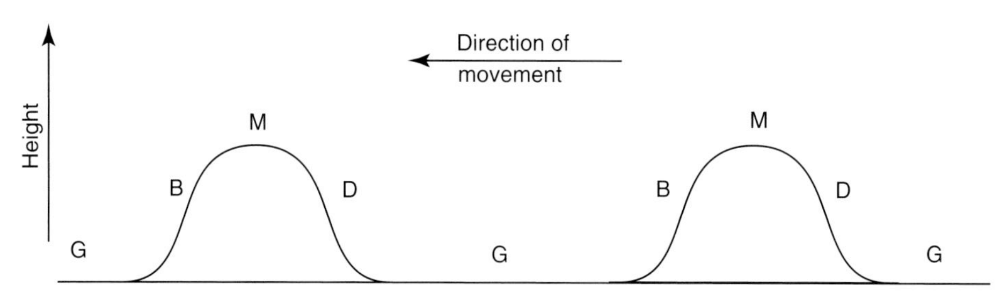


Figure 6.1 The relationship between pattern and process: the alternation between the mature hummock phase (M) and the hollow or gap phase (G) (e.g. in *Festuca ovina* dominated sites) proceeds through building phase (B) from hollow to mature and through the degenerate phase (D) from mature to hollow. In this abstract example, building occurs on one side and degeneration on the other causing directional movement of the system. The same general scheme may apply in *Sphagnum*-dominated wetland hummock–hollow systems, brousse tigrée vegetation stripes, or wave-regenerating forest systems (based on Watt 1947).

Our search of the literature in preparing this chapter was itself an interesting process. Many papers that have 'spatio-temporal analysis' or 'space–time analysis' in the title or key words, are using the term in a somewhat more general way than we have in mind, providing summaries of information rather than a detailed analysis (e.g. Knapp 1998; Civerolo & Rao 2001). The methods we include in this chapter will be somewhat more technical, as will become apparent. We have, however, limited the discussion almost entirely to statistical analysis, avoiding extensive discussions of modelling because the topic of spatially explicit dynamic models, for example, is both outside the domain of this book and sufficiently rich and complex to deserve a book of its own (see Dieckmann *et al.* 2000, among others).

The data used for spatio-temporal analysis can be classified in a number of different ways, but an important criterion will be whether they are continuous or discontinuous in time or in space. For example, if we are monitoring the environmental conditions in a nature reserve, using an array of hygrothermographs placed throughout the reserve, the data are continuous in time but discontinuous in space. On the other hand, if we are studying the movements of animals using global positioning system (GPS) radio collars, which provide a report on position once every few hours, the data will be continuous in space but discontinuous in time. In both those examples, we may make assumptions about interpolation: what the conditions are between sites with hygrothermographs or what the animals were doing between position reports. In cases such as permanent sample plots, in which tree stems are mapped and re-mapped at intervals, no interpolation may be necessary: stem No. 23 was alive in 1970, standing dead in 1978 and a downed log in 1985. In such cases, the observation and analysis of spatio-temporal pattern brings us very close to observing the processes that contribute to the pattern, because there is sufficient data to recover all the important events and changes.

To complete the classification of data types, it is possible to have data that are more-or-less continuous in both space and time, for example the flight of a butterfly or dragonfly (observed from a suitable distance), although the observer might want to divide the movement into temporal units. Similarly, the tracks of animals in the snow are also of this kind. On the other hand, mark-recapture data (which are records of animals caught, labelled or tagged somehow, released and then caught again) are discontinuous both in space, trap locations, and in time, trapping session or date.

A concept that is basic to our discussions in much of this chapter is that of spatio-temporal autocorrelation. This refers to the lack of independence between objects, events, observations or measurements due to their positions in space and in time. The simplest kind is the case of short-range positive spatio-temporal autocorrelation where samples are more similar when they are closer together in space or in time (as discussed for space alone in the previous chapters). For example, Setzer (1985) used a Mantel test (see Chapter 3) on spatial and temporal distances between aphid galls on cottonwoods (*Populus deltoides*), and found that galls close in space were likely to suffer mortality close in time. In comparison, more complexity will be found in cases involving cyclic behaviour such as diurnal migration, such as the vertical migration of zooplankton, where autocorrelation will be positive at short space and time lags, becoming negative over short space and longer time lags, and then positive again over even longer time lags (cf. Ohman 1990).

Just as there is a variety of measures for spatial autocorrelation (e.g. Moran's $I$; Geary's $c$), a number of different indices of spatio-temporal autocorrelation is possible. One such is Griffith's space–time index (Griffith 1981; Henebry 1995):

$$I_{s-t} = (nT - n)\frac{\sum_{t=2}^{T}\sum_{i=1}^{n}\sum_{j=1}^{n} w_{ijt-1} z_{it} z_{jt-1}}{\sum_{t=2}^{T}\sum_{i=1}^{n}\sum_{j=1}^{n} w_{ijt-1} \sum_{t=1}^{T}\sum_{i=1}^{n} z_{it}^2}. \tag{6.1}$$

There are $T$ temporal units and $n$ spatial units, with $w_{ijt}$ as the weights and the $z$s are the deviations from the overall mean of the observations. Clearly, this measure combines an evaluation of temporal autocorrelation (of $z_{it}$, say, with $z_{i,t-1}$) with an evaluation of spatial autocorrelation at individual times (of $z_{it}$ with $z_{jt}$).

If we are measuring autocorrelation for a particular separation or 'distance' class, $d$, the weights might be more precisely written as $w_{ijt}(d)$. As with Moran's $I$, the expected value is a function of the negative reciprocal of the number of samples:

$$E(I_{s-t}) = \frac{-(T-1)}{T(nT-1)}, \tag{6.2}$$

which is approximately $-1/nT$ for large values of $T$. For large sample sizes, the assumption of convergence to normality is justified (Henebry 1995). Figure 6.2

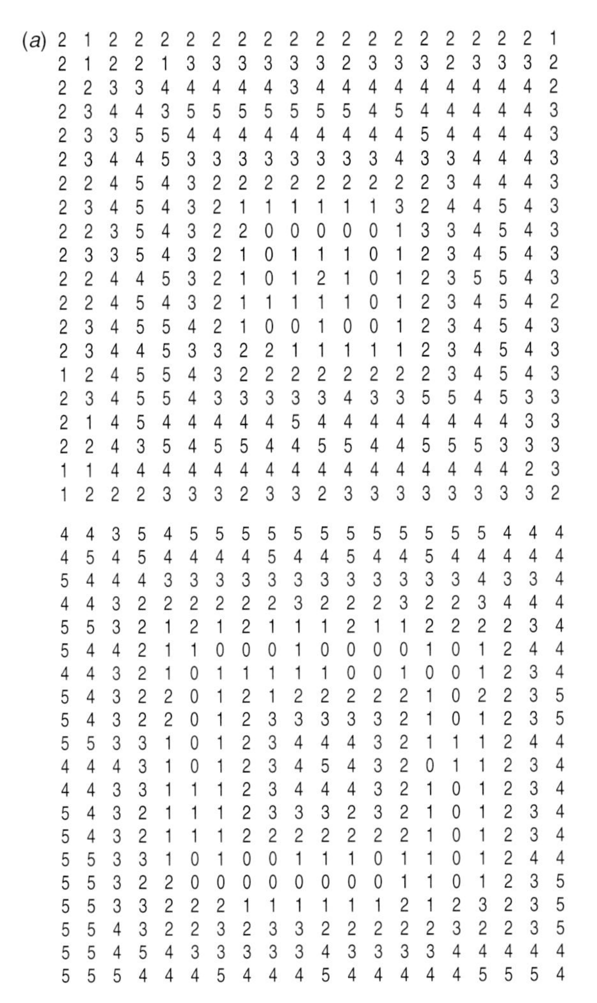


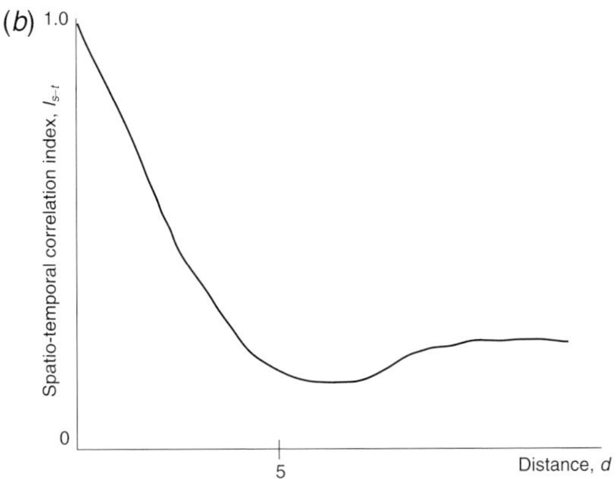


Figure 6.2 (*a*) Artificial data for spatio-temporal analysis, with waves of higher and then lower density moving out from the centre. (*b*) The analysis of the artificial data using Griffith's space-time correlation index. (*c*) Field data (*Nardus stricta*) from Law *et al.* (1997); higher values represent higher density. Each side of the 15 × 15 grid measures 50 cm. (*d*) The analysis of the field data using Griffith's space-time correlation index.

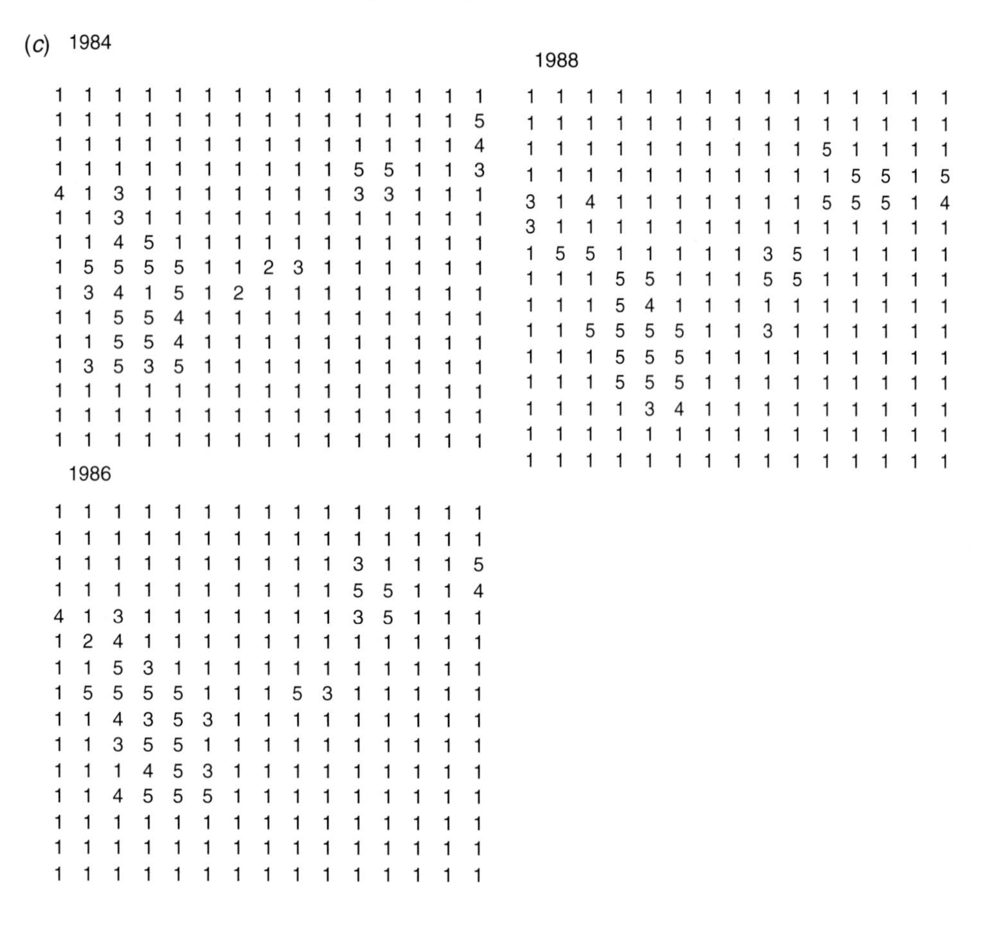


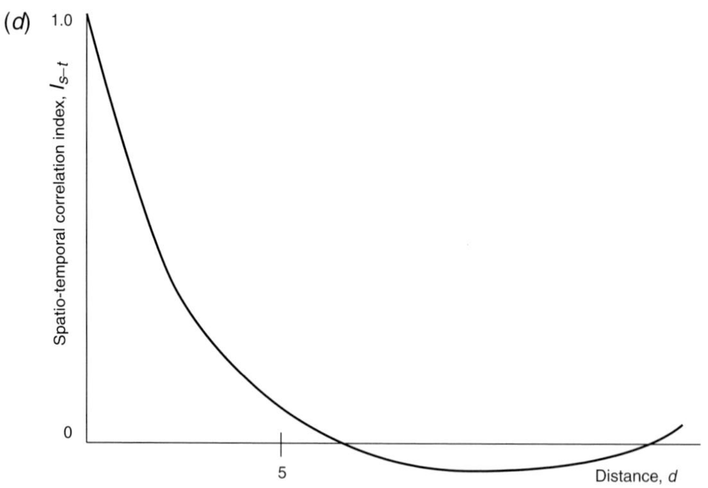


Figure 6.2 *(cont.)*

shows two examples of the application of this index to artificial data (Figure 6.2*a*) of a travelling wave (Figure 6.2*b*) and to *Nardus stricta* L. data presented in Law *et al.* (1997), three years of which are shown in Figure 6.2*c*, with the analysis in Figure 6.2*d*. One problem with this approach is the possibility of spatio-temporal 'anisotropy' introduced by the fact that, in ecological examples, the units of measurement in space may not be comparable to the unit used to measure time. In some examples in physics, there may be a natural spatio-temporal equivalence due to relativity, such as a second in time and a light-second in space. Analogously, there may be some ecological examples in which there is a natural equivalency between spatial and temporal measures, based on the speed of dispersal or travel, but there will be many cases in which such an equivalence is not available.

In this chapter, we begin by examining spatio-temporal analysis by discussing studies that look at change in some of the spatial statistics already described in this book. We then proceed to talk about four kinds of truly spatio-temporal analysis, based on join counts, cluster change detection, polygon change and movement through space. The section following the group of methods attempts to tie together process and pattern by looking at situations in which we can analyse some kind of record created by the processes themselves, in particular, tree establishment, growth and mortality, clonal plant mobility and the boundaries of crustose lichen colonies.

## 6.1 Change in spatial statistics

One of the most simple kinds of spatio-temporal analysis is to examine changes in almost any of the spatial statistics described so far, as a function of time. Small changes can be accounted for by variation in the same underlying process, but large changes may suggest a change in the process itself. There are many examples of this approach in the literature. For example, Wu *et al.* (2000) studied the 'progress' of fragmentation in tiger bush (brousse tigrée) landscapes by comparing the curves of lacunarity as a function of scale (see Chapter 2), observed in different years. Brousse tigrée is a phenomenon of arid regions in which there are bands of woody vegetation, running across the direction of water flow, alternating with bare ground. It was described in more detail in Chapter 1 in relation to the concept of anisotropy. In their study of these landscapes in southwest Niger, Wu *et al.* (2000) found that lacunarity increased between 1960 and 1992 (Figure 6.3), indicating the continuing fragmentation of the woody vegetation in the landscape.

Dale & Zbigniewicz (1997) studied a somewhat similar phenomenon at a smaller scale. They used 100 m transects of 10 × 10 cm contiguous quadrats in shrub-dominated communities near Kluane Lake in the Yukon to examine the effect of a peak in snowshoe hare density on two of its winter food plants, *Salix* and *Betula*.

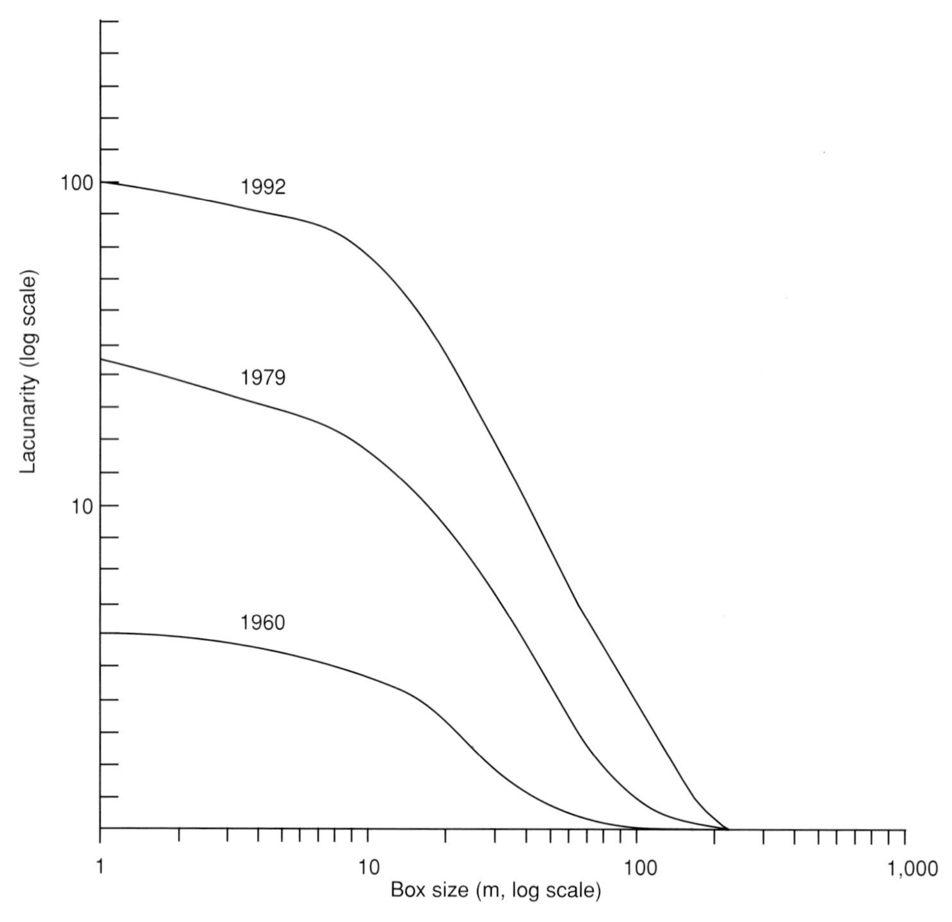


Figure 6.3 Changes in lacunarity curves for brousse tigre sites. The lacunarity maximum increases markedly from 1960 to 1992, indicating more open ground, but the basic shape does not change noticeably, indicating little change in the other basic characteristics of the spatial pattern.

The snowshoe hare (*Lepus americanus*) populations have cyclic fluctuations in density, with an amplitude of a factor of 5–25, and a period of about 10 years (cf. Krebs *et al.* 2001). They sampled the shrub vegetation in 1988, just before the peak in density, and in 1993, just after it. They used 3TLQV and Galiano's (1983) NQV (see Chapter 2) to detect the scale and patch size of the shrubs' patterns, expecting that the herbivore population peak would lead to fragmentation of the patches, leading to reduced scales of pattern and smaller patches. What they found, however, was that despite the intensive and extensive browsing of the twigs of these plants, the basic characteristics of the pattern recovered very quickly. This suggests that the basic characteristics of the spatial pattern of these shrub-dominated areas may persist for decades.

Vacek & Lepš (1996) studied the effect of neighbouring trees on tree vitality and mortality in five permanent plots in the mountains of the Czech Republic, monitored for 18 years. They used Ripley's *K*-function analysis (Chapter 2) to examine the spatial pattern of *Picea abies* trees in 50 × 50 m plots. They found that although most of the plots started with a slight tendency to aggregation in 1976, by 1993, the pattern had changed to one of overdispersion at scales of 2–5 m, attributable to the higher mortality rates of trees with large and close neighbours. Kenkel *et al.* (1997) report a similar trend for *Pinus banksiana* in Manitoba: a clumped pattern in the early stages of stand development gave way to a more-or-less random pattern at intermediate ages (30–40 years), with a tendency toward overdispersion at later stages. We will return to a more detailed analysis of tree establishment and mortality in Section 6.6.1.

As a last example in this section, we cite Nestel & Klein (1995) who used Moran's *I* (see Chapter 2) to study the spatio-temporal patterns of dispersion of adult leaf-hoppers (Homoptera: Cicadellidae) through the growing season of fruit orchards in Israel. The spatial pattern of one species, *Asymetrasca decedens* (Paoli), began as aggregated and changed from aggregated to random with each successive generation. The pattern of a second species, *Edwardsiana rosae* L., remained aggregated throughout the entire season. The authors suggest that knowledge of the early season aggregation of both species could be used to develop an integrated management strategy for these insect pests.

This general approach to spatio-temporal analysis, using basic spatial statistics at two or more times is straightforward and clearly can be very informative and useful. It will become more important as an analytical approach as more long-term studies continue or are established. The main disadvantage is that any summary statistics and changes in them may miss some of the important details of the actual changes to individual units in the spatial structure. One of many themes running through this book is the thought that methods that merely detect departures from randomness, such as spatial clustering, may not be telling us enough; for spatio-temporal analysis, for example, it may be important to know the size, spacing, compactness and positions of the clusters (perhaps especially in pest management applications) in addition to the general degree and scale of aggregation.

## 6.2 Spatio-temporal join count

In plant ecology, the concept of spatio-temporal pattern goes back at least to Watt (1947), who described how certain vegetation types tended to occur close together in space and time. One factor that contributes to this phenomenon is the clonal nature of some plants. Clonal growth forms are often described based on the spatial pattern of the ramets, which is related to patterns of establishment in space and time. The 'phalanx' form is characterized by a compact spatial structure of the ramets and the 'guerrilla' form is characterized by much more loosely arranged ramets (Lovett Doust 1981). These two forms are endpoints of a continuum, and a general growth form can be described on the basis of the spatial arrangement of ramets, and the predictability of ramet establishment in space and in time.

Methods used in elucidating spatio-temporal pattern have been developed for single factor autocorrelation analysis to consider factors separated by two 'lags', i.e. intervals along axes of space and time. One such method, Griffith's space-time index, given above (Griffith 1981; cf. Henebry 1995), is related to Moran's $I$, but examines the values of a variable at different places and times. This index is appropriate to the analysis of continuous variables such as plant density, but if a binary variable such as presence : absence of individuals is considered, then join count statistics can be used.

Join count statistics are a way of measuring association in nominal data distributed on a lattice or grid (Cliff & Ord 1981; see Chapter 3). Here we consider the case of binary data where each cell of a lattice can take only one of two values such as black or white. A join is defined as a connection of a particular lag between pairs of defined cell types, for example black to black. Figure 6.4*a* shows a simple example where joins of contiguous black cells are counted in a $7 \times 7$ square. In the spatio-temporal approach, a two-dimensional lattice represents a single spatial dimension, such as a transect of $n$ quadrats and $m$ intervals of time (cf. Little & Dale 1999). Suppose the black cells represent either plant establishment or the presence of at least one individual of a particular species. If it represents establishment, then the occurrence of plants of different ages in the same quadrat can then be distinguished by placing several black cells in that column of the lattice. Join lengths are specified by combination of the intervals along the axes: $(d = 2, t = 3)$ signifies joins of length two on spatial axis and three in time. Spatio-temporal association can be determined by comparing the observed number in a class to the number expected from randomness. The simple null hypothesis ($H_{01}$) is that the observed number can be accounted for by random occurrence. If more joins of a particular class $(d, t)$ occur than are expected, this indicates a tendency for stems to be separated by a distance of $d$ in space and $t$ in time.

As a further refinement to this approach, Little & Dale (1999) were interested in the first stem that established in each quadrat; indicated by black cells in a lattice with only one black cell per column. Therefore, two further null models can be used, in which black cells are randomly arranged but with no more than one in a column. In the uniform row model ($H_{02}$), the probability that each of the $n$ columns contains a black cell is $r$ and the probability of a black cell in any column is distributed uniformly among its rows. In the 'top black' model ($H_{03}$), black cells may occur in

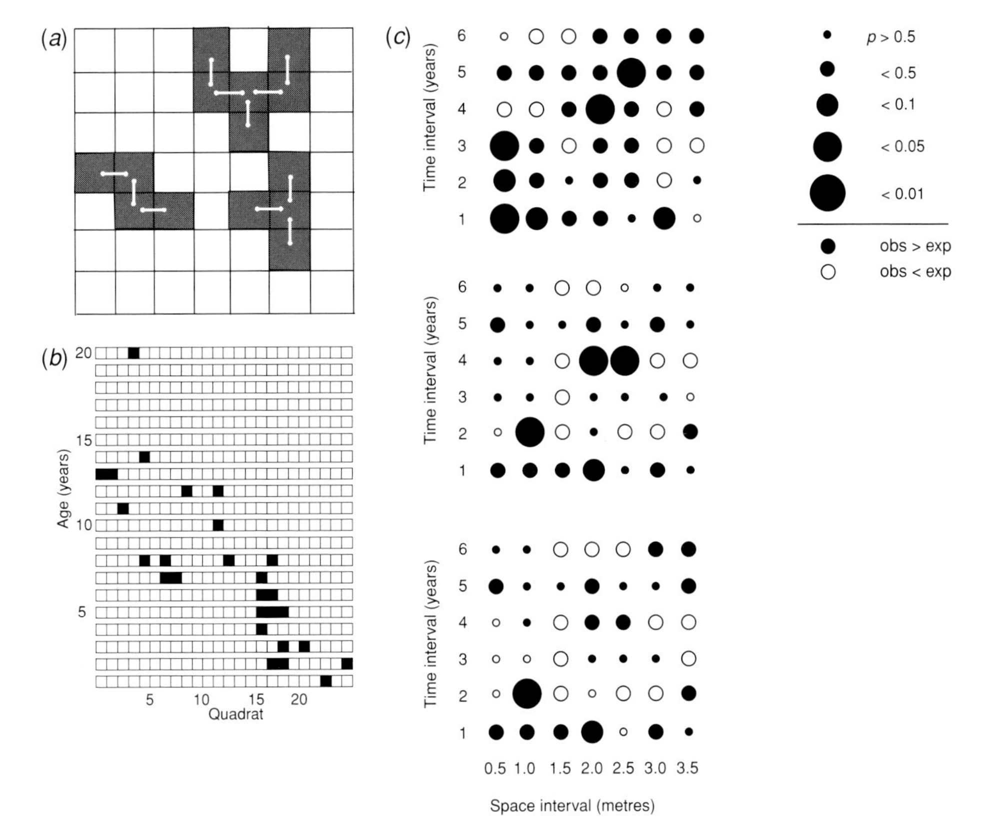


Figure 6.4 (*a*) Join counts of black-to-black contiguities in a 7 × 7 square; there are 11, as indicated by the dumb-bell 'simple join' symbols. (*b*) Space and time data for the establishment of *Populus balsamifera* at Ministik, Alberta. (*c*) Spatio-temporal join count analysis of the *Populus balsamifera* data. At each space and time lag combination, the size of the symbol represents the significance level and the colour codes (○, •) represent the sign of the difference between observed and expected values. The three null hypotheses are fully randomized (upper), uniform row (middle) and 'top black' (lower).

any cell of the lattice with probability $p$, but only those in the highest row in each column (representing the earliest to establish) are retained.

For each join class, 10,000 'data' lattices were generated for each of three random models, which were used to calculate a reference distribution and expectation. The association statistic of each join class was assessed based on the deviation of observed from expected join frequencies. The results are displayed as two factor 'correlograms', using circles, the sizes of which represent the probability and the positions of which represent the join classes. Figure 6.4*b* presents the *Populus balsamifera* establishment data from a transect at Ministik, Alberta, and Figure 6.4*c* shows the results of this analysis using three different null models. It shows that the fully randomized model has high values at (0.5, 1), (0.5, 3), (2.0, 4) and (2.5, 4). The other two models have high values at (1.0, 2) and (2.0, 4), the latter probably being a resonance peak. Overall, the results suggest that it is common for stems to be separated by two years in age and by 1 m along the transect, indicating that the clone advances in biennial pulses.

This technique is essentially a temporal adaptation of the lichen mosaic method described by Dale (1995; 1999, see Chapter 2) but the important difference in this case is the use of several null models, rather than just one. As we commented in Chapter 2, one way in which our spatial analyses can become more sophisticated, and thus more useful, is to use a range of null models rather than just simple 'randomness' for comparison with the outcomes we observe.

## 6.3 Spatio-temporal analysis of clusters and contagion

The kinds of analyses described in this section originated in the context of the study of clusters and spread of incidences of human disease, and some of the vocabulary will persist in our discussion. The methods, however, will translate well into many ecological areas, not just the incidence of disease, pathogens or parasites, but also the locations of rare epiphytes, nitrogen fixing symbionts and so on. The non-temporal version of the basic approach is called 'cluster detection', which should not be confused with the multivariate technique known as 'cluster analysis'. Methods for detecting disease clusters for which the 'at risk' population is unknown are essentially versions of univariate point pattern analysis. The epidemiological literature tends to emphasize approaches in which the locations (or at least the number in a given area) of individuals at risk are known, as well as affected individuals. The question is then whether the diseased cases are more clustered than can be explained by local variations in the at risk population. This approach is a version of bivariate point pattern analysis, and strongly resembles other ecological questions of a similar nature, such as 'Are the *Solidago* plants with galls more aggregated than can be explained by the overall patchiness of the plants?' (Dale & Powell 1994).

Wakefield *et al.* (2000) provide a comprehensive review of the methods for cluster detection in the general area of spatial epidemiology (see also Fotheringham & Zhan 1996; Jacquez 1996). The methods fall into several categories. 'Traditional' methods include a simple comparison of the numbers of cases observed in different areal units (townships, counties) with the expected number (based on population and global disease rate), using a goodness-of-fit test. 'Distance : adjacency methods' include Moran's $I$ for rates in contiguous areal units and Diggle & Chetwynd's (1991) variant of Ripley's bivariate $K$-function analysis for point data. Locally specific methods include the moving window approach and risk surface estimation

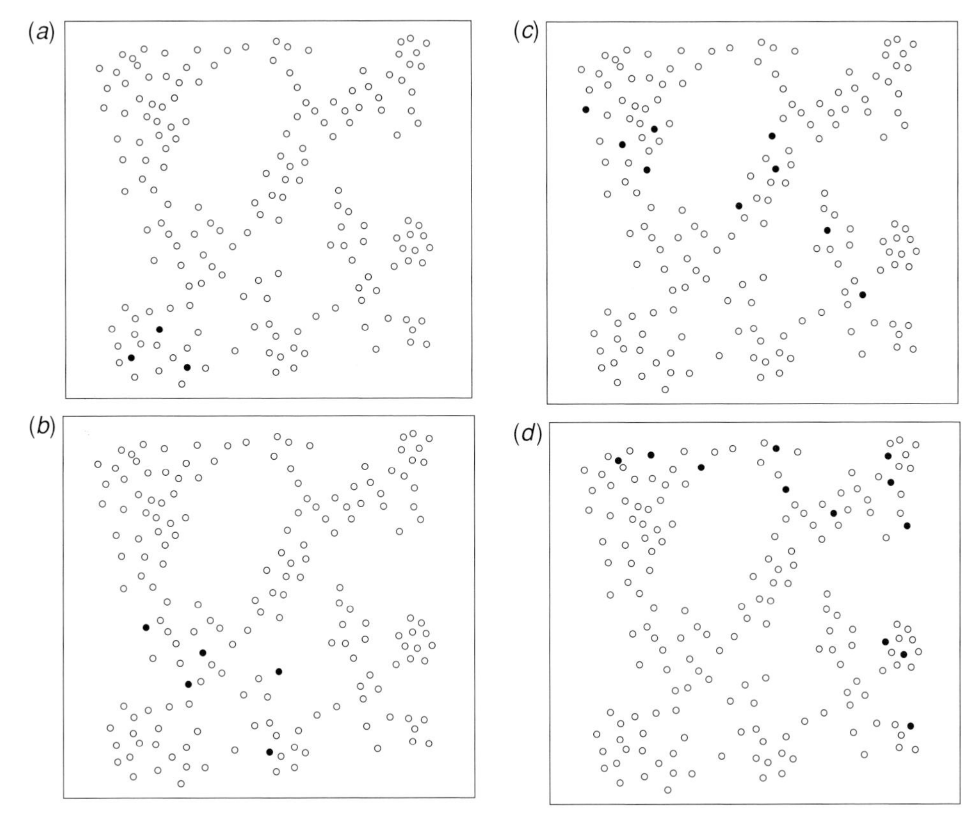


Figure 6.5 (*a*)–(*d*) The incidence of newly diseased organisms (filled) in a population at four different times. The spread of disease is obvious.

(see Wakefield *et al.* 2000 for details.) Our intention here is merely to describe a couple of approaches suitable for spatio-temporal analysis.

The first set of methods is best appreciated by looking at the four parts of Figure 6.5, which shows the progression of a disease (black dots) through a population (all dots) at four time periods. Clearly, the disease is spreading out from one corner of the figure, and the black colouring indicates a case that is new in that time period, rather than being a cumulative record.

A simple approach is to define a threshold value for 'near' vs. 'far' in time and a threshold value for 'near' vs. 'far' in space. Pairs of disease incidences are then categorized as 'near' or 'far' in time and 'near' or 'far' in space to produce a $2 \times 2$ contingency table. The table can then be tested with the usual statistics to determine whether incidences close in time also tend to be close in space (Knox 1964). For example, from Figure 6.5, using thresholds of 2 time steps and of 40% of the side of the sample area, we get these counts:

| | Time | |
|---|---|---|
| | Near | Far |
| Space | | |
| Near | 114 | 2 |
| Far | 125 | 110 |

The goodness-of-fit statistics are highly significant indicating spatio-temporal association among the disease incidences, in spite of the large number of pairs that are near in time but separated in space (chiefly in time period 4, Figure 6.5*d*).

In the introduction to this chapter, we mentioned the use of the Mantel test to determine spatio-temporal clustering in the example of aphid galls on cottonwood trees. The Mantel test compares two 'distance' matrices (see Chapter 3), and those distances can be of the spatial and temporal separations of events. In this context the Mantel test can be seen as an extension of Knox's approach, using the measured distances, rather than just 'near' and 'far' (Bailey & Gatrell 1995). Where $d_{ij}$ is the separation of two events in space and $s_{ij}$ is the separation in time, the basic Mantel statistic is:

$$Z_M = \sum_{i-1}^{n-1} \sum_{j=i+1}^{n} d_{ij} s_{ij}. \tag{6.3}$$

Evaluation of the test statistic is most easily accomplished by a randomization test. In the example shown in Figure 6.5, the observed Mantel statistic is $Z_M = 1219.14$ which was found to be highly significant by a randomization test (of re-labelling with set numbers of each kind of event), once again indicating an association between temporal and spatial proximity.

Notice that, in spite of our earlier promise, the approaches described so far do not use any of the information of the 'at risk', but disease-free, population. In fact, the next method we describe does not use that information either. That method is the spatio-temporal version of Ripley's *K*-function analysis, using only focus events (e.g. disease incidences) and counting the number of focus events within distance $t$ and time $\tau$ of each event. Observed and expected values are compared in the usual way and plotted as a function of $t$ and $\tau$. There are many potential problems with this approach, the first being the possible incommensurability of time and space units already discussed. The second is that unless there is a very long time series of observations, temporal edge effects can be an important factor. The third question is whether, because time is directional, a one-sided search template should be used rather than the two-sided '$t$-bar' template depicted in Figure 2.24.

It seems that in the literature, a truly bivariate spatio-temporal analysis based on Ripley's $K$-function is missing. This is an area where further developments are needed and should be expected.

## 6.4 Polygon change analysis

In Chapter 2, concerned with the basic methods of spatial analysis, there was an extensive discussion on the analysis of points or events in the plane, and considerably less on the analysis of irregular polygons in the plane; this reflects the emphasis in the literature. Similarly, published studies on spatio-temporal analysis also emphasize point events rather than polygons, quite naturally when the subject is the dynamics of tree stems, for example, or the locations of disease foci. The analysis of polygons offers considerably greater complexity than the analysis of point patterns, because of the greater complexity of the data themselves. The difficulties of dealing with polygons can sometimes be avoided by representing the centre of the polygon by a point with associated size and shape characteristics, and with the adjacency of polygons represented by lines between the points. These networks can then be analysed for autocorrelation in the neighbour network using statistics such as Moran's $I$ or Geary's $c$ (see Chapter 3). For example, the polygons in Figure 6.6*a* can be reduced to a graph of their connections to neighbours by contiguity and their areas (Figure 6.6*b*). Autocorrelation analysis shows significant similarity among neighbours, based on their areas, but information about other characteristics of the polygons is lost.

When we now consider analysing a dynamic system of polygons, the situation becomes even more complex, because not only do the characteristics of the polygons change (position, size, shape), as do their connections to neighbours, but also old polygons may disappear and new ones develop. One approach would be to calculate summary statistics for each of several observation times and then examine changes in those summaries; for example, Peralta & Mather (2000), in a study of deforestation in Amazonia, used indices of lacunarity, patchiness and area–perimeter fractal exponent to summarize the changes. There is an advantage in using more than one summary statistic, obviously, but even so, a lot of detail on the characteristics of the polygons will be missed, entailing the loss of potentially important information.

Sadahiro & Umemura (2002) have developed a sophisticated approach to the analysis of changing polygons. The conditions of their scheme are that the polygons are immobile, so that individual animals, or herds or flocks, are not eligible, and that change occurs in a discontinuous fashion, so that objects like temporary pools that shrink and expand in a continuous way are also excluded. With those exclusions aside, their treatment seems flexible and useful. They divide the stepwise behaviour of an individual polygon into six primitive events, illustrated in Figure 6.7:

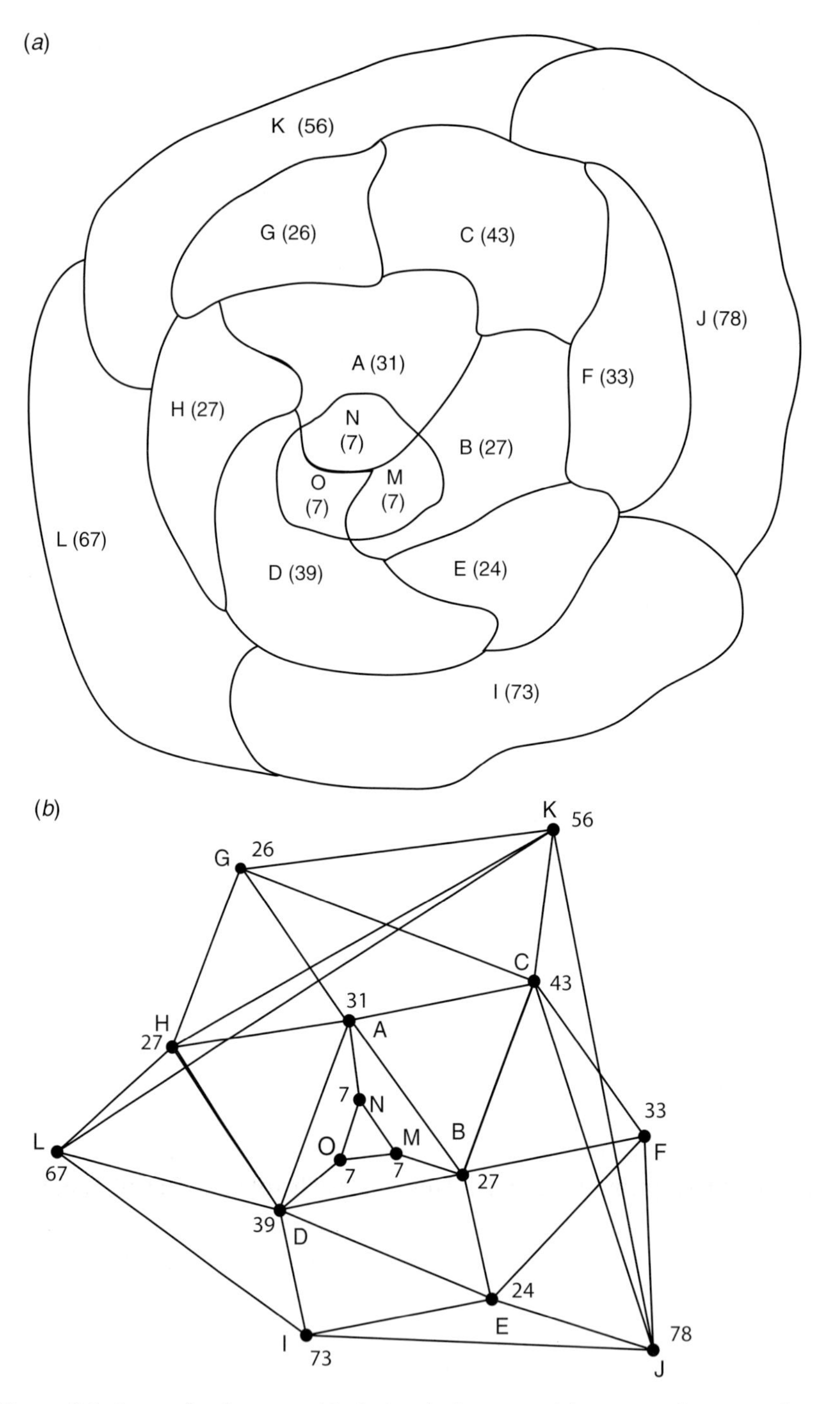


Figure 6.6 A set of polygons with their relative areas (*a*) portrayed as a graph (*b*) with the connections in the graph representing the contiguity (shared edge) of polygons. The nodes of the graph retain the areas of the polygons for further analysis.

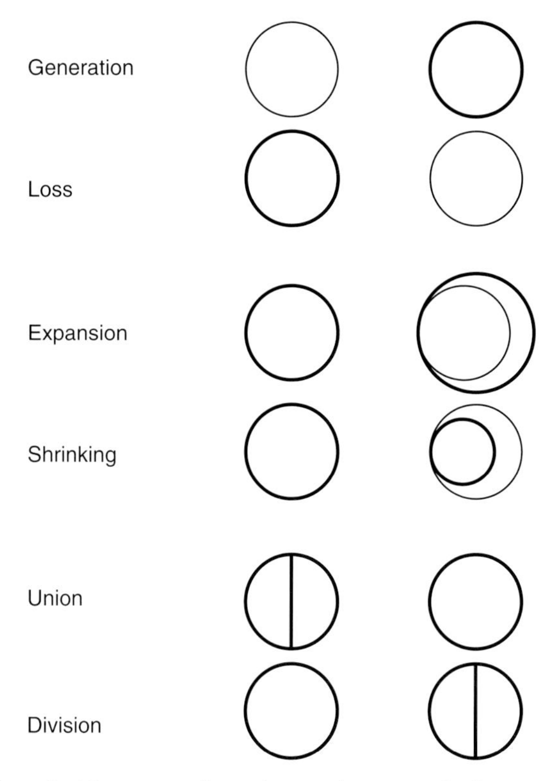


Figure 6.7 The primitive events for polygon change analysis: generation, loss, expansion, shrinking, union and division.

(1) generation, i.e. the appearance of a new polygon;
(2) disappearance, i.e. the loss of a polygon;
(3) expansion, i.e. the increase in area occupied;
(4) shrinkage, i.e. the loss of area;
(5) union, i.e. two polygons merging; and
(6) division, i.e. a polygon splitting into two.

The changes observed between two different times can then be described by combinations of these primitive events. This approach follows that of Claramunt & Thériault (1997) who proposed a more complicated scheme of sixteen primitive events, in part to be able to deal with possibly mobile polygons. An additional assumption is that there are no problems in tracing the identity of polygons; the time interval is assumed to be sufficiently short that polygons that overlap are two temporal versions of the same polygon.

The two sets of polygons from the two observation times, $\Gamma_1$ and $\Gamma_2$, are overlaid to create a new set of polygons, $\Gamma_u$ (Figure 6.8). These new polygons can then be classified into three groups, $\Omega_0$ which existed at both observation times,

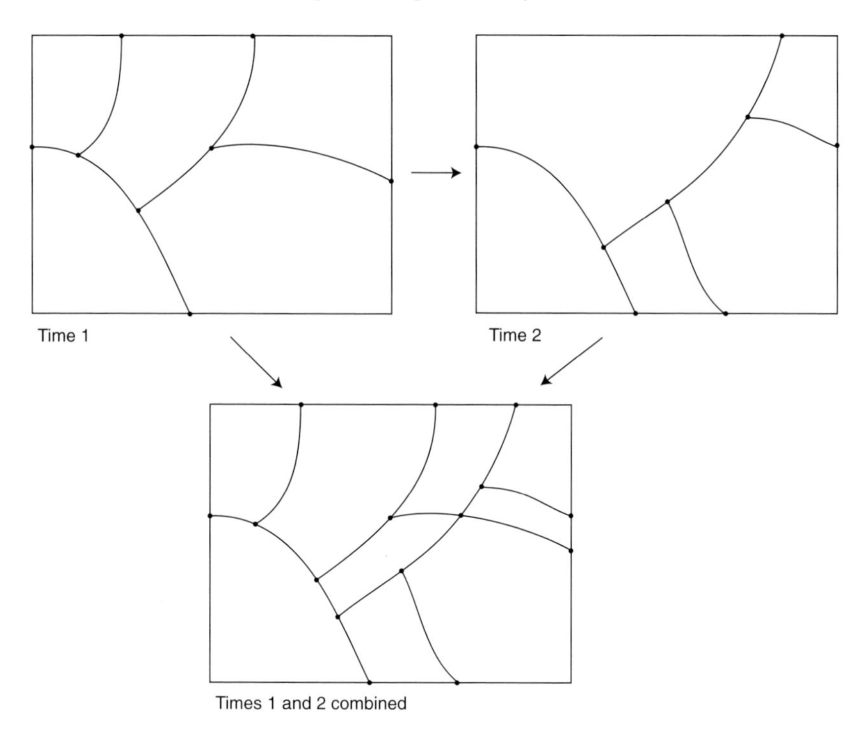


Figure 6.8 The two sets of polygons from time 1 and time 2 ($\Gamma_1$ and $\Gamma_2$, upper diagrams) are overlaid to create a combined set ($\Gamma_u$, lower diagram). There are five polygons in each of the first two sets and nine in the combined set.

$\Omega_1$ which existed at time 1 but not time 2 and $\Omega_2$ which existed at time 2 but not time 1. The arcs in the diagrams are classified into twelve groups based on the combinations of four possible states in which they existed at times 1 and 2: boundary, partition, internal to polygon and absent. Figure 6.9 illustrates these classes. The two classifications are then used to deduce possible sequences of primitive events that gave rise to the observed changes. The changes between two observation times are decomposed into the smallest number of primitive events possible, and one useful statistic is then the number of such primitive changes, $M_e$, which can be standardized to the total number of polygons:

$$m_e = \frac{M_e}{|\Gamma_1| + |\Gamma_2|}. \tag{6.4}$$

The authors suggest that, as a refinement, instead of raw event counts, the events could be weighted by some function specific to the kind of event, $f_X$. For example, generation and expansion events could be weighted by the area gained and shrinking and disappearance events by area lost, with partition and union events having

Time 1

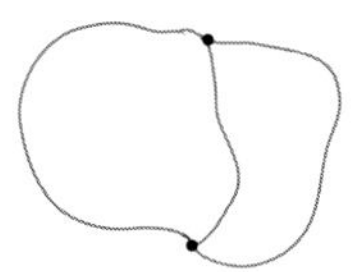

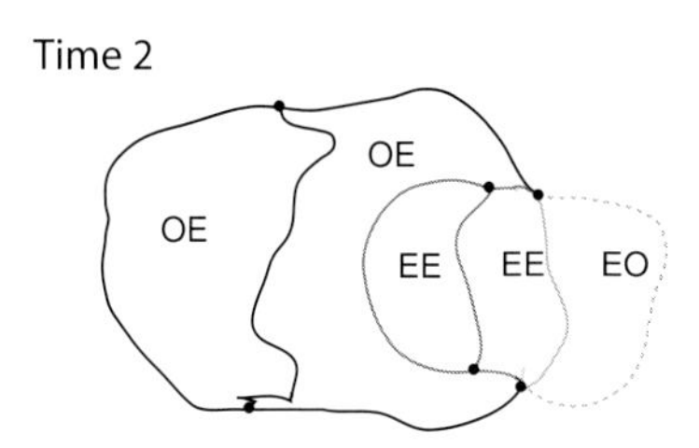


Times 1 and 2 combined

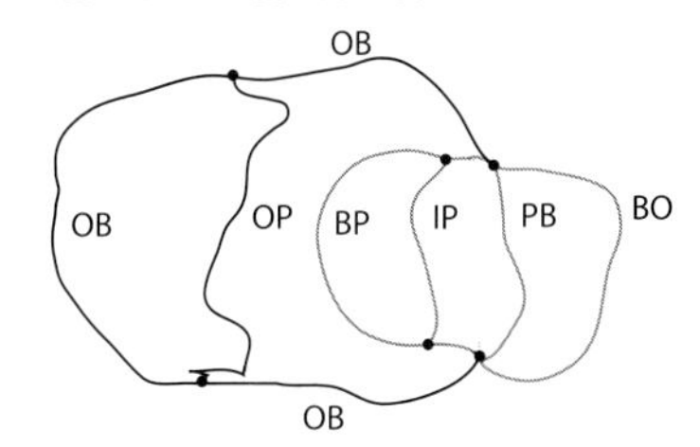


Figure 6.9 In this example, there are two polygons in the first set, four in the second and five in the combined set. The polygons can be classified as 'EO', extant at time 1, but not at time 2; 'OE', extant at time 2, but not at time 1; and 'EE', extant at both times. The edges that define the polygons can be classified as 'BO', boundary at time 1, but extinct at time 2; 'PB', partition then boundary; 'IP', interior then partition; 'BP', boundary then partition; 'OP', partition new at time 2; and 'OB', new boundary.

weight 0. In that case, the area index would be:

$$m_A = \frac{A(\Omega_1) + A(\Omega_2)}{A(\Gamma_1) + A(\Gamma_2)}. \tag{6.5}$$

This approach is new and has few examples of application, but it clearly presents a great deal of promise for future analysis. While it does deal with the dynamics of a set of polygons, it does not include characteristics such as their shape in the analysis. In addition, in this treatment, if the polygons being studied form a complete mosaic filling the study region, the kinds of transitions that can occur become much more limited. The analysis of full mosaics (in which the polygons form a complete tiling of the plane) is another related area that deserves further work and exploration.

## 6.5 Analysis of movement

There are many circumstances in ecological studies in which the movement of individual organisms is of crucial interest: the spread of disease organisms or vectors in epidemiology, the identification of home ranges in wildlife ecology or the spread of a clonal plant into new habitat, to suggest a few examples. The quantitative analysis of such movements will depend in part on whether the movement itself (or its record) is more or less continuous, like a beetle wandering through a grassland, or it occurs as discrete units, such as a butterfly's foraging journey stopping at individual inflorescences, or the identifiable ramets of a growing clone. If the movement is continuous, it is often divided up into units, using some criterion (possibly arbitrary) such as the positions observed every five minutes. The concept of units of movement will be discussed further, later in this section.

When the movement occurs in units, or can be divided up into units, analysis usually proceeds based on the lengths of the units of movement and the angles between them (see Figure 6.10). Crucial to an understanding of the correct way to analyse this kind of data is an understanding of the calculation of the average of a set of angles. Cain (1989) points out that many authors in ecology have made mistakes in this process. We cannot just take a simple average the way we do for a linear scale like distance. For example, if measured on a scale from $0^\circ$ to $360^\circ$, the angles 90 and $270^\circ$ have an average of $180^\circ$; measured on a scale of $-180^\circ$ to $+180^\circ$, the same angles, now $90^\circ$ and $-90^\circ$, have an average of $0^\circ$ (see Upton & Fingleton 1989: Chapter 9).

For any set of angles, whether the absolute angles of the steps, labelled $\alpha_i$ in Figure 6.10, or the 'turning angles', $\delta_i$ in the same diagram, the angles are represented by vectors of unit length and coordinates $(x_i, y_i)$, as in Figure 6.11. The coordinates of the mean vector are:

$$\bar{x} = \frac{1}{n}\sum_{i=1}^{n} \cos(\alpha_i) \quad \text{and} \quad \bar{y} = \frac{1}{n}\sum_{i=1}^{n} \sin(\alpha_i). \tag{6.6}$$

In polar coordinates, it is

$$(\bar{x}, \bar{y}) = (r_a \cos\varphi, r_a \sin\varphi), \tag{6.7}$$

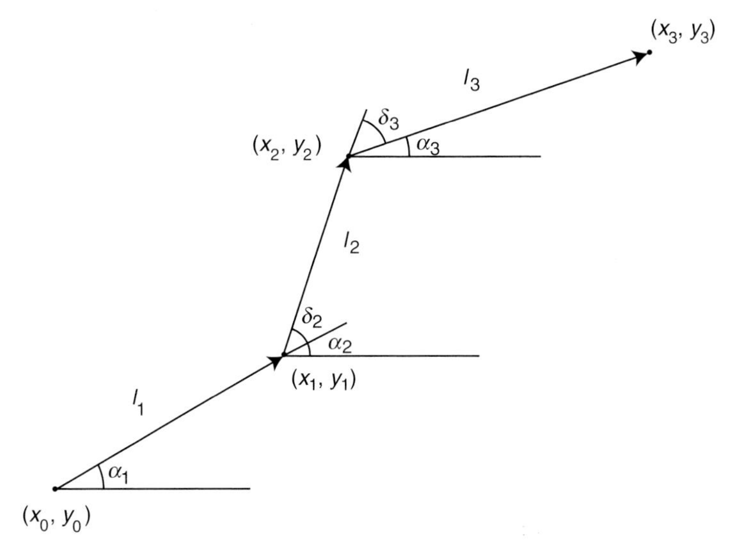


Figure 6.10 Movement portrayed as a series of straight-line units, $l_1$ to $l_3$, with known start- and endpoints. The $\alpha$s are the absolute angles and the $\delta$s are the changes in angle.

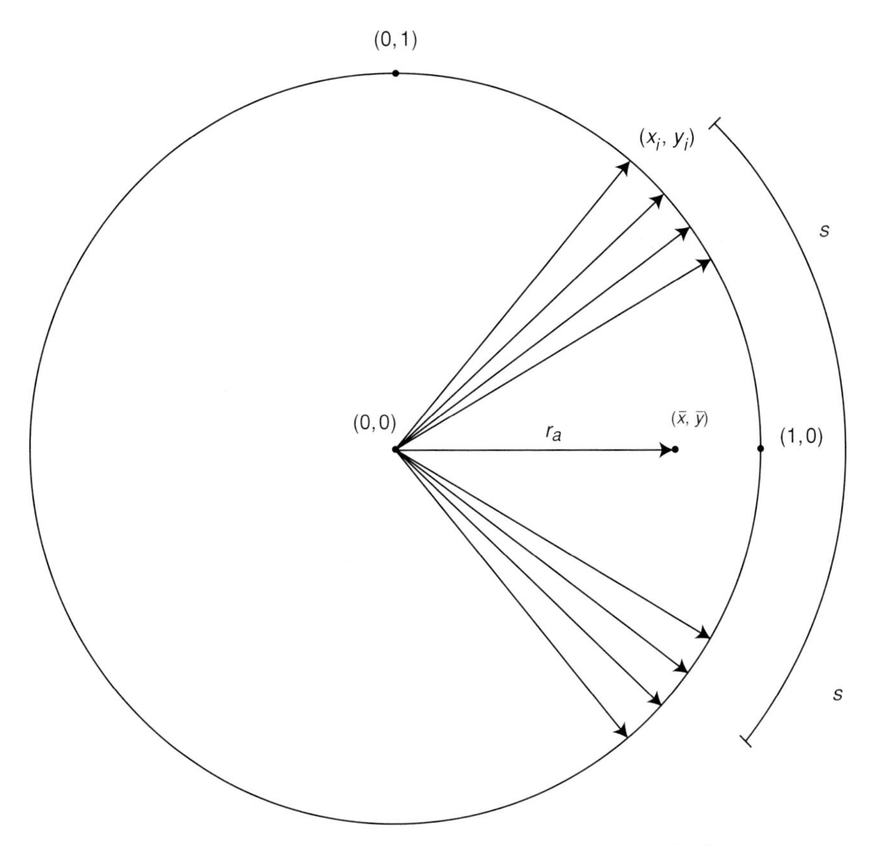


Figure 6.11 Calculation of the average of a set of eight angles depicted as eight unit vectors, $(x_i, y_i)$. The angular concordance is $r_a$ and the angular deviation is $s$.

where $\varphi = \tan^{-1}(\bar{y}/\bar{x})$ and

$$r_a = \sqrt{\bar{x}^2 + \bar{y}^2} = \frac{1}{n}\sqrt{n^2\bar{x}^2 + n^2\bar{y}^2} = \frac{1}{n}\sqrt{\left(\sum_{i=1}^{n} \cos\alpha_i\right)^2 + \left(\sum_{i=1}^{n} \sin\alpha_i\right)^2}. \tag{6.8}$$

The measure $r_a$ should not be referred to as a measure of angular correlation because it cannot take negative values; it is more appropriately referred to as a measure of angular concordance or angular concentration (Zar 1984). Alt (1990) referred to it as 'parallelicity'! It takes value 1.0 when all the angles are the same and value 0 when the vectors cancel each other out (Upton & Fingleton 1989). For example, in Figure 6.11 there is a good agreement in the direction of the angles and $r_a$ is 0.75. The circular equivalent of the standard deviation for linear data is $s$, the angular deviation (Batschelet 1981):

$$s = \sqrt{2(1 - r_a)}. \tag{6.9}$$

In our example, $s = 0.71$, which can be converted to degrees by multiplying by $180°/\pi$ (Batschelet 1981), here 41° (see Figure 6.11), showing again that there is a relatively small variance in the set of angles.

In considering a path such as that depicted in Figure 6.10, and seeking a measure of the angular autocorrelation, for lag 1, the average of the cosines of the turning angles is a good candidate (Batschelet 1981). Figure 6.12 illustrates this concept. When the two steps of the path are aligned in the same direction, correlation is 1.0; when they are at right angles, the correlation is 0; and when they are directly opposite, it is −1.0. Turchin (1998) suggested that this statistic should not be used for turning angles because 'turning angles are typically concentrated around zero (so that) the difference between two successive turning angles is likely to be near zero, even if there is no autocorrelation. This would result in a significantly positive, but spurious, angular autocorrelation'. We disagree and would argue that turning angles concentrated near zero, indicating a tendency for motion to continue in the same direction, is real autocorrelation as a characteristic of the data.

As a preliminary analysis of movement data, we suggest calculating the radial (distance) and angular correlation for a range of lags. Figure 6.13 shows the results of this kind of analysis for five sets of artificial data with different kinds of autocorrelation. (We could also look at the autocorrelation of net displacement as a function of lag.) Turchin (1998) discussed the possibility of problems associated with 'oversampled' data sets; i.e. if an animal's location is recorded every second, there is so much autocorrelation in the data, that we may seem to have too much data for the information we can get out of it. The exploratory analysis recommended here, however, will allow a simple evaluation of the 'resolution' of the data.

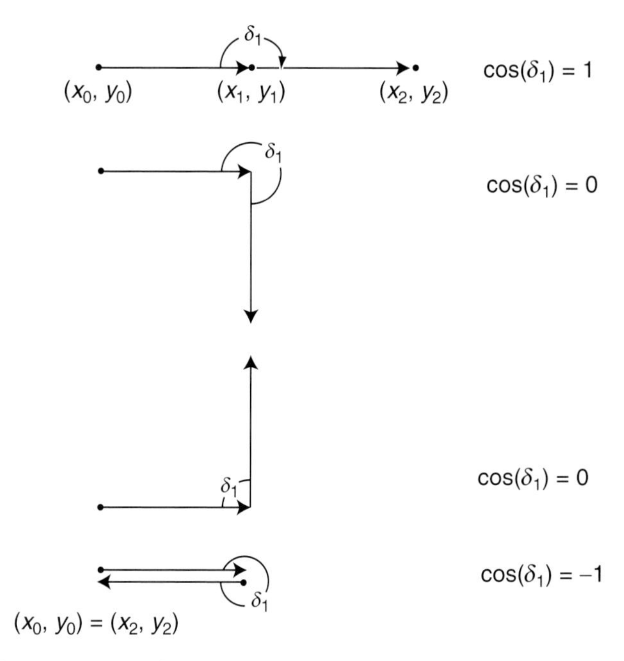


Figure 6.12 The cosine of the turning angle as a measure of angular autocorrelation.

Similarly, Turchin (1998) also discussed the aggregation of individual steps (the finest scale of resolution of the movement) into 'moves', with more recognizable breaks or turns between them (Figure 6.14*a*). This may not be necessary or desirable; the evaluation of radial and angular correlation as a function of lag should provide the information needed. In addition, the angular correlation graph will allow an assessment of whether there tends to be an alternation of right and left turns, without the need for a separate runs test (cf. Sokal & Rohlf 1995). If that question is of particular interest, that approach can also be used to provide an actual statistical test rather than just an exploratory indication of tendency.

The most difficult problem to deal with in real field data may lie not in the resolution of the data, or in over- or under-sampling; it may lie in the lack of stationarity of the process. As an example, a colleague has provided us with about 3,000 locations of an individual elk taken every 2 hours with a GPS radio-collar (Merrill unpublished). Even one-tenth of the data represents about a month of time, and it is easy to imagine how an elk's behaviour might change from the first half of a month, such as July, to the second half, as the montane vegetation changes rapidly. This possible departure from the underlying assumptions must be taken into account both in analysis and in interpretation.

As a simple null hypothesis, with which to compare the observed characteristics, we might consider the well-known 'random walk' (cf. Turchin 1998).

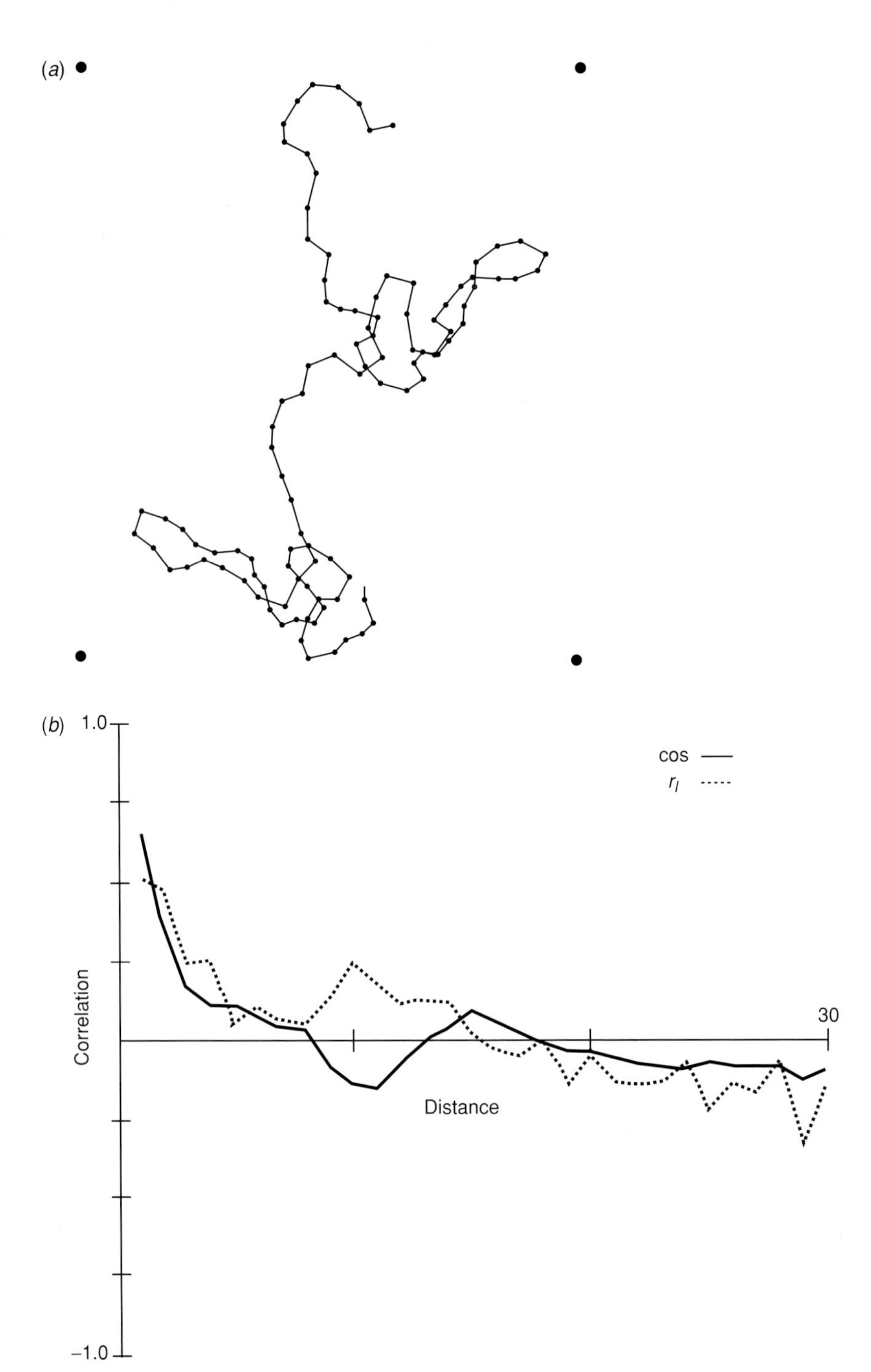


Figure 6.13 Pairs of paths and the autocorrelations of the units' lengths and angles as a function of distance or lag: (*a*), (*b*) short-range positive autocorrelation in both length and angle; (*c*), (*d*) angles and lengths both independent; (*e*), (*f*) directional bias, but lengths independent; (*g*), (*h*) directional cycles, lengths independent; and (*i*), (*j*) cyclic behaviour in lengths, angles independent. The dots in the illustrations of paths are the corners of the study area.

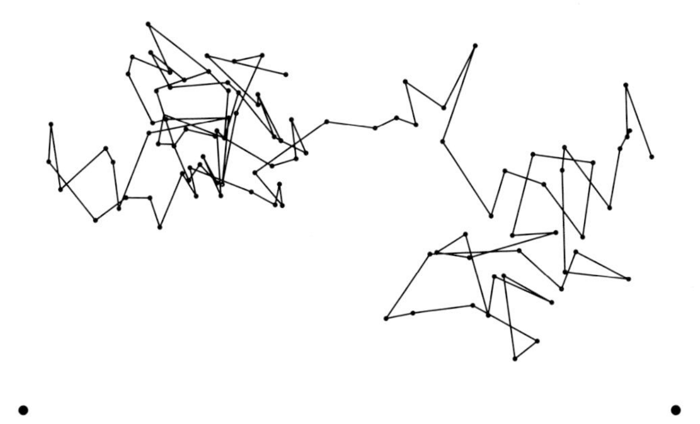


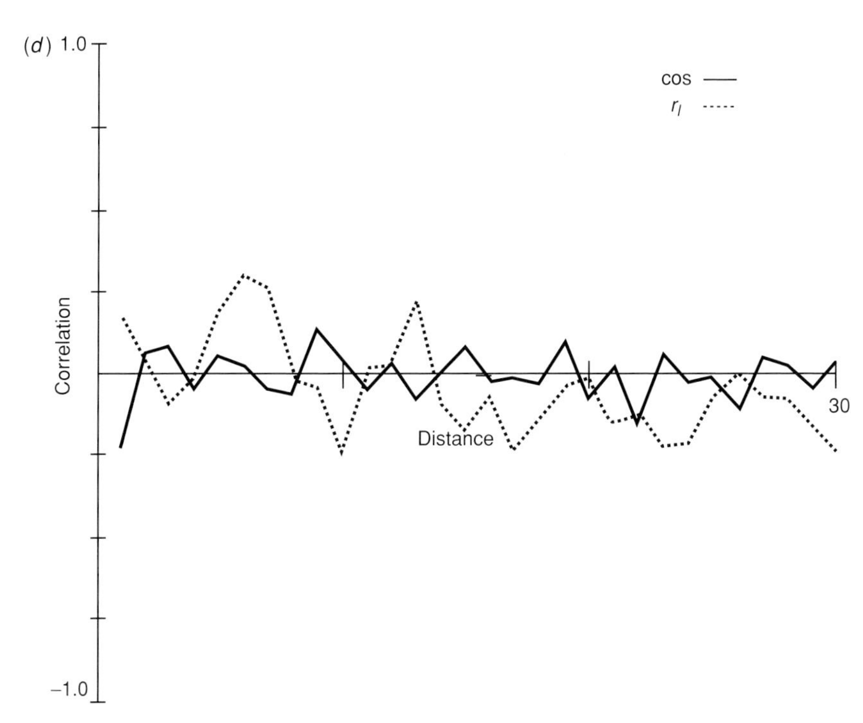


Figure 6.13 (*cont.*)

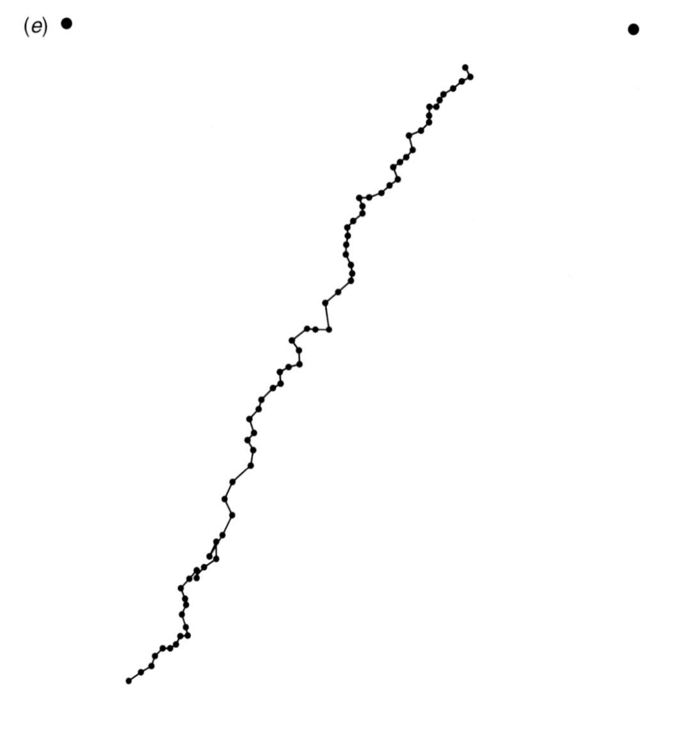


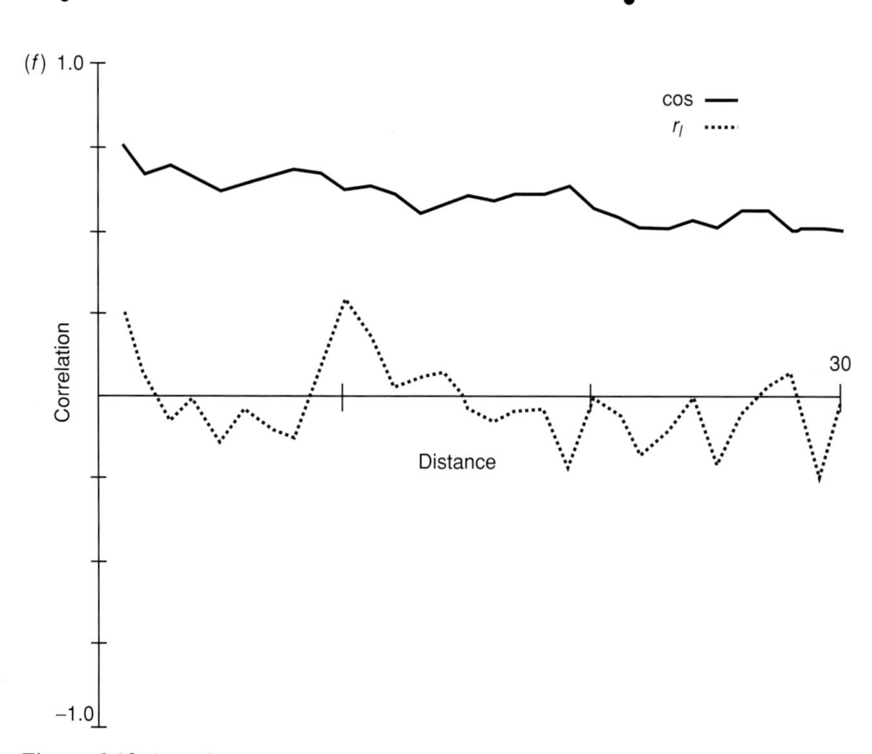


Figure 6.13 (*cont.*)

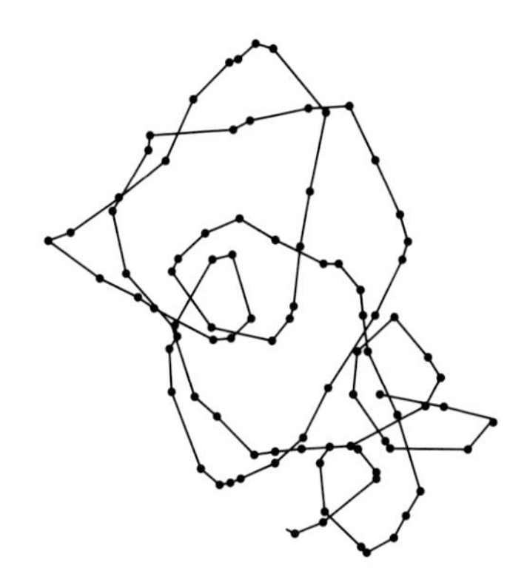

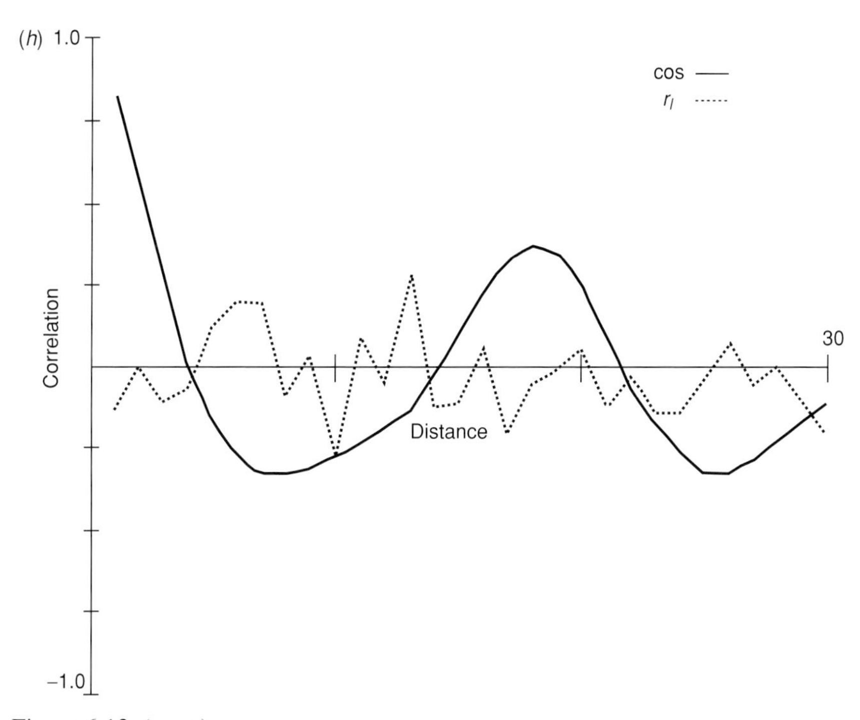


Figure 6.13 (*cont.*)

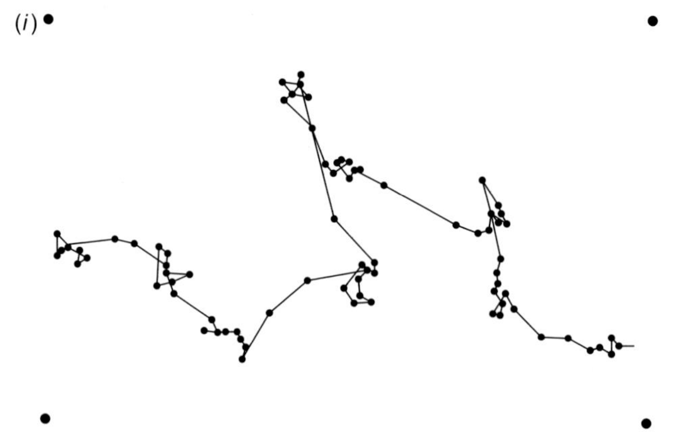


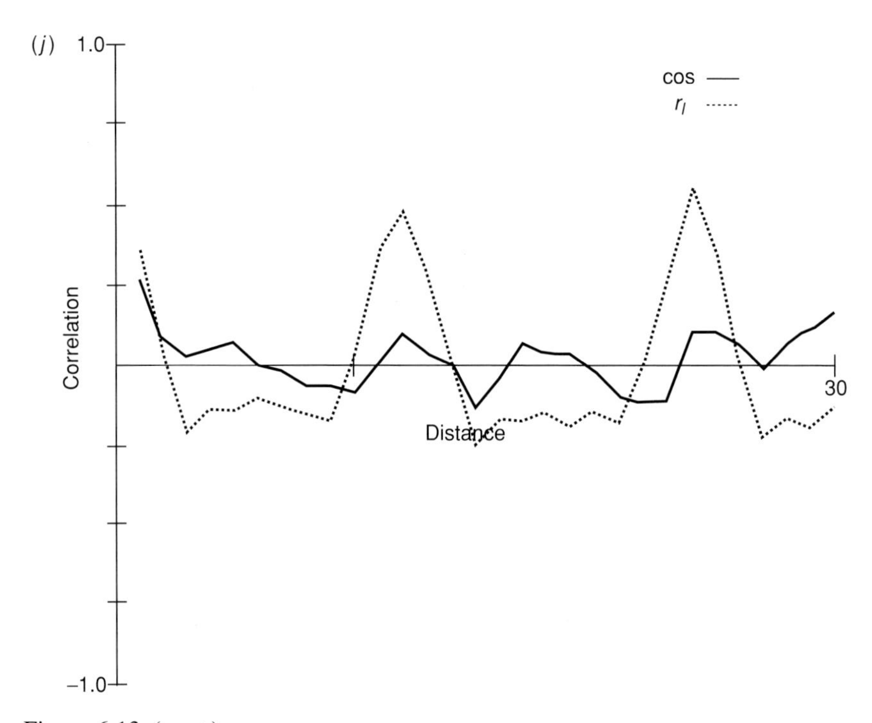


Figure 6.13 (*cont.*)

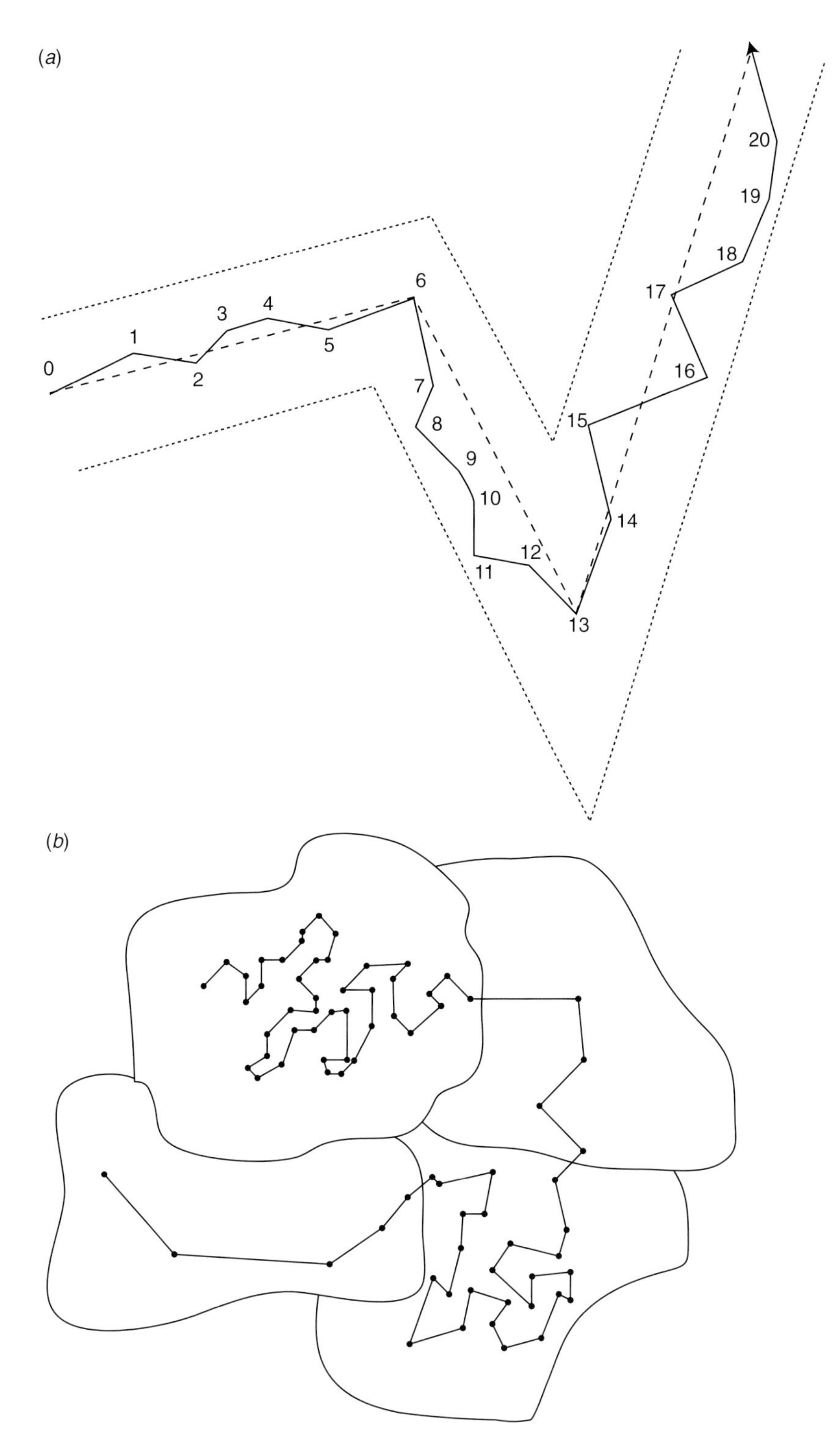


Figure 6.14 (*a*) The proposed aggregation of unit steps into 'moves'. (*b*) Movement has more turns and shorter lengths in preferred habitat. (The curved boundaries delineate different habitat types.)

The null model of the random walk is that the direction of each unit of movement is randomly chosen from the full circle of possible directions and that the length of the unit is drawn from a distribution of unit lengths, which can be estimated from the data. It is easy to calculate the expected displacement of the individual (the expected net displacement is 0), and other characteristics such as the distributions of angles, resulting from a random walk, and in general the paths of clonal plants and mobile animals do not match those well. Angles near 180° are less common and displacement tends to be greater than in this model (Turchin 1998).

As an alternative to the random walk model, comparisons with a 'correlated random walk' (CRW) model are popular (Kareiva & Shigesada 1983). In this model, the expected net displacement is still 0 as in the simple random walk, but its other properties are different because of autocorrelation of the units that make up the path. The model usually looks at the first-order autocorrelation of length and direction of the units of movement and compares the observed and expected net squared displacement and path 'tortuosity' (Wiens *et al.* 1993) or 'sinuosity' (Williams 1992; Sanuy & Bovet 1997). Kareiva & Shigesada (1983) examined this model and provided details on the derivation of the expected values. Another approach to the evaluation of the appropriateness of the CRW model is to compare the observed net displacement after $n$ steps (or the net squared displacement) with that predicted from the model. Many studies have used one or other of these approaches to evaluate the movement of caribou, clover, golden rod, toads and caterpillars (Cain 1990; Cain *et al.* 1995; Sanuy & Bovet 1997; Bergman *et al.* 2000; Doak 2000). Kareiva & Shigesada (1983) recommend that we should go beyond the simpler first-order models and look at more complicated examples using higher-order Markov models. That is, the length and direction of a movement unit can depend not just on the one immediately preceding it, but perhaps on the characteristics of the last two, three or more units of movement. For example, Figure 6.13*f* illustrates behaviour that has strict cycles, but a more realistic version of the same cyclic behaviour but with a random component could be modelled by a high-order Markov model.

Another alternative is to examine 'biased correlated random walks' in which a general directional tendency, either absolute or relative to some habitat element, is included (Turchin 1998), so that the expected net displacement is not 0. This approach is similar to adding 'drift' to the model, by including a set absolute directional component to each unit of movement (Wiens *et al.* 1993). For example, Schultz & Crone (2001) looked at the tendency of the 'Fender's blue' butterfly (*Icaria icarioides* ssp. *fenderi*) to move toward its host plant (*Lupinus sulphureus* ssp. *kincaidii*) by including a bias component of movement toward a nearby lupin patch in a correlated random walk model, whatever its absolute direction. The bias component was significant for a range of distances from a patch. Whichever kind of model is used, randomization techniques based on re-ordering the individual units of the movement or Monte Carlo procedures using parameters determined from the observed characteristics of the path can be used to evaluate the results (cf. Manly 1997), but we will describe some general concerns about randomization tests in Chapter 7.

In characterizing the spatial complexity of movement, a number of authors have advocated measuring the fractal dimension of the path, either using the 'dividers' method (see Chapter 3) or based on re-normalization procedures (Wiens *et al.* 1993). Turchin (1996) advised that this is a risky procedure unless the path is truly self-similar so that the fractal dimension remains constant over a range of spatial scales. This seems to be good advice, and this method should be used only with an understanding of that risk of misinterpretation. In addition, Schultz & Crone (2001) pointed out that descriptive measures of fractal dimension (or tortuosity) are difficult (impossible?) to translate into useful movement parameters that can then be used to predict the distributions and dynamics of populations. This is an important deficiency because for the wildlife biologists who use these methods, prediction is what it is all about.

Another important theme in the analysis of this kind of data is the search to associate the locations and the characteristics of movement paths with the types of habitat through which the animal moves and thus to evaluate differential habitat preference and habitat use (see Figure 6.14*b*). A difficult and controversial aspect of this analysis is how to carry out statistical tests relating animal movement to habitat type, and Manly *et al.* (2003) provided a useful review. Before we proceed to discuss this topic we suggest that you . . . STOP.

Stop reading this book for a few minutes and go to your stack of reprints (or 'e-prints'!) and re-read Hurlbert's (1984) discussion of pseudo-replication and the design of ecological field experiments. We need to make sure that our understanding of 'units' and 'replication' is clear and thorough. An important concept that is brought out in Hurlbert's discussion begins with a description of an experiment to compare the rates of decomposition at 1 and 10 m depth in water. Eight netting bags filled with leaf material are placed at 1 m depth and eight at 10 m depth. The important concept here is that, in this case, the leaf bags are not the experimental units; the locations at which they are placed are, and the bags are really just measuring devices. If the eight bags for each depth are placed at only one location for each depth, we cannot really test for differences between depths. We can only test for differences between two locations, one of which happens to be at 1 m and the other at 10 m; i.e. we have simple pseudo-replication. In order to examine the effects of depth, the leaf bags must be dispersed to allow us to make inferences more general than the results of just two locations. Sacrificial pseudo-replication occurs when there are true replicates, but the data for replicates are pooled before analysis or where two or more samples taken from each experimental unit are treated as independent samples. Hurlbert's discussion reminds us that we need to consider the identity of the experimental unit and the identity of a replicate in proceeding to design statistical tests for locational data. It also reminds us that we need to approach the practice of pooling data with great caution . . .

NOW, . . . to return to the topic of testing the relationship between animal location and habitat type . . .

The simplest approach is to look only at the locations of the animals, without considering the properties of the path of movement. Neu *et al.* (1974) advocated a simple goodness-of-fit test to compare the frequency of animals' presences in particular habitat types with the availability of those habitat types in the landscape. For example, with only two habitat types, A and B, known to be present in a ratio of 2 : 1 in the landscape, the observed data of 175 animal locations in type A and 125 in type B would be compared with the expected values of $\frac{2}{3} \times 300 = 200$ for A and $\frac{1}{3} \times 300 = 100$ for B by calculating:

$$X^2 = \frac{(175 - 200)^2}{200} + \frac{(125 - 100)^2}{100} = 9.375. \qquad (6.10)$$

The test statistic $X^2$ is compared with the $\chi^2$ distribution with one degree of freedom and in this example the result is highly significant. The conclusion is that the animal is preferentially using habitat B. It is worth reminding ourselves that this goodness-of-fit test is sensitive to the overall sample size. Given the same proportions but only 60 observations, 35 and 25 in the two habitat types, the result would not be significant. With 10,000 observations, the result of this test is almost certain to be significant if there is any tendency at all to depart from non-preferential habitat use.

If more than two habitat types are being considered, following the overall test, a Freeman–Tukey standardized residual could be calculated, comparing the observed and expected value for each habitat type, to determine which contribute most to the overall significance. Values of magnitude greater than 2.0 indicate important contributions to the overall significance, but we cannot ascribe a particular significance value to individual cells (cf. Bishop *et al.* 1975). Table 6.1, following, provides an example.

The table shows an apparently significant departure of the observed from the expected ($X^2 = 71.3$ on 4 d.f.; $G = 58.5$; compare with $\Sigma z^2 = 54.9$) with the avoidance of habitat type C and the preferential use of D and E contributing most to the overall significance.

The table below is based on the situation in which the proportions of the five habitat types can be treated **known**, perhaps from airphoto interpretation or GIS analysis. If, on the other hand, the values in the third column represent the frequencies of the habitat types in 400 random samples, the analysis is different. The null hypothesis now is that columns 2 and 3 in the table are both estimates of a common

Table 6.1 *A comparison of habitat use with habitat availability*

| Habitat type | Observed count | Expected count | Freeman–Tukey $z^a$ |
|---|---|---|---|
| A | 180 | 200 | −1.43 |
| B | 110 | 100 | 1.00 |
| C | 30 | 60 | −4.48 |
| D | 50 | 30 | 3.21 |
| E | 30 | 10 | 4.64 |
| Total | 400 | 400 | $\Sigma z^2 = 54.9$ |

[a] The Freeman–Tukey statistic is a standardized residual that can be compared with $N(0, 1)$ for assessment.

frequency distribution and so expected values are calculated for both. For example, the expected value for both columns and type A is 190. Now, $X^2 = 26.5$ on 4 d.f. ($G = 27.2$) and only habitat type E has a Freeman–Tukey standardized residual with an absolute value greater than 2. The fact that the evaluation of habitat availability is based on a sample has greatly lowered the value of the test statistic and has changed the interpretation to one that would concentrate on the disproportionate overuse of habitat type E.

This contingency table approach, as proposed by Neu *et al.* (1974), has some advantages, including the fact that it can be used for one or for several animals (with suitable caution in how the results for several animals are combined; White & Garrott 1990). Remembering that Hurlbert (1984) pointed out that this kind of goodness-of-fit test is the most misapplied statistical procedure, we might have some concerns about what is being implicitly defined as the experimental unit and as the replicate if we include data from several animals. Thomas & Taylor (1990) shared the concerns about the possible misuse of the goodness-of-fit test and point out a number of other problems including tests that do not control the 'experiment-wise' error and the sensitivity of tests to the subjective inclusion or exclusion of resources. Even leaving those concerns aside, this approach cannot be used as described because it has one truly fatal flaw.

The problem is, of course, that in using the standard goodness-of-fit test, autocorrelation in the data has not been accounted for. Positive autocorrelation makes statistical tests too liberal, giving more apparently significant results than the data actually justify. We have discussed this general problem at length in the previous chapter, in the context of the effect of spatial autocorrelation on statistical testing. The same sorts of considerations will apply, however, to the spatio-temporal autocorrelation inherent in animal movement data. It is a bit strange that in recommending Neu's approach, White & Garrott (1990) mentioned the problem of autocorrelation without any suggestion of how it can be addressed.

The optimistic view is that if you thin out the data so that the observations used in the analysis are further and further apart in time and space, at some point they will become independent of each other. This is the concept of 'time to independence', which has received some attention in the behavioural ecology literature (Swihart & Slade 1985, 1986; Solow 1989; Salvatori *et al.* 1999). It seems unlikely to apply well in the case of studying a single animal, the behaviour of which may be more consistent than comparisons among animals, just because it is the same animal with its own idiosyncrasies, memories and so on. Differences in the approaches used depend on whether the focus is on a single animal or on a population (Manly *et al.* 2003). Millspaugh *et al.* (1998) emphasized the importance of considering the biological characteristics of the organisms in attempting to assess autocorrelation and the independence of observations. Second, behaviour may exhibit cyclic patterns, which may introduce negative autocorrelation at some lags (temporal or spatial), and does not provide independent samples or observations either.

Without repeating too much of the material of the previous chapter, it is sufficient to say that there is no easy solution to this problem and no simple 'time to independence' trick that will make it disappear (see Rooney *et al.* 1998). The fact that autocorrelation is not significantly different from zero (cf. Swihart & Slade 1985) does not mean that it has been shown to **be** zero and can therefore be treated as such. In the preceding chapter, we gave a good example of how 'insignificant' autocorrelations can have a large cumulative effect. Minta (1992) concluded that time to independence may be 'practically unachievable' for many species, and Rooney *et al.* (1998) recommended the use of short temporal sampling intervals to produce a rich data set. Otis & White (1999) pointed out that there seems to be disagreement in the literature concerning the importance of autocorrelation in the analysis of habitat selection. (Surprisingly, Otis' 1997 paper on habitat selection with multiple patches does not discuss problems arising from autocorrelation.) They recommended using tests based on the variation among individual animals for which the number of degrees of freedom is not affected by the number of locations for each animal. For example, if the habitat use by each of 10 animals can be used to create 10 rankings of habitat preference, the rankings can be compared using a non-parametric rank test (Friedman 1937; Conover 1980). What would be the experimental unit and what is a replicate here? Of course, if some of the animals used are part of the same family or the same herd, problems with pseudo-replication can still arise because of lack of independence among the units treated as replicates (cf. Weber *et al.* 2001). Manly *et al.* (2003) provided detailed advice on designing studies depending on whether an individual or a population is the subject of interest. They also described the use of a resource selection function, which is based on the ratios of observed to expected sample counts in the different habitat categories.

The question remains, however, whether there is a way, using the characteristics of the path of a single animal, to look at preferential habitat use. We will now describe one approach. It is based on the fact that in favoured habitat, the movement will have greater tortuosity with more frequent and tighter turns; this leads to less net displacement and greater residence time in those patches (Turchin 1998). Any set of $k$ steps in a movement path can be assigned an index of tortuosity, or as we express it 'compactness'. A number of different measures could be used for this purpose, but we will describe one that we believe is new to the literature. The index of compactness can be based on the 'convex hull' of that portion of the path, which is the smallest convex polygon that contains it. Where $m$ is the diameter of the convex hull, the largest node-to-node distance in the convex hull, and $L$ is the total path length within it, then a simple measure of path tortuosity is $L/m$ (Claussen *et al.* 1997). The same authors also discussed a measure based on turning angles but those can be used only when the total path is available for analysis, not just a sample of it, such as spaced observations. Figure 6.15 shows two examples of this measure; a number of other measures are also possible for this purpose.

For each point $i$ in the path and integer $k$, we can calculate the compactness for a subpath of length $k$ centred on $i$. Then, those scores can be compared for different habitat or vegetation types in which the centre points of the subpaths occur, either by averaging or in some spatially explicit way (e.g. contour maps) for a range of subpath lengths. Figure 6.16 gives an example. Statistical testing can be carried out by superimposing the path itself, after random translation, rotation and reflection on the habitat map, say 1,000 times, and recalculating the scores. The scores from the original position of the path can then be compared with the values from the randomizations and thus evaluated for significance. As always, the assumptions that seem reasonable and the questions being asked will affect the randomization procedure chosen. The usefulness of compactness as a measure will depend on the pattern of behaviour.

As with the discussion of using position data to evaluate habitat use, there is much discussion in the literature of the proper evaluation of an animal's home range or of an animal's territory. Again, the presence of autocorrelation in the data is an important feature that must be considered. We are not going to comment on the technical aspects of home range evaluation, but we will point out that spatial and temporal autocorrelation seem to be implicit in the concept of an animal's home range. We agree with the overall recommendations of de Solla *et al.* (1999) and Rooney *et al.* (1998) that more data will provide clearer answers, but that autocorrelation needs to be evaluated in the determination of those answers.

There are other topics related to the analysis of the movement of individuals that we have not addressed here. For example, we have not discussed the

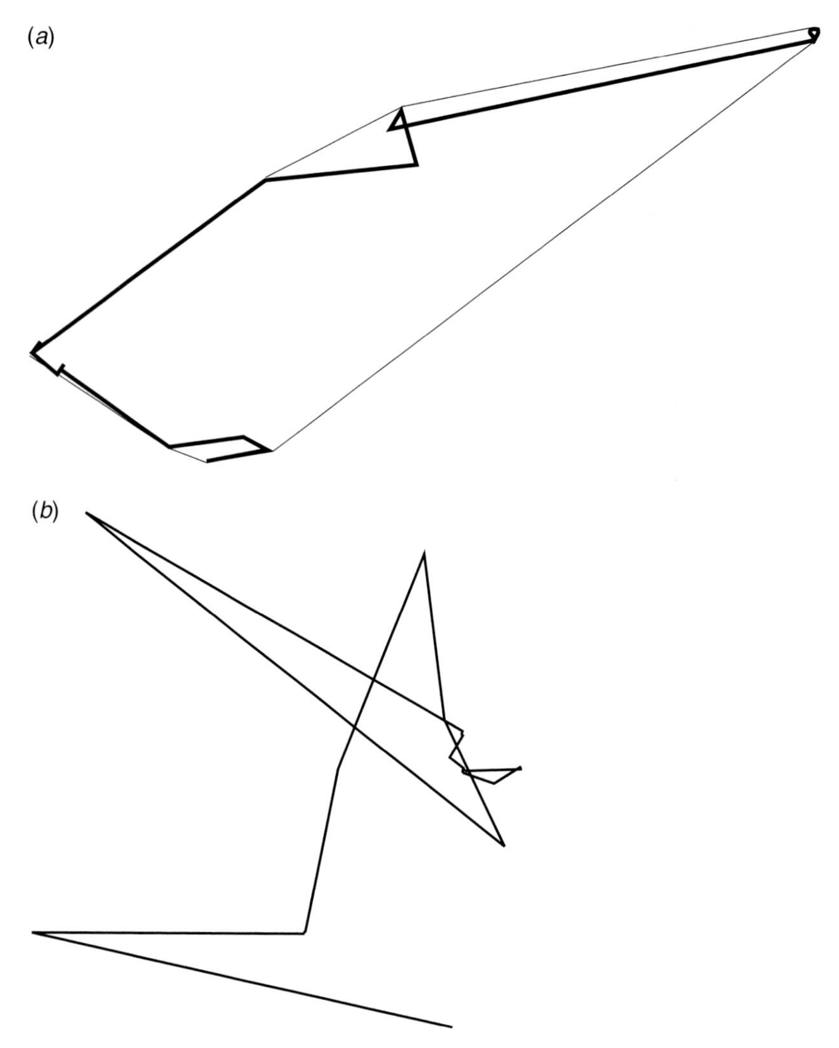


Figure 6.15 Measures of path compactness or tortuosity. (*a*) A sample of elk movement of low tortuosity: $L/m = 1.56$. The bold lines are the path and the fine lines complete the convex hull. (*b*) A sample of elk movement with higher tortuosity: $L/m = 4.12$.

mark-recapture methods associated with trapping session on a grid of traps. Usually the aim is an estimate of density and other characteristics related to the temporal dynamics of the population being studied, but it would be easy and straightforward to extract some spatial information from such data. As a general comment, however, it is clear that the topic of the spatio-temporal analysis of the movement of individuals is both important and worth further research on methods and their application.

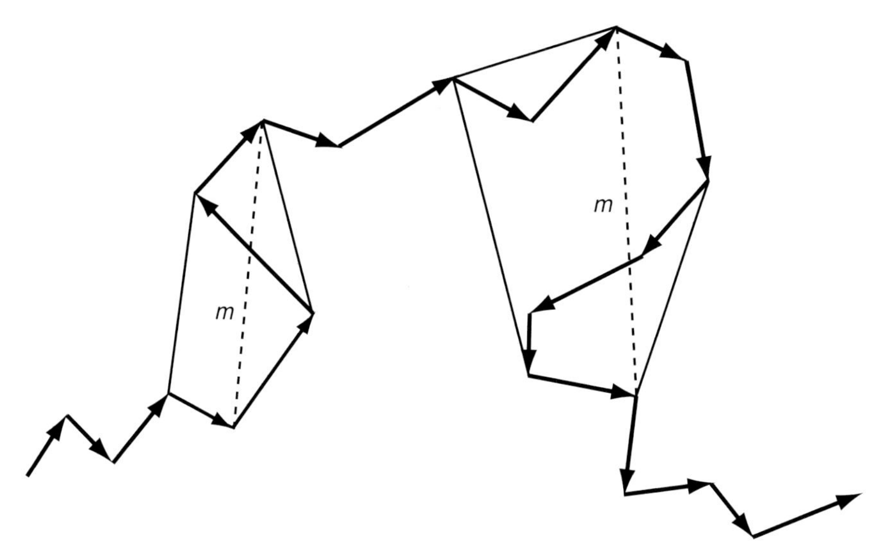


Figure 6.16 Measures of path compactness or tortuosity can be localized for given lengths of subpath, here 4 or 8 units; the bold lines are the path and the fine lines complete the convex hulls in the two locations. The dashed lines are the diameters, $m$. The measure of compactness is 1.64 for the first subpath and 2.31 for the second.

## 6.6 Process and pattern

### *6.6.1 Tree regeneration, growth and mortality*

In both population ecology and its application in forestry, there has been an abiding interest in the fates of trees after the regeneration of a forest following a major disturbance such as a fire or harvesting. From the point of view of population ecology, the emphasis is on the density of stems and the 'self-thinning' process, as density declines with ongoing tree mortality (Kenkel *et al.* 1997). The question for naturally regenerating stands is to what extent the size and proximity of neighbours determine the probability of mortality. From the point of view of forest managers, the emphasis is on the appropriate densities for planting and the size and age at which thinning should be carried out to maximize economic yield. In mixed forest (e.g. deciduous and conifer) there may also be questions about the timing and intensities of interventions to adjust the ratios of different species. For either kind of question, there is value in the long-term study of permanent plots in which the establishment, growth and mortality of identified individual stems are recorded at regular intervals over many years. Data on tree growth increments are typically analysed by multiple regression procedures, with the spatial factors being included as the distances to or densities of neighbouring trees, together with their heights and canopy volumes. Mortality can be analysed in a similar way, but using logistic regression to evaluate the probability of mortality as a function of those kinds of variables (Woollons 1998).

In the absence of established permanent plots, it may sometimes be possible to obtain some of the same kinds of information by reconstructing the history of a forest stand based on tree rings, the sizes and positions of living trees and the sizes and positions of recent and not-so-recent mortalities. For example, Carrer & Urbinati (2001) analysed structural and tree-ring variables of a timberline forest (Italian Alps) at fine spatial and temporal scales, and found that the positive autocorrelations they observed could be attributed to microsite differences, but that the spatial structure of radial growth was sensitive to extreme meteorological events. They concluded that succession seemed to be moving the forest toward a system governed by gap regeneration dynamics that would maintain the coexistence of the two main tree species, *Larix decidua* and *Pinus cembra*. Brodie *et al.* (1995) reconstructed the history of a clone of poplar (*Populus balsamifera*) in northern Quebec from the age, diameter and positions of all living and dead stems. Partial Mantel tests suggested that the clone developed in three phases: post-fire colonization, consolidation and then, directional expansion. The numerous small-diameter dead stems were aggregated and the mortality seemed to be density dependent, probably due to intracohort competition. These two examples (of the many available) illustrate the wealth of information that can be obtained from such retrospective spatial analyses of trees and the insight they can provide in the ecological processes that they document.

### 6.6.2 Plant mobility

The interest in the spatial dynamics of forest stands is driven at least in part by their economic importance, and the persistence of tree rings, as records of past processes, provides a useful insight to the past. An obvious disadvantage of trees as study organisms is that the processes take such a long time. Many authors have looked at the spatio-temporal structure of plant populations and communities by using herbaceous plants for which the processes are considerably more rapid. Here the rhizomes or runners, rather than annual growth rings, provide the record of past processes, such as clonal expansion.

For example, Evans & Cain (1995) studied the ‘foraging behaviour’ of a clonal plant, *Hydrocotyle bonariensis* (Apiaceae), in response to patches of grass. Instead of calculating a ‘bias’ component from growth angles, as described for Schultz & Crone (2001) above, they classified rhizome growth in the vicinity of grass patches into three categories: veers toward patch, veers away from patch and no change in direction. The numbers in the three categories were then subjected to contingency table analysis which showed that where the grass was patchy, the rhizomes tended to veer away from it. Where the grass was uniform or absent, no significant veering behaviour was observed. We might be tempted to use this approach to study animal movements, rather than using the biased random walk approach described above, but we would have to be very concerned in the statistical analysis about the large amount of positive spatial autocorrelation in the data (from using one or only a few animals), as we have already discussed.

Cain & Damman (1997) studied the patterns of reproduction and clonal growth in the woodland herb, *Asarum canadense*, comparing the characteristics in an early successional forest with those in a late successional forest. They used the Pearson correlation coefficient to examine the autocorrelation of rhizome length and branching angles over time. Rhizome lengths were autocorrelated (first order, we assume, but it is not stated) in both ages of forest, but apical angles were not. Lateral branching angles, however, were negatively autocorrelated (first order again), with a tendency to alternate branching to the left and to the right of the parent rhizome. This study provides a relatively rare ecological example of the spatial analysis of branching linear structures.

This topic again leads us back in the direction (!) of a discussion of spatial autocorrelation and its effect on our evaluations. For the purposes of the current topic, we can end by pointing out only that it is an important first step to assess and quantify the autocorrelation in the data. Even that first step may provide some insight into the characteristics of the processes underlying the observed spatio-temporal patterns.

### 6.6.3 Lichen boundaries

Just as it may be possible to reconstruct past history from tree rings, persistent rhizomes and other similar data for a forest stand, in other systems there may be other obvious traces of the past that can be used to reconstruct process from pattern. One such example is the boundaries of crustose lichens growing on rocks. Lichens are composite organisms, consisting of a specialized kind of fungus with one or more photosynthetic algae, and have an amazing range of growth forms, colours and ecological situations (documented in a recently published and wonderful book by Brodo *et al.* 2001). Crustose lichens are those that have no lower cortex and are therefore so intimately attached to their substrate that they cannot be removed from it; many of them have their thallus fringed by a region of unlichenized (fungus only) tissue called the prothallus (Brodo *et al.* 2001). The thallus itself is often divided into coloured photosynthetic patches called aureoles, as in *Rhizocarpon* species (Figure 6.17). Although it is possible for one crustose lichen to invade another's thallus (Brodo *et al.* 2001, Plate 4), it also seems common for a 'truce' boundary to be established between two colonies, particularly if they are of the same species, often resulting in a complete mosaic of colourful patches separated by the black lines of the prothallus (which is why the *Rhizocarpons* are commonly known as 'map lichens'). If the observed interthalline boundaries represent a truce

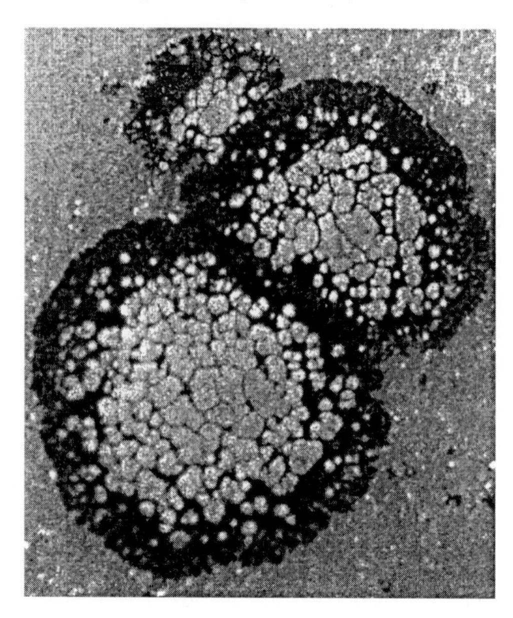

Figure 6.17 Three small thalli of yellow *Rhizocarpon*. The black prothallus and coloured aureoles are obvious. There appears to be a 'truce' boundary between the two larger thalli.

line, established and maintained from the time of contact, then we can deduce the relative rates of thallus growth from the shape of the boundary as we will now describe, following Dale (1985).

We can begin by considering a number of different models of how the radial growth rate of a crustose lichen thallus (pictured as a perfect circle) might depend on the radius of the thallus.

(1) Model A: growth independent of radius, which gives linear growth at a constant rate $a$:

$$\frac{\mathrm{d}r}{\mathrm{d}t} = a. \tag{6.11}$$

(2) Model B: growth rate proportional to radius, giving exponential growth:

$$\frac{\mathrm{d}r}{\mathrm{d}t} = ar. \tag{6.12}$$

(3) Model C: growth rate proportional to radius, up to radius $s$ and constant thereafter, because only those parts of the thallus closer than $s$ to the margin can contribute to radial growth:

$$\frac{\mathrm{d}r}{\mathrm{d}t} = ar|r < s;$$
$$\frac{\mathrm{d}r}{\mathrm{d}t} = as|r \geq s. \tag{6.13}$$

(4) Model D: growth rate at first proportional to radius but declining after reaching radius $s$ (Benedict 1967):

$$\frac{dr}{dt} = ar|r < s;$$
$$\frac{dr}{dt} = q < as|r \geq s. \tag{6.14}$$

(5) Model E: growth rate changes very slowly from exponential to linear (Hill 1981):

$$\frac{dr}{dt} = \frac{ars}{r + 2s}. \tag{6.15}$$

If the two thalli are the same size when they meet, under any of the five models just described the resulting boundary will be a straight line. For thalli of unequal sizes, Model A produces a boundary that is a hyperbola (concave toward the smaller thallus, of course) and Model B produces a boundary that is a circle (see Figure 6.18*a*, *b*). Model C gives a boundary of two parts with an obvious transition (Figure 6.18*c*), and Model D produces one that folds back on itself, as in Figure 6.18*d*. Hill's model results in an oval-shaped boundary (Figure 6.18*e*) in which it is possible that the smaller thallus becomes completely surrounded by its larger and faster-growing neighbour.

Based on the preliminary investigation of saxicolous lichens in the Canadian Rockies, described in Dale (1985), it was reported that straight boundaries between thalli of approximately equal size were common, boundaries were usually smooth with no obvious 'break' points and curving boundaries were almost always concave toward the smaller thallus. Further research confirms these features, suggesting that the growth rates tend to follow a smooth transition from exponential to linear growth, as in Hill's equation, Model E above (Armstrong 1992; Armstrong & Smith 1996). There is no evidence in the boundaries between thalli we have examined that radial growth actually slows down at larger sizes as suggested by Armstrong's data for *Rhizocarpon geographicum* (Armstrong 1983).

We admit that the shape of the growth curves of lichens as a function of radius may not be of broad ecological interest, but there are closely related topics in the general area of spatio-temporal analysis that are. In discussing the five models above, we made the comment that, no matter what the shape of the growth curve, provided it was the same for all thalli, the boundary would be a straight line if the thalli were the same size when they met. We can describe the same fact with another image: a number of propagules colonize a plane surface at the same time, producing circular colonies that grow outward at the same rate, with growth ceasing at any points of contact; this creates boundaries that are straight lines (Figure 6.19). If the process continues until the plane is filled, the result is the familiar tessellation

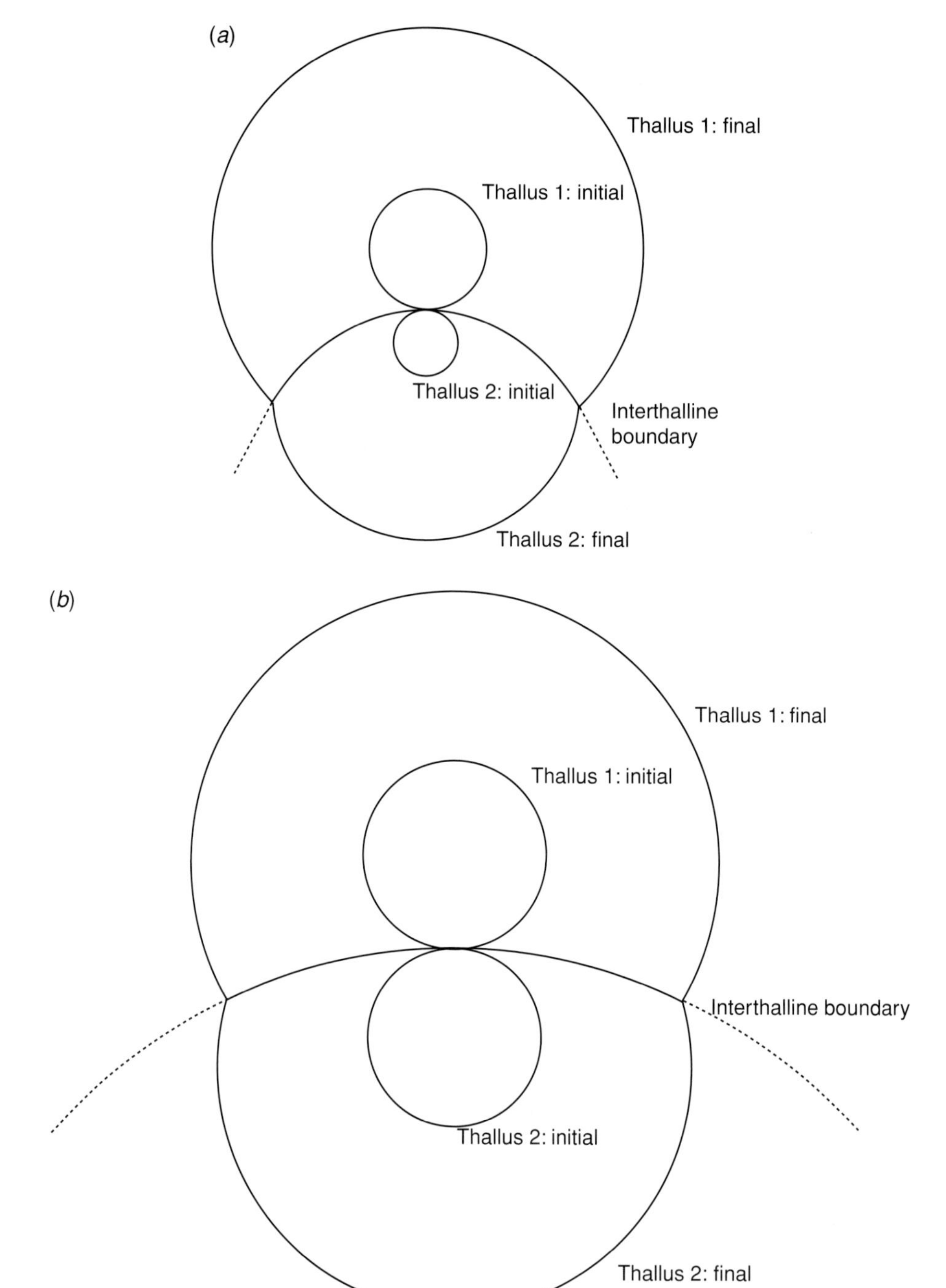


Figure 6.18 The predicted interthalline boundary shapes for different models of growth: (*a*) Model A, constant; (*b*) Model B, exponential; (*c*) Model C, exponential and then constant (redrawn from Dale, 1995); (*d*) Model D, exponential and then declining; and (*e*) Model E, exponential declining gradually to constant.

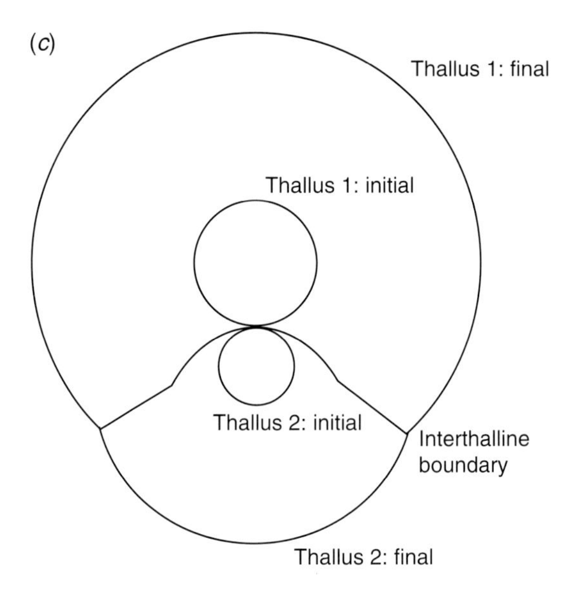


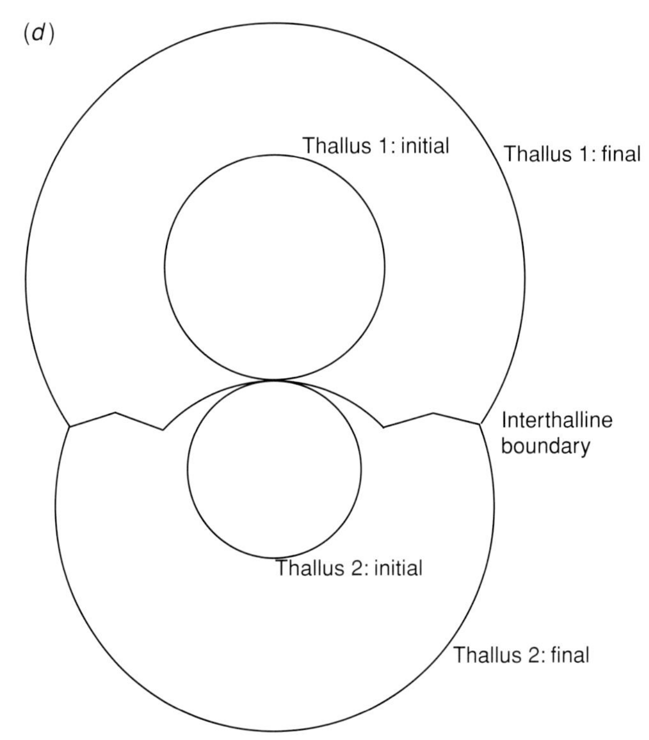


Figure 6.18 (*cont.*)

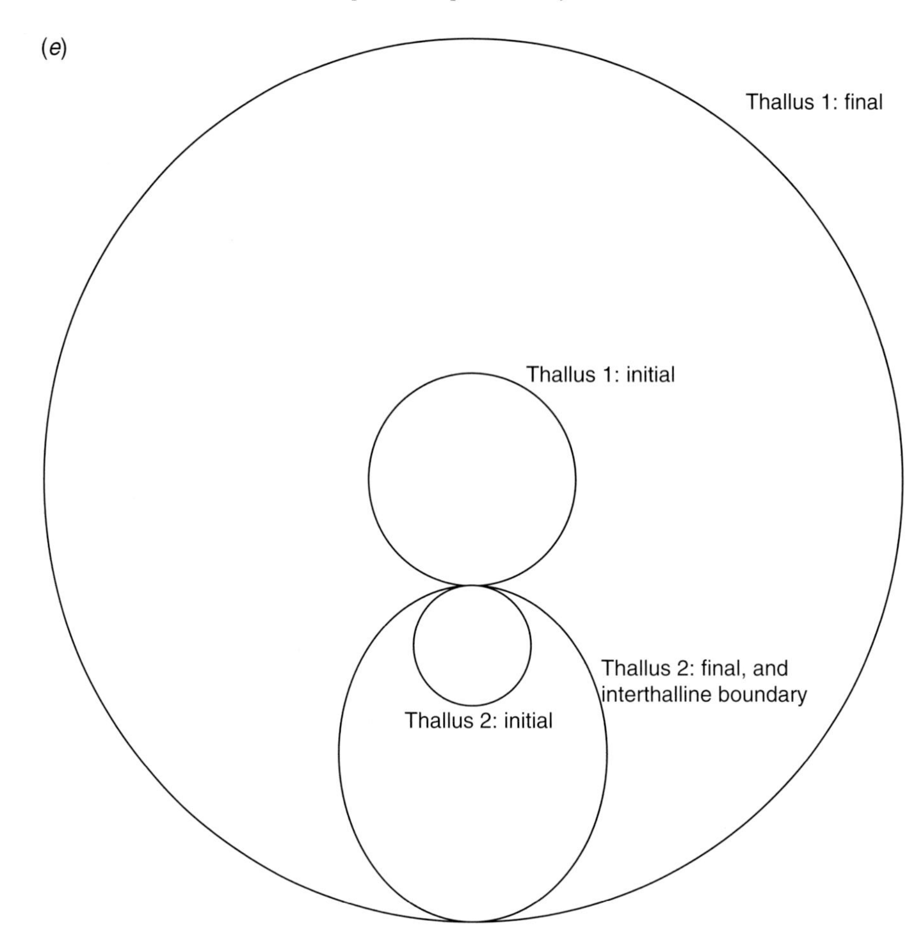


Figure 6.18 *(cont.)*

known variously as Dirichlet, Voronoi or Thiessen polygons (cf. Okabe *et al.* 1992 and Chapter 2 of this book). This is often alluded to in the study of plant population ecology, particularly for processes like competition and self-thinning, because the polygon associated with a plant (or event or propagule) is the area and the resources that it can pre-empt because they are closer to it than to any other (Mead 1966). Those seedlings, for example, with the smaller polygons have fewer resources easily available and are more likely to suffer mortality (Watkinson *et al.* 1983; Mithen *et al.* 1984; Owens & Norton 1989; cf. Kenkel 1991).

Given the same picture, of propagules arriving on a plane surface and growing into colonies that stop where they meet, if they do not arrive simultaneously, the resulting boundaries will not be straight and the shape will now depend on the growth rate model (Figure 6.20). Boots (1980) and Frost & Thompson (1988) have examined a number of different curved boundary tessellations that might arise in

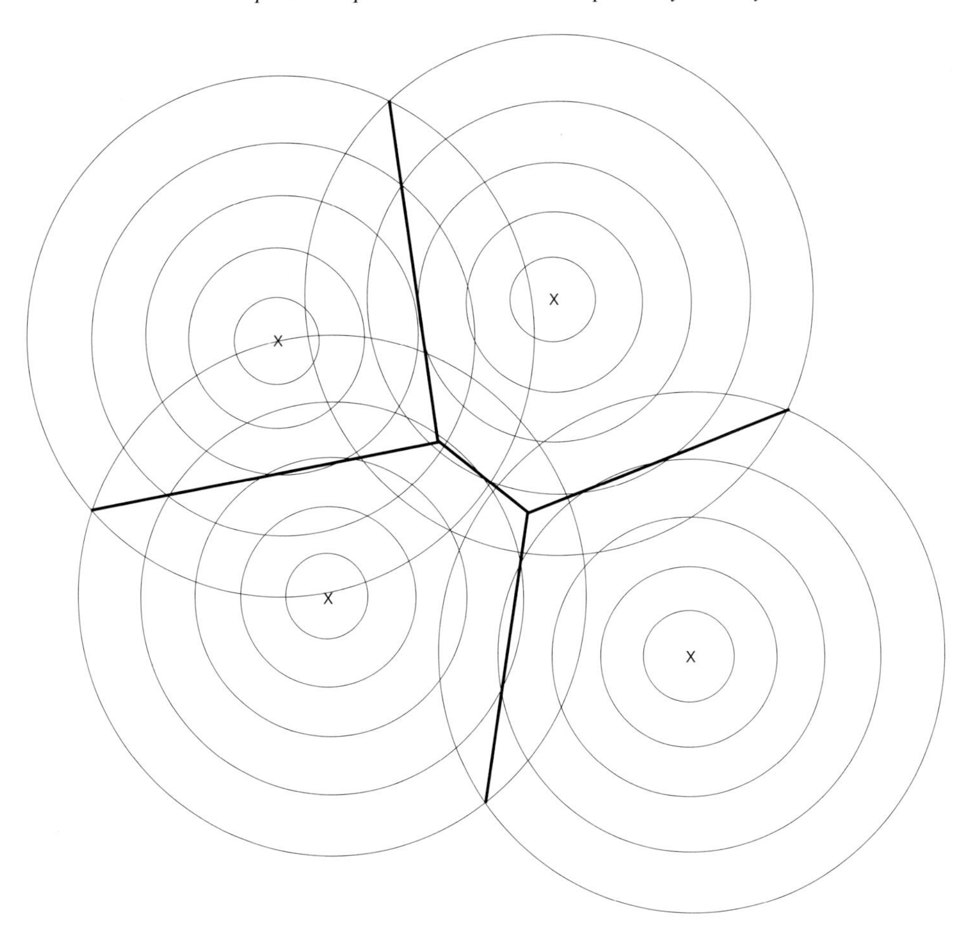

Figure 6.19 Synchronous colonization produces straight interthalline boundaries.

this way, but there are few applications of this interesting concept in an ecological context (cf. Kenkel 1991). This is another area in the general field of spatial ecology that deserves and would reward further investigation.

## 6.7 Spatio-temporal orderliness and spatial synchrony

The basic concept of spatio-temporal autocorrelation analysis is that samples that are closer together in space or time probably are more similar than samples that are taken further apart. In a simple world that might be true, but in both space and time, it is also possible for more widely spaced samples to be similar. In space, a simple alternation of patches of high density with gaps of low density will produce cyclic behaviour for almost any measure of spatial autocorrelation as a function of distance. Temporal cycles will produce the same behaviour as a function of time and

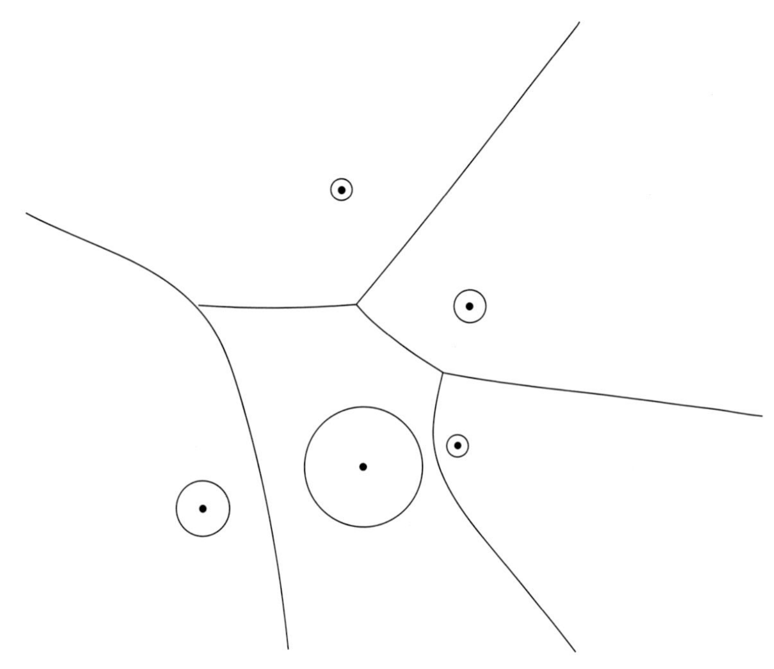

Figure 6.20 Asynchronous colonization can produce curved interthalline boundaries, with the shape depending on how the growth rate changes with size. The circles are boundaries at time 1 and the other lines are boundaries at time 2.

we are familiar with many examples in ecology, such as masting by trees and the famous population cycles of snowshoe hares and lynx, or small mammals such as lemmings and voles. Often these cyclic phenomena are more-or-less synchronized over large areas, and a number of studies has examined the relative effects of the characteristics of population dynamics and dispersal and of external forces, such as climatic events, in causing large-scale synchrony. Synchronous cycles are not the only form of spatio-temporal organization observed in natural systems or in the models used to investigate their properties. Expanding circles, travelling waves and spirals of high density are other possibilities, somewhat like the wave-regeneration in some forests or the development of banded vegetation in 'brousse tigrée' described above. Much of the literature on spatio-temporal organization has concentrated on the obvious cycling of certain populations because they are widespread (in more ways than one!) and provide a challenge to our understanding of the factors that determine the abundances and distributions of organisms (cf. Krebs 2002).

Given a cyclic system that is found over a wide geographic area, it is of interest to ask about the relationship between the cycles at different locations, as a function of distance. Do the cycles exhibit the same periodicity or does the period change with location? If the cycle lengths are more or less the same, are the cycles in synchrony over the geographic range or does synchrony decline with distance?

There are several different, but closely related, ways of analysing data from a set of spatial locations in order to answer these kinds of questions. In general, the data for any particular location, $i$, will consist of a time series of population densities, $\boldsymbol{N}_i = N_{i1}, \ldots, N_{it}, \ldots, N_{iT}$. Density data are often log-transformed before analysis:

$$X_t = \log(N_t + 1). \tag{6.16}$$

The method described by Hanski & Woiwod (1993) uses these log-transformed data, but removes the first-order temporal autocorrelation effects by fitting the equation:

$$X_{t+1} = a + bX_t, \tag{6.17}$$

and then using the residuals for further analysis:

$$R_t = X_t - (\hat{a} + \hat{b}X_{t-1}). \tag{6.18}$$

One general approach is the calculation of the cross-correlation coefficient between the two series with a time lag of zero. Where there are two time series, $x_1$ and $x_2$, both of length $T$, their cross correlation is:

$$r_{12}(0) = \frac{\sum_{t=1}^{T}(x_{1t} - \bar{x}_1)(x_{2t} - \bar{x}_2)}{\sqrt{\sum_{t=1}^{T}(x_{1t} - \bar{x}_1)^2(x_{2t} - \bar{x}_2)^2}}. \tag{6.19}$$

Where the data consist of two time series for each of a number of locations, it is also possible to examine the effect of spatial distance on the cross correlation coefficient of the two time series. Tobin & Bjørnstad (2003) provided an interesting example of the application of this approach to a study of the spatio-temporal relations between a prey species, the house fly *Musca domestica*, which is a serious pest in commercial hen houses, and a predatory beetle, *Carcinops pumilio*. Given the cross correlation of the two time series for each of a number of locations (108 in one large hen house and 162 in another), a kernel function can be used to give an estimate of the cross correlation at any given distance. They found that during the exponential growth phase of the fly population, the beetles were strongly negatively cross correlated with their prey at local spatial scales.

Hanski & Woiwod (1993) used a related approach to examine the spatial synchrony of populations of single species. Studying the densities of different kinds of insects (moths and aphids) in England, they used the cross correlation between the residuals of conspecific population densities for each pair of sites, $r_{ij}$, and plotted them as a function of distance between sites, $d_{ij}$. The $y$ intercept of the linear regression of correlation as a function of distance was used as a species-specific measure of synchrony to be compared with a measure of population variability through time. They found a positive relationship between synchrony and temporal variability for aphids and noctuid moths, but not for geometrid moths.

Another method is to make a direct comparison of the $n \times n$ matrix of correlation coefficients, $r_{ij}$, with the $n \times n$ matrix of inter-site distances, $d_{ij}$. The usual technique for comparing such a pair of matrices is the Mantel test (Chapter 3), which uses a randomization procedure to test for the significance of the relationship between the two. Koenig & Knops (1998) cautioned that because there are more pairwise correlation coefficients than there are sites, 'the potential for pseudo-replication biasing the results of statistical tests cannot be ignored'. They suggested that, because autocorrelation is expected to decline with distance, the Mantel correlogram may not produce results of ecological interest, when it detects only a decline in autocorrelation as a function of distance. Figure 6.21*a* gives an example of the correlation of tree-ring widths in *Picea glauca* as a function of geographic distance in northern Alberta (Peters 2003). The results for two trees per site and five sites show little evidence of a systematic decline in correlation with distance, indicating that, in this case at least, the situation is not the simple one that Koening & Knops (1998) describe. In fact, the simple decline with distance may not be as common as those authors suggest, as we described in Chapter 5.

The same authors (Koenig & Knops 1998) recommended the use of a 'modified correlogram' to display the results graphically, by plotting the mean correlation coefficient between the time series of randomly chosen pairs of sites within specified distance classes; to avoid pseudo-replication problems, each site is used only once. Ranta *et al.* (1997) used a randomization technique to compare the level of synchrony at any particular site with others, by choosing other sites at random, thus avoiding the same problem. Figure 6.21*b* shows a re-analysis of the same *Picea glauca* tree-ring data of Figure 6.21*a*, but using randomly chosen pairs of trees that are then not re-used (sampling without replacement). The conclusions drawn would be the same. Where each site has two time series of data, for example acorn production and annual growth in oak trees, the cross-correlation coefficient between the two series can be used in the same way (Koenig & Knops 1998). It is not clear to us how great a problem this re-use of data really is; as described in Chapter 2, many exploratory analysis techniques are based on the repeated use of the same data; TTLQV being an extreme example (Chapter 2). There is the usual

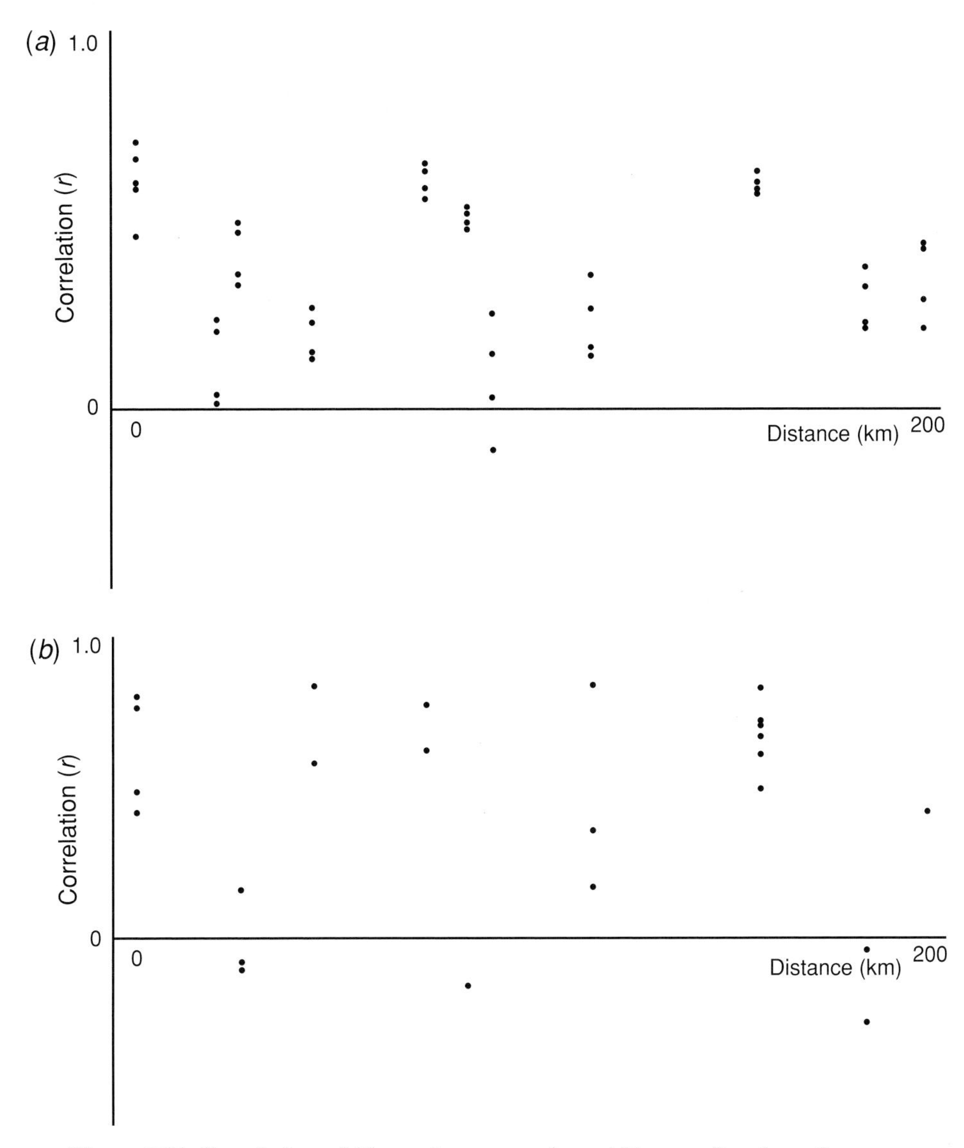


Figure 6.21 Correlation of *Picea glauca* tree-ring widths as a function of geographical distance in central Alberta. (*a*) Five sites, the same two trees per site. (*b*) The same five sites, two trees per site, but the trees not re-used (sampling without replacement).

trade-off between independence and the effective use of the information available. This comment applies to a wide range of methods not just to the correlation techniques being discussed here.

There is a number of variations on these basic methods. Sutcliffe *et al.* (1996) used the non-parametric Spearman's correlation to evaluate the synchrony of butterfly population dynamics. This approach uses the ranks of the abundances rather than the values themselves and may be less affected by extreme values (cf. Conover 1980). As with other correlation statistics, it applies to pairs of series and for more than two series in a region, the measures for all pairs could be averaged to give a regional measure of synchrony. An alternative is to use Friedman's measure of concordance for several series (Conover 1980). Bjørnstad *et al.* (1999) averaged the pairwise cross correlations to get a regional measure of synchrony but based them on the year-to-year changes in density, not on the densities themselves.

Bascompte & Solé (1998) describe how the analysis of two data sets confirmed the predictions from spatially explicit dynamic population models of spontaneous self-organization in the form of spirals or travelling waves of population density. Since then, a number of studies have demonstrated the phenomenon of travelling waves of density, e.g. in field voles (*Microtis agrestis*, MacKinnon *et al.* 2001) and in red grouse (*Lagopus lagopus* ssp. *scoticus*, Moss *et al.* 2000). At a larger scale (the entire Canadian boreal forest, of the order of 5000 km across), Viljugrein *et al.* (2001) found that although there is broad-scale synchrony in the population cycles of mink (*Mustela vison*) and muskrat (*Odatra zibethicus*), peaks and troughs in these cycles generally appear first in the Athabasca basin and spread from this epicentre. Similarly, while the basic oscillation of spruce budworm (*Choristoneura fumiferana* Clem.) is the same across all of Ontario (about 1000 km), large outbreaks appear first in the eastern zone, followed by the central zone, and then 5 or 6 years later in the western zone (Candeau *et al.* 1998). It is not our purpose to review the mechanisms believed responsible for this travelling wave phenomenon, but see Sherratt *et al.* (2000) and Sherratt (2001) for interesting discussions. Travelling waves are detected in the spatio-temporal data by looking for anisotropy in the spatial covariance (Bjørnstad *et al.* 1999a, b; Lambin *et al.* 1998). If there is a travelling wave, cross correlation declines markedly with distance in directions perpendicular to the wavefront, but does not decline with distance parallel to the wavefront. The plot of cross correlation as a function of distance, described above, is merely divided into a few direction classes and examined for differences. Lambin *et al.* (1998) and colleagues have described a modelling method to help estimate the speed and direction of such a travelling wave and to determine its statistical significance.

There are two related questions of the consistency of spatial pattern through time and of the synchrony of temporal patterns in space. This section has attempted to provide a description of the various methods that can be used to answer these two questions. Answering these questions merely leads (of course) to more questions, now concerning the ecological processes that lead to the spatio-temporal patterns we detect.

## 6.8 Chaos

In the preceding section, we discussed some of the spatial aspects of population dynamics, or the temporal aspects of patterns of population density, but we avoided the topic of spatial (or spatio-temporal) chaos, because it is sufficiently interesting and important to merit a section of its own. Used in a technical sense, 'chaos' refers to a system's behaviour in time or space that is irregular and possibly very complex but that is strictly deterministic. Chaos is not the same as randomness or stochastic behaviour, although it may appear to be unpredictable and aperiodic. One distinctive feature of chaos is that the overall behaviour can be very sensitive to very small changes in conditions (the so-called 'butterfly effect', Schroeder 1991). We will begin our discussion of chaos as it relates to spatio-temporal analysis with a review of an example that is probably very familiar to the reader.

Consider the following difference equation that describes the dynamics of the population density at a particular location and time, $N_t$, as a function of the density at the preceding time, the growth rate of the population, $r$, and the carrying capacity for the population, $K$:

$$N_t = rN_{t-1}(K - N_{t-1})/K,$$

which can be rewritten as

$$n_t = rn_{t-1}(1 - n_{t-1}). \tag{6.20}$$

This equation describes logistic population growth, which is almost exponential when the population is well below the carrying capacity, with the growth rate being slower at higher densities and decreasing to zero at the carrying capacity. The growth rate is negative when the carrying capacity is exceeded and the population declines. The behaviour of the population that is derived from the application of this equation depends on the intrinsic growth rate, $r$. When $r$ takes the value 2.5, the equilibrium value of $n_t$ is 0.6, and the population converges to this value, no matter what the starting density. If we plot a two-dimensional diagram of $n_t$ versus $n_{t-1}$, any trajectory will converge to the point (0.6, 0.6), which can be thought of as an 'attractor' under these conditions (Figure 6.22*a*). When $r$ is 3.2, while there is an equilibrium value of 0.6875, it is almost never reached, because the equilibrium is unstable, and $n_t$ alternates between two values, 0.513 and 0.7995 (Figure 6.22*b*). When $r$ is increased to 3.4, $n_t$ cycles among four values, approximately 0.875, 0.383, 0.827 and 0.501 (Figure 6.22*c*). Further increases cause doubling of the lengths of the cycles, but the behaviour soon becomes aperiodic (chaos!). What is most fascinating about this simple system is that as $r$ continues to increase, the behaviour returns to simple cycles, then back to chaos and so on (usually illustrated with the 'bifurcation to chaos' figure that appears in many

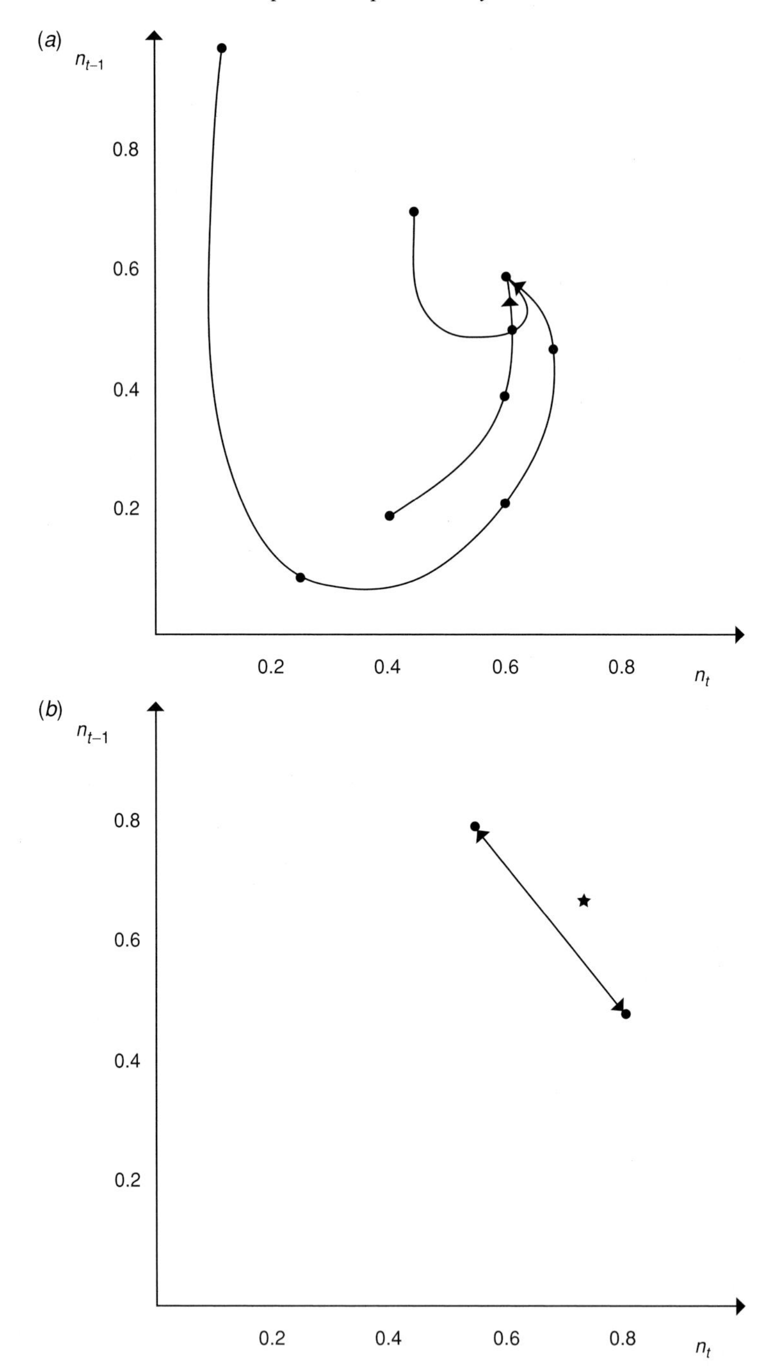


Figure 6.22 (*a*) When $r = 2.5$, the different trajectories in the phase space converge to the equilibrium point of (0.6, 0.6). (*b*) When $r = 3.2$, the trajectories do not converge to the unstable equilibrium value of 0.6875 (the star), but converge to an alternation between two other values. (*c*) When $r = 3.4$, the 'equilibrium' is a cycle among four different densities.

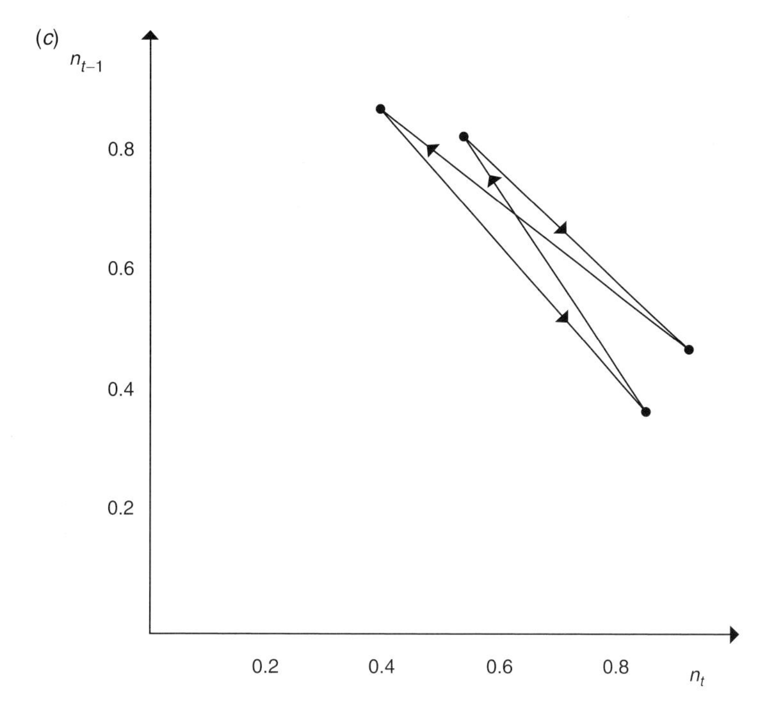


Figure 6.22 (*cont.*)

places, e.g. Schroeder 1991, Figure 12.11). While as a simple model of population behaviour, this approach to chaos seems unrealistic because the values of $r$ required to achieve chaos are unnaturally high, it provides an important lesson on the potential importance of non-linear dynamics.

How does this relate to spatio-temporal analysis? The first question is how to detect chaos in time series from individual locations and to determine whether natural systems indeed can be chaotic. In 1995, Solé & Bascompte wrote:

> Detection of chaos in ecological data is one of the most challenging problems in contemporary ecology. It is not enough to certify that time series are complex, we need a comprehensive approach to that complexity. . . . [T]he question of chaos in nature is still an open one.

In that decade, much effort was devoted to trying to meet that challenge, but without complete success, some of which we shall describe below (cf. Stone & Ezrati 1996; Perry *et al.* 2000).

If chaos does exist in ecological systems, a second question then is what are the relationships among chaotic data series at a number of locations and to find whether the relationships have a spatial component. Another question is: if there is such a thing as spatial chaos, how do we detect it and what are the characteristics

Table 6.2 *Two pairs of time series with slightly different initial values*

| | $r = 3.95$ | | | $r = 3.25$ | | |
|---|---|---|---|---|---|---|
| $t$ | $n_i$ | $m_i$ | $\Delta_i$ | $n_i$ | $m_i$ | $\Delta_i$ |
| 0 | 20.00 | 21.00 | 1.00 | 20.00 | 21.00 | 1.00 |
| 1 | 63.20 | 65.53 | 2.33 | 52.00 | 53.92 | 1.91 |
| 2 | 98.22 | 89.22 | 2.65 | 81.12 | 80.75 | 0.37 |
| 3 | 29.51 | 37.98 | 8.47 | 49.87 | 50.52 | 0.74 |
| 4 | 82.16 | 93.05 | 10.88 | 81.25 | 81.24 | 0.01 |

of its dynamics? We might also ask about the relationship between space and time in chaotic systems and how we might determine that relationship.

We will begin by considering the first question, concerning the detection of chaos in a single time series. At first glance, a chaotic time series seems indistinguishable from a noisy stochastic series, and it may seem to be a challenging problem to detect chaos: actually it is not, at least in theory. The basis for detecting chaos goes back to a characteristic of chaos described above: the fact that small differences become amplified. Let us consider an example based on Eqn (6.20) above. Consider two cases, both with starting values for two series of $n = 20$ and $m = 21$, and examine $\Delta_i = |m_i - n_i|$ for increasing or decreasing differences; in the first case $r = 3.95$ gives chaos and the differences increase, and in the second, $r = 3.25$ does not, and differences tend to decline (Table 6.2).

The diagram plotting $n_t$ against $n_{t+1}$, as in Figure 6.22, is called a phase space diagram, in which time is not represented as an axis, but is included explicitly in the drawing of the trajectories, through time, of the combinations of values. In complex situations, we may plot only the 'attractor' to which the trajectories converge (Figure 6.22*b*). The attractor may be a single point, as mentioned above, a finite loop of 2, 4, 8, ..., points, or in the case of chaos, a 'strange' attractor that is infinite but bounded and often (always?) fractal (of fractional dimension).

Where there is chaos, the trajectories that started close together diverged in the phase space; where the behaviour was not chaotic they converged. To put it in more mathematical terms, consider two trajectories in a phase space that are separated by a small amount, $\varepsilon$, at time $t$; at time $t + \tau$, they are separated by $\varepsilon_\tau = \varepsilon e^{\lambda\tau}$, where $\lambda$ is a constant that is characteristic of the system, known as the Lyapunov exponent. From the first four or five values of the two cases described in Table 6.2, above, in the chaotic example, $\lambda$ is about 0.6 (calculated from $\ln(10.88)/4$) and in the cyclic example, it is about $-1.15$ (calculated from $\ln(0.01)/4$). In fact, it is the Lyapunov exponent that will allow us to detect chaos: if it is greater than zero, there is chaos. If it is less than zero, there is convergence. An exponent of zero indicates cyclic

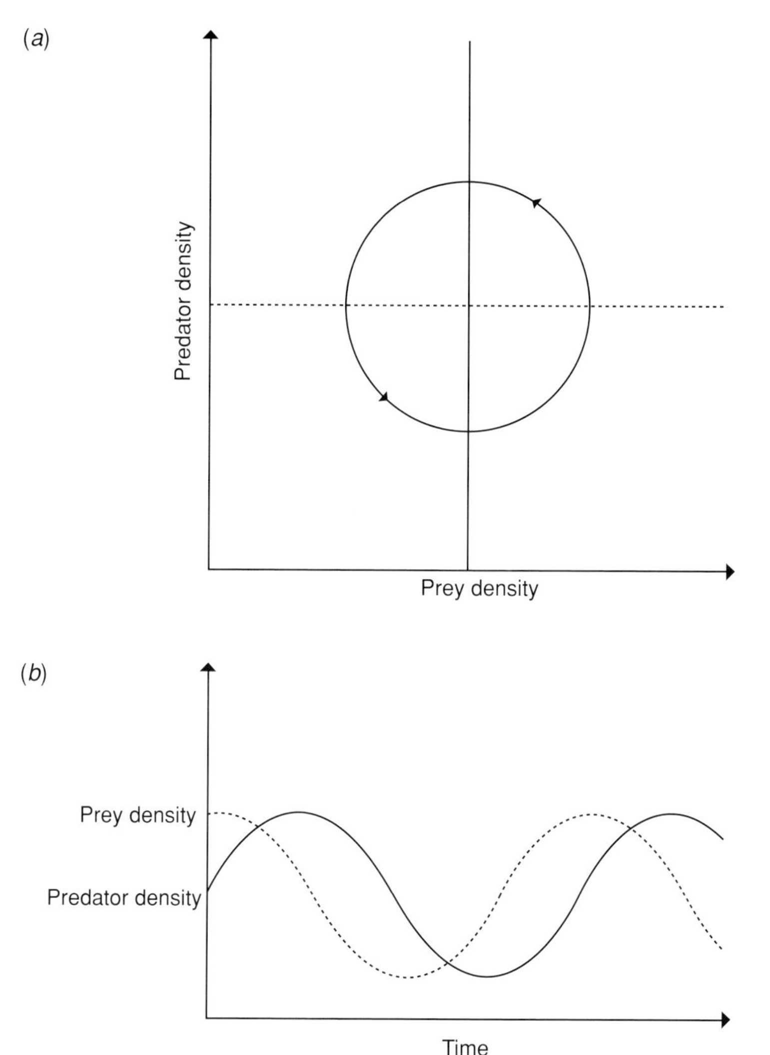


Figure 6.23 The cyclic behaviour predicted by simple predator–prey models, depicted in phase space (*a*) and as relative densities as a function of time (*b*). The predator density increases when prey are plentiful, but high predator densities drive prey density down. Predator density declines when prey are scarce.

behaviour such as the limit cycles predicted from simple predator–prey models, familiar from ecology textbooks (Figure 6.23). It seems that all that is required to detect chaos, then, is to determine the Lyapunov exponent from the data and then draw a conclusion based on its value.

Problem solved . . .? Well, no, not exactly. The difficulty is that to get a reliable estimate of $\lambda$ from real data, long data series may be needed, and those are seldom available in ecological studies. One feature of ecological data is that even if there is a chaotic deterministic foundation to the data, we would expect a stochastic component of exogenous factors to be added on somehow (Dennis *et al.* 2001), and some methods of analysis may be affected by this 'noise' in the data (cf. Ellner & Turchin 1995). Even if long data series are available, the characteristics of the population, such as the intrinsic growth rate, may change over time, so that the Lyapunov exponent we are trying to estimate is not constant. As Ellner & Turchin (1995) concluded, it may be more sensible to ask of any particular system 'When and how often is it chaotic?' rather than 'Is it chaotic?' This suggestion fits with one theme in this book, the distinction between global and local evaluation of the characteristics of the phenomena being studied.

In addition to the technique of estimating the Lyapunov exponent, a number of other methods to detect chaos have been proposed; see Stone & Ezrati (1996) and Perry *et al.* (2000). Using response surface methods, Turchin & Taylor (1992) found a number of different dynamics in natural populations of insects and vertebrates, but only one, the aphid *Phyllaphis fagi*, was thought to be chaotic. Perry *et al.* (1993) included further data from the same population of *Phyllaphis fagi*, and from other populations of the same species, and decided that its dynamics were actually stable, not chaotic. Dennis *et al.* (2001) in studies of laboratory populations of flour beetle (*Tribolium* sp.) found that high period cycles were more likely to be a good description of the population dynamics than chaos. Knowledge of the biology of the system being studied seems to be essential for interpreting the dynamics, whatever techniques are used to investigate them. From Perry *et al.* (2000), one might conclude that many systems are on the 'edge of chaos', with Lyapunov exponents around zero. While chaos may turn out to be rare in nature, it is important when studying dynamics and spatio-temporal systems to keep in mind that it is possible.

Given that chaos is possible in purely temporal systems, what is possible when a spatial structure is included? The easy answer is that, in theory, almost anything can happen. A straightforward way to include a spatial element in simple dynamic models is to have two or more metapopulations, each governed by the same underlying model, but linked by dispersal between adjacent subpopulations. Gonzalez-Andujar & Perry (1993) investigated such linked populations and concluded that linking the populations reduced the occurrence of chaos. Ruxton (1993) responded pointing out that linked populations may still be chaotic. Doebeli & Ruxton (1998) extended this work on metapopulation dynamics, showing that long-range dispersal can stabilize otherwise complex dynamics and that short-range dispersal can destabilize otherwise stable dynamics.

To illustrate these effects, we will use a model of two populations, both governed by Eqn (6.20), and either linked by dispersal or isolated from each other. We will consider several different situations, all with $r = 3.58$. With starting densities of $n = 0.50$ and $m = 0.80$, the behaviour is chaotic, with the attractor as shown in

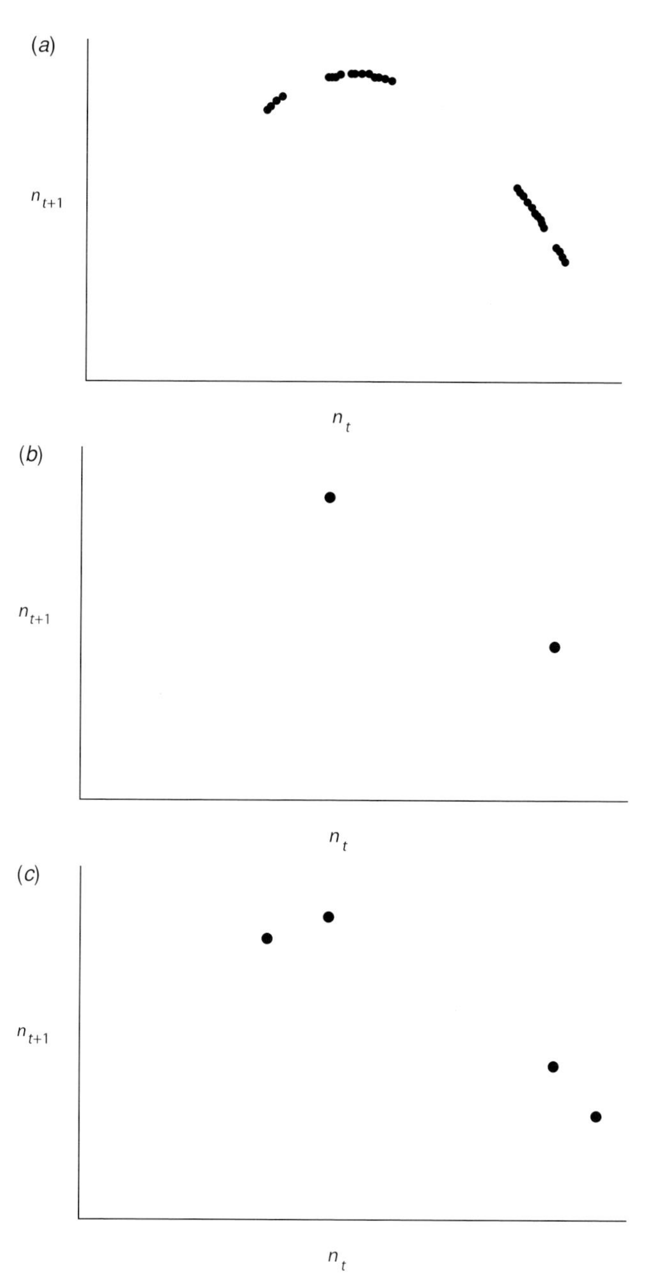


Figure 6.24 The effects of interchange between populations on the edge of chaos, with $r = 3.58$. (*a*) $n = 0.5$, $m = 0.8$: chaotic behaviour; (*b*) 5% interchange: chaos suppressed, a cycle of period 2. (*c*) $n = 0.9$, $m = 0.8$, cycle period 4 with 5% exchange; but with $n = 0.85$, $m = 0.8$, the chaotic behaviour returns as in (*a*).

Figure 6.25 Something like spatial chaos in a cellular automata system, Wolfram's (2002) 'rule 150'.

Figure 6.24*a*. When the same starting values are used, but the populations are linked by an exchange (dispersal in each direction) of 5% of each population, the behaviour becomes a cycle of period 2, as shown in Figure 6.24*b*. Because the system is 'on the edge of chaos', the outcome can be changed merely by changing the starting densities, while the degree of linkage is unchanged. Starting with $n = 0.90$ and $m = 0.80$, the result is a cycle of period 4 (Figure 6.24*c*). Starting with densities of 0.80 and 0.85, the two subpopulations essentially act as one and chaos returns, and the situation reverts to that shown in Figure 6.24*a*. This example, based on a very simple system of two subpopulations, illustrates well the potential complexities of the behaviour of chaotic systems with spatial structure. Think of the possibilities with greater spatial complexity! This kind of interaction between subpopulations

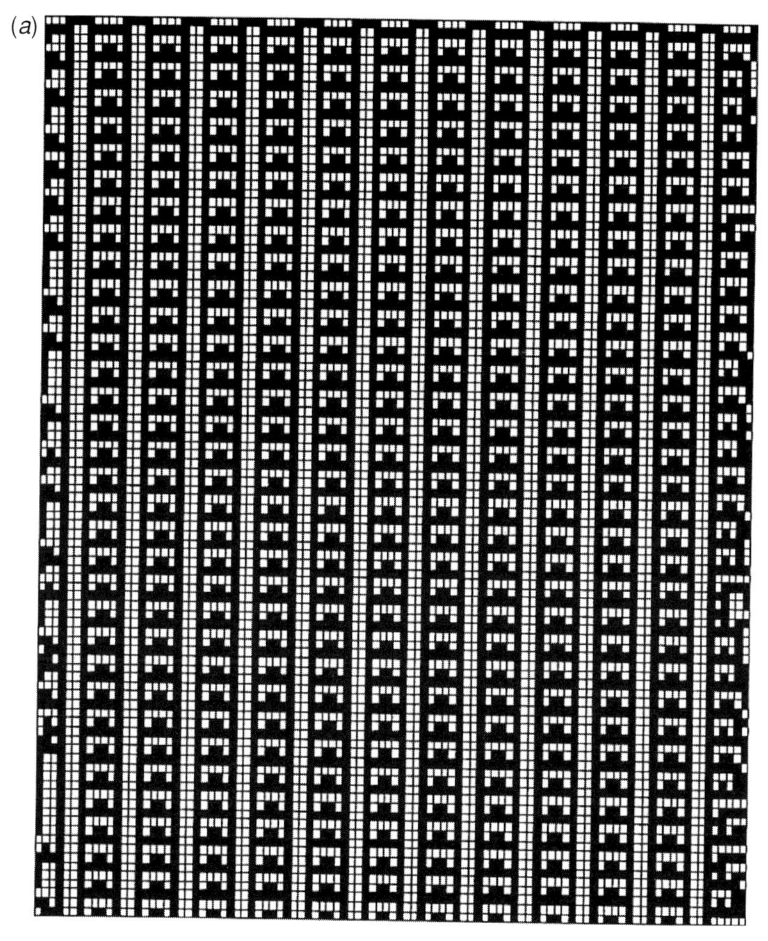


Figure 6.26 Sensitivity to starting conditions in cellular automata: three outcomes of Wolfram's (2002) 'rule 73' with different initial patterns (top row).

should be considered in interpreting spatial dynamics or the spatio-temporal patterns of density.

Spatial chaos? Why not? At least in theory... Any model that can generate chaos in time can generate chaos in space, but it is not clear how applicable such models would be in describing real systems. On the other hand, Petrovskii & Malchow (2001) described a spatially explicit predator–prey model in which spatio-temporal chaos appears in a subdomain of the system and then spreads to take over the whole space. A kind of purely spatial chaos can be found in the development of some cellular automata models. For example, Wolfram (2002) displayed a number of examples which seem to have complex aperiodic behaviour, which he refers to as 'random', but which, because they result from deterministic rules, should more properly be referred to as chaotic (e.g. p. 227, 'rule number 150',

(*b*)

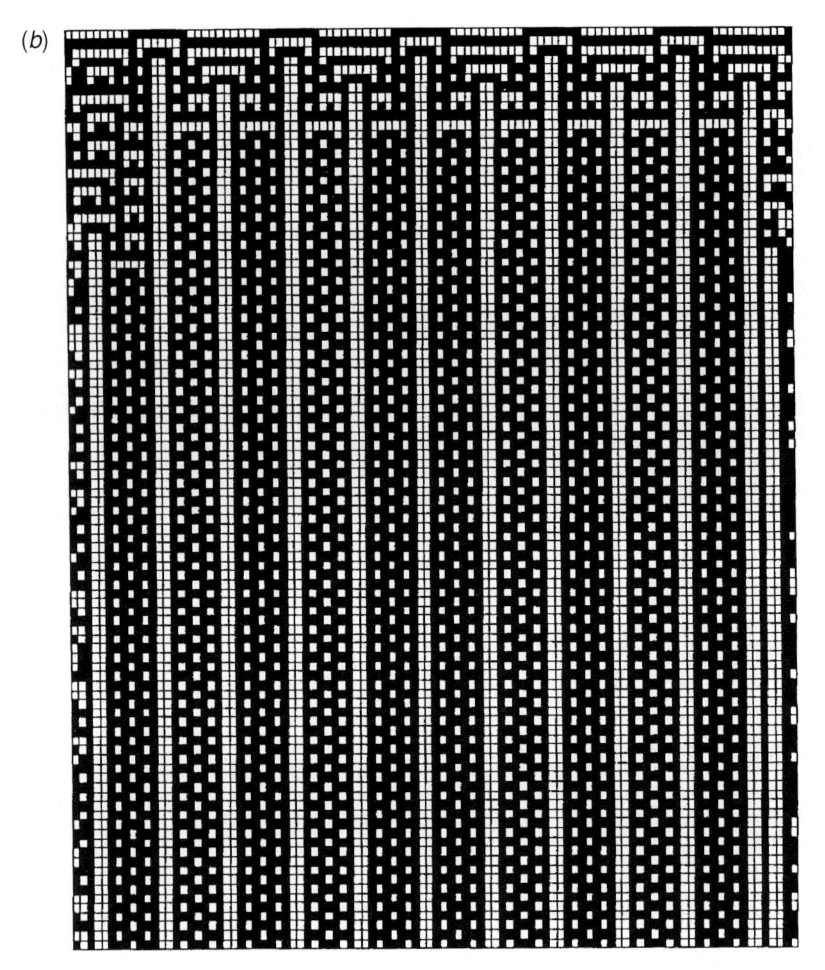

Figure 6.26 (*cont.*)

Figure 6.25). Some cellular automata resemble chaos in their sensitivity to initial conditions; Wolfram's 'rule 73' produces qualitatively different patterns for different starting conditions (Figure 6.26*a*, *b*, *c*). These cellular automaton systems resemble temporal systems more than spatial chaos, in some ways, because there is a strict directionality in them for cause and effect (top to bottom in our figures).

Given the interaction of deterministic effects, stochastic factors and the underlying spatial structure, it will probably be even more difficult to demonstrate true spatial chaos in nature than it has been to demonstrate true temporal chaos. Diks *et al.* (1997) warned that if spatio-temporal chaos exists and we study it only through time or in space, we may be misled because we need to look at both space and time together.

For the purposes of this book and of this chapter on spatio-temporal analysis, we have probably said enough on the fascinating topic of chaos. The main point

Figure 6.26 (*cont.*)

is to be aware that in biological systems, non-linear dynamics should always be considered as a possibility, and that chaos and the near approaches to chaos cannot be totally ruled out as reasonable explanations of the observed behaviour.

## 6.9 Concluding remarks

The area of spatio-temporal analysis, and the phenomena with which it deals, is definitely one of the most fascinating and rapidly developing in ecology today. It is critical to a mature ecological understanding, not just in the sense of spatial pattern and temporal process, but also in the sense of spatial dynamics and temporal pattern. For example, the work on spatial synchrony and asynchrony described in the section above is providing important insights into the basic drivers of population dynamics and community interactions. The analysis of animal movement through its habitat, and how that is related to the habitat structure is an area of active research where we expect to see rapid developments. The analysis of polygon change is also one that deserves further work and effort.

### *6.9.1 Recommendations*

Spatio-temporal analysis is a field which can handle, or perhaps requires, rich data sets; many of the techniques described here will be most rewarding with detailed spatial information and many times of observations.

In conducting statistical tests for this sort of study, we need to be constantly aware of spatial and temporal autocorrelation, their effects and the processes that give rise to them. In many instances, we also need to be concerned about the possible pitfalls of pseudo-replication. A constant theme throughout this book is the lack of independence between observations. While that lack of independence causes problems for statistical tests through autocorrelation, it is also the property that makes prediction possible, and prediction (of many kinds, including interpolation as well as extrapolation) is crucial to the scientific value of ecology. Predictions, in general, are more powerful and potentially more useful when they are quantitative, and so our advice is to have lots of high-quality numerical data. We discussed the problems and interesting qualities of spatial and temporal autocorrelation in the previous chapter, and it is an important consideration in analysing spatio-temporal data of all kinds. We do not agree with the concept of somehow thinning out data in order to get 'independent' observations because it is wasteful and it may not work, as discussed in Chapter 5! The concept of time-to-independence or distance-to-independence is mistaken. We need to learn to take advantage of that lack of independence in the data and use it for our own purposes. Therefore, it is much better to use all the information available and to evaluate the characteristics of autocorrelation in the data to be used in later analysis.

# 7
# Closing comments and future directions

## Back to basics

Both authors of this book are very visual in their approach to problems, as is evident in the number of figures we have used. It is not surprising, therefore, that we advocate the visual evaluation of every step of the analysis process. Plot the data, plot the results of analysis and, when fitting a model, plot the residuals. There are insights to be gained throughout the process.

The first step in any analysis is to plot the data. Many problems can be avoided by this simple step, combined with an awareness of potential problems and some thought. A common mistake in dealing with spatial data is that the $x$- and the $y$-coordinate axes (the columns in a data file) are not used in the right order during the analysis. Indeed, the input format needed varies with the statistical software or GIS package being used; for example, the location of the origin (0, 0) can be either in the upper left corner (as in a matrix array) or in the lower left corner (as sampled in the field). This difference in the location of the origin can result in analysing the mirror image of the data, which is not always an important problem but in some applications may be critical. Furthermore, plotting the data can provide obvious and useful information to guide the choice of which spatial statistics to use and which methods will detect the spatial structure of the data. This early informal evaluation of the spatial behaviour of the data is also useful later in interpreting the results of an analysis and in understanding the spatial patterns and the processes that generated them.

Having analysed the data, plot the results. It is quite possible for a global analysis to detect very little pattern in non-stationary data, but the non-stationarity could be revealed by spatially explicit local analysis. For example, in point pattern analysis, Ripley's $K$-function may show little departure from CSR, where Getis' method of plotting the scores may reveal spatial trends (see Chapter 2). In our analysis of published variograms of ecological data, described in Chapter 5, we found many

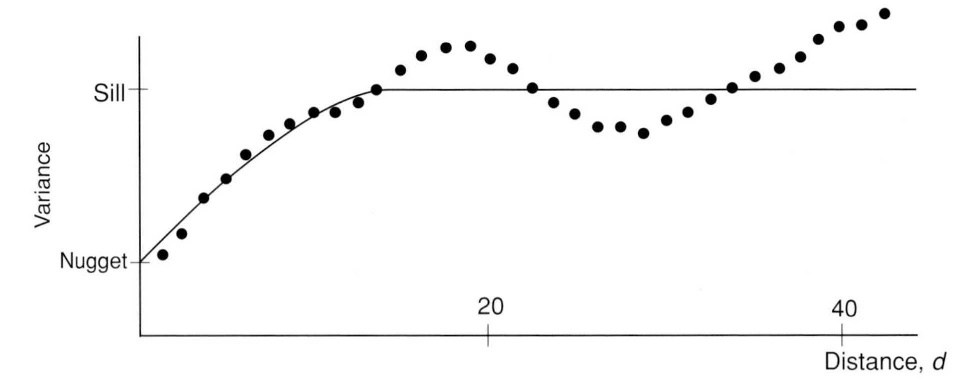


Figure 7.1 A variogram model (line) fit to observed variances as a function of distance (points). In some senses, the model explains much of the characteristics of the observed values but, in other ways, it is not a good fit.

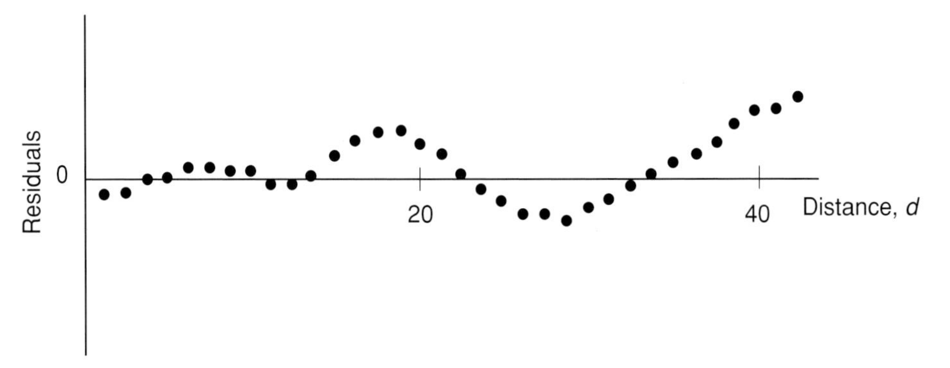


Figure 7.2 The residuals of the model fitted in Figure 7.1. They are clearly not independent.

instances of variogram models used for data that obviously were not good descriptions. Figure 7.1 gives an example, in which the model explains a large proportion of the variability in the data and yet is not really a good fit in other senses. Plotting the results and some thinking about what they show might persuade us to change the approach we use.

When fitting models, plot the residuals. The last example can also illustrate the importance of plotting the residuals of a fitted model. The assumption is often that the residuals have a normal distribution with a constant variance and are independent of each other. The residuals plotted in Figure 7.2 clearly do not meet those conditions.

In the same spirit of understanding how the data are analysed, some detailed knowledge of data storage and the analysis algorithm is always beneficial. Some statistical and GIS software packages are too user-friendly and the user is kept unaware of the details of analyses actually performed. Understanding the programs will make you more alert to potential miscoding that could have taken place in some, especially when using shared code from website sources. The spatial analyses for this book were performed using both software packages (Passage, BoundarySeer, ClusterSeer, Splus+Spatial, GS+, Surfer, CANOCO, IDRISI and ArcGIS) and specifically written programs in computer languages (Fortran, VisualBasic, QuickBasic).

The selection of the appropriate spatial extent and grain at which to study the process of interest can be tricky when prior knowledge is not available. To perform a meaningful study, information about the spatial and temporal domains of the process, as well as spatial and temporal response scales of the patterns, is needed. Such information can be obtained by carrying out a pilot study (see Legendre *et al.* 2002, among others). To facilitate the sampling design of a pilot study or the actual field work, when there is no time available for a pilot study, a variety of other sources may be useful. These include aerial photographs, remotely sensed images, vegetation maps, digital elevation maps, hydrologic maps and bathymetric charts, as well as knowledge from previous studies on the variables, species and systems of interest from reports, papers and colleagues' expertise.

Another important, and also simple, suggestion is that larger sample sizes often offer more options for analysis, as well as more power. This may be particularly important when the particularities of the data do not permit parametric analyses, and randomization methods seem to be the best approach to analysis. We know that in many areas of ecology it is rare to have sample sizes of $n = 30$ or more. From examining the problems of fitting models to data (of known structure) described in Chapter 5, it is clear that in some circumstances, a sample size of even $n = 100$ is really too small. To model the spatial structure of the data may require much larger samples than we are used to having available. Similarly, for the detection of spatial pattern, a transect of 40 contiguous quadrats is definitely too short, because of the very limited number of lags or block sizes that can be examined. Under those circumstances, smaller and more numerous sample units will be more effective. For studies in two dimensions, the same considerations apply, but there is the important issue of the trade-off between extent and grain of sampling when resources are finite. As for many of these issues, a balance of considerations and limitations is required.

More data does not always mean better data. Indeed, more and more ecological studies are carried out at the landscape level at which novel and challenging questions can be investigated. Usually, such studies use either aerial photographs or remotely sensed data, which can provide very large data sets (e.g. tens of thousands of pixels). Beside the obvious problems with remotely sensed data, such as spatial accuracy, image distortion and misclassification (see Burrough & McDonnell 1998), the data usually cover a large area where several environmental factors and ecological processes can occur and therefore non-stationarity, rather than stationarity, can be assumed (see discussion in Section 7.2). In such a context, global spatial analyses (Chapters 2 and 3) of the data for the entire area should not be performed unless the data are first spatially stratified and partitioned into spatially homogeneous subareas (see Chapter 4). We recommend using local spatial statistics (Chapter 3) under these circumstances.

## 7.1 Programming skills

In offering advice to graduate students in almost any branch of ecology, one of the most important recommendations is to acquire at least some programming skills. This may sound like an old-fashioned suggestion, given the wealth of software packages available, but that wealth of software is itself part of the reason for the advice. It is very dangerous and potentially very misleading to use software programs when the details of their calculations are not made explicit. As a single example, one popular analysis package, in calculating the variogram for a range of spatial displacements ($\mathbf{h}$), uses divisor $n$ for all spatial intervals rather than $n_h$ (i.e. the number of pairs at the spatial displacement $\mathbf{h}$). The user needs to know this difference in the divisor used in order to interpret the results correctly. The second reason for the advice 'learn to program' is that it opens the door for the researcher to explore methods or variants of methods of their own devising without relying on others for help, providing greater flexibility. In addition, there may be a time lag between the creation and publication of a new method of analysis and its general availability in popular software packages. The ability to write or to modify analysis programs will allow the researcher to implement the most up-to-date methods.

## 7.2 Stationarity

In thinking about the assumption of stationarity and the detection of non-stationarity, it is important to acknowledge that, depending on the relative scale used and 'the luck of the draw' (i.e. the characteristics of a particular realization of an underlying process or model), a stationary process can give rise to an inhomogeneous or apparently non-stationary pattern. For example, a Poisson–Poisson or Neyman Type A process gives rise to clumps of events and, given only a few clumps, it is possible they will occur in the same part of the plane. For example, Figure 7.3 shows a randomly generated pattern, which appears to exhibit non-stationarity. There are five clumps of events, but they are all on one-half of the plane even though all parts of the plane have an equal probability of a clump occurring in it. The probability

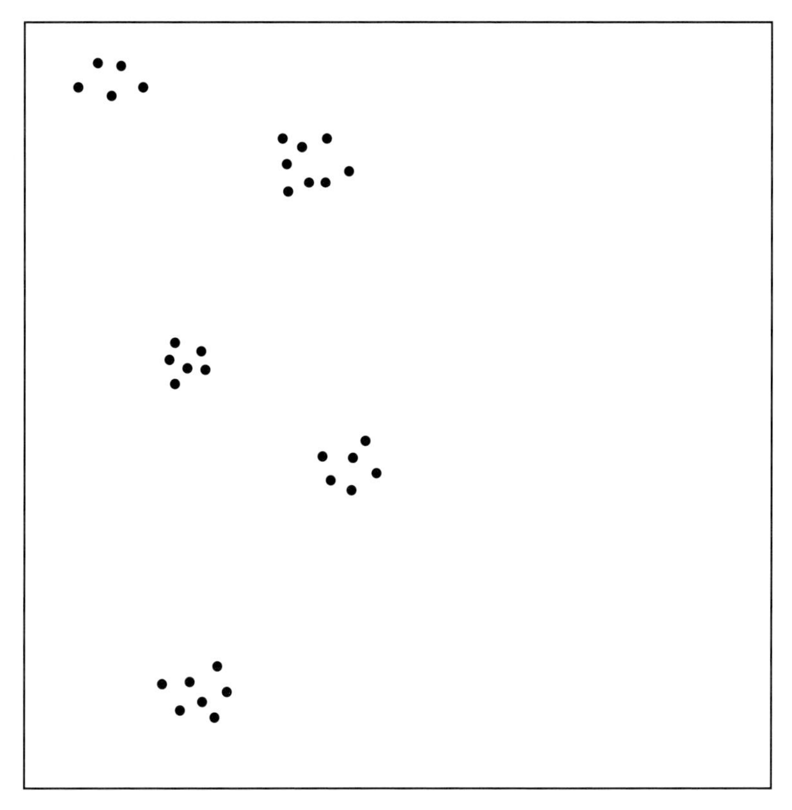

Figure 7.3 An apparently non-stationary pattern of events (dots) resulting from a stationary underlying process.

that all the clumps occur in one-half of the plane is something like $1/2^4 = 0.0625$; so that while this kind of pattern would be somewhat rare in randomly generated patterns, it is not unexpected.

As we stressed before, the data set collected from field sampling is, at best, very much like a single realization of an underlying model, and apparent inhomogeneity may be the result of processes that are, in fact, stationary. A related comment is that we need to be concerned about the power of our procedures to reject the null hypothesis of interest, especially the null hypothesis of stationarity. In many instances, a useful exercise is to determine in advance the strength of the spatial pattern that would be required to reject the null hypothesis, given the sample size or effort expended (i.e. the power of the test). Combined with a pilot study to determine some of the characteristics of the spatial structure of the system being studied, this prior knowledge of the magnitude of sample size needed can provide important guidance in the study design and in subsequent analysis.

In trying to determine the fit of different kinds of models to data in spatial series, such as the AR or MA models described in Chapter 5 or some combination of them, the same proviso applies. Short series may not provide sufficient data to allow us to determine correctly the underlying model. In Chapter 5 we gave artificial examples in which the series of data with $n$ equal to 100 were not best fit by the model that gave rise to them. With field data, the same concern will apply, with the added potential problem that to have longer data series may require greater spatial extent, increasing the risk of encountering true non-stationarity in the underlying processes. Larger data sets may be necessary to detect weak patterns.

## 7.3 Null hypotheses

One of the themes of this book is the suggestion that a single simple null hypothesis (e.g. the complete spatial randomness of events) may not be particularly interesting or particularly useful in an ecological context. In many instances, a hierarchy of null hypotheses of increasing restriction and sophistication will be much more informative. In parallel, when we consider the use of randomization techniques for testing these hypotheses, a series of increasingly restricted randomization schemes will tell us a lot more that is meaningful than a single unrestricted randomization which can only test the simplest null hypothesis. In designing randomization procedures, it is also important to think through the worst possible case, the 'pathological' data set and how the proposed procedure would respond to it. We may be able to resolve our concerns about a particular randomization procedure by finding a counter-example that shows up its faults.

As an example, consider a sampling design of eight transects of four sample plots each, with three in unburned forest and one in a burned area. The abundance of a particular species was determined in each plot and the question of interest is whether the abundance is different between the burned and unburned areas. The analysis first proposed was to use 1000 iterations to create a reference distribution of the density in the unburned forest, by taking only one of the three 'unburned' values in each transect and calculating the average. These averages were then to be used to create a distribution of mean densities and the mean density of the burned area plots (calculated once) was to be compared to the distribution. This mean was to be declared significantly different from the unburned area mean if it fell above the 97.5% value of the distribution or below 2.5%. This initial proposal seems straightforward, but Table 7.1 below gives a counter-example of why the proposed analysis is not the best.

The mean value for the intact forest is 6.0 in every iteration, and so the distribution is very narrow. The burned area has a mean of 5.0, which is outside the distribution, leading to the conclusion that it is significantly low. An examination

Table 7.1 *The abundance of a given species in 32 plots arranged in 8 transects*

| | Transect | | | | | | | |
|---|---|---|---|---|---|---|---|---|
| | 1 | 2 | 3 | 4 | 5 | 6 | 7 | 8 |
| Condition | 10 | 10 | 6 | 6 | 6 | 6 | 2 | 2 |
| Unburned | 10 | 10 | 6 | 6 | 6 | 6 | 2 | 2 |
| | 10 | 10 | 6 | 6 | 6 | 6 | 2 | 2 |
| Burned | 7 | 7 | 5 | 5 | 5 | 5 | 3 | 3 |

of the distributions of the two sources of data suggests that they overlap greatly, and that the conclusion may be mistaken.

This example provides a good illustration of the close relationship between the null hypothesis and the randomization method used to test it. What was actually tested here was the hypothesis: 'The burned area mean is not different from the unburned area mean (which can be treated as a given)'. A better version of the null hypothesis is that the densities of the two areas are the same, so that the observed values in the two areas provide an estimate of a common mean. This logic leads to a different randomization method: the observed difference in mean density between the two areas ($6.0 - 5.0 = 1.0$) is compared to the distribution of the difference when all 32 observations are randomized. When that approach is used, the difference is not significant: 155 of 1,000 trials have an observed difference greater than 1.0. That is the result of a complete randomization, which destroys all the spatial structure in the data. Given the obvious trend in overall densities in the transects, a restricted randomization, within transects, should be considered. Then the null hypothesis is: 'Given the overall density within each transect, the burned and unburned densities do not differ.' Randomizing within transects produces a significant result, with only 4 of 1,000 trials giving a greater difference between the means than that observed.

In addition to showing the close relationship between the hypothesis being tested and the randomization technique used, this example also shows the potentially very important difference between complete and restricted randomizations (see also the discussion in Section 7.6).

## 7.4 Numerical solutions

Another theme for this book, which was not as well developed or as obvious as some of the others, is the usefulness of numerical, as opposed to analytical, solutions to methodological problems. As a single example, consider the details of edge correction for a technique such as Ripley's $K$-function analysis. For a simple square or rectangular plot, there are edge corrections available based on the position of the index point and the radius of the circle being used, $t$. If, however, the study region is not a simple rectangle, or worse yet, has curving boundaries, formulae for edge correction would be very complicated. A numerical solution such as described in Chapter 2 seems like a sensible alternative, particularly as the power of personal computers continues to improve at an impressive rate. The same increase in computing power makes possible a range of computer-based solutions to problems, such as the 'model and Monte Carlo' approach to dealing with spatial autocorrelation in statistical testing. It also changes the way in which models can be used by ecologists. We can use them to explore the effects of particular structures on our understanding of what is occurring (cf. Legendre *et al.* 2002), rather than trusting models as reasonable representations of the data themselves, and then basing the analysis on that trust. (Watch the assumptions!)

We have not spent much of this book on discussions of 'classical' questions of experimental design and the analysis of variance (ANOVA), except to provide some thoughts on the relationship of design and analysis to the spatial structure (spatial autocorrelation) of the environment in which the experiment is carried out (Chapter 5). Without citing specific examples, the reader will probably not need much convincing that this can be a rather confusing area, particularly with complex designs, even before spatial considerations are included. Statistical textbooks may sometimes seem to disagree, or are not always clearly in agreement, which makes life difficult for ecologists, trying to analyse and interpret their data correctly. (One graduate student, after a seminar on some tricky statistical concepts, asked, 'Does this mean we have to be statisticians as well as ecologists?' The answer was, 'No, but you do need to know where to get reliable advice.') Faced with several possible alternatives for ANOVA (often all apparently equally justifiable), one approach is to use artificial data to provide guidance. Generate several sets of 1,000 (say) iterations of data equivalent to those you wish to analyse with small to large treatment effects (3, 5, 8, 12, 20, 30%,...) and examine the behaviour of Type II error. That should provide a realistic guideline for the choice of analysis. As a more general suggestion, it may be useful in many circumstances, particularly when beginning a new type of project, to create and analyse artificial data of the form expected to be produced, to anticipate analytical problems in advance.

In the same way that the researchers have to be precise about the null hypothesis tested, they need to be certain that the spatial statistic they use is indeed doing what they want. For example, Mantel tests, as well as partial Mantel tests, estimate the degree of relationship between coefficients of similarity (distances) between pairs of observations, rather than working with the original data while considering the relative spatial arrangement of the sampling locations. Therefore, Mantel tests do not take into consideration the spatial autocorrelation itself, but rather the relative spatial arrangement of the samples.

## 7.5 Statistical difficulties

There is a variability in geostatisticians' abilities to determine the most appropriate geostatistics (Englund 1990), and also among spatial statisticians and statisticians per se. Hence, ecologists (like all biologists) need to be responsible about their data analyses and subsequent ecological interpretation. Statistical significance does not always result from a significant ecological process. For example, there could be a significant difference between the degree of spatial patchiness between two populations due to the spatial structure of the habitats rather than to the species' ability to move in a fragmented landscape. The reverse is also true; a non-significant statistical result can still have an important ecological meaning. This is especially true for the detection of spatial pattern in the presence of positive spatial autocorrelation: each sampling unit does not contribute a full degree of freedom to the statistical tests (Chapter 5), but we actually want to learn about the pattern itself. In the case of adjacent sampling units in the presence of spatial structure, each sampling unit does not bring a full degree of freedom but the similarity of adjacent samples tells us something else about the size of the spatial structure: that it is larger than the sample units. Consequently, when adjacent units have similar values, this 'redundant' information is informative about the scale of pattern.

Some parametric tests are more sensitive, less robust, than others to the presence of spatial structure in the data. For example, the $t$ distribution does not actually change much in shape between an effective sample size of 20 and one of 50. Therefore, tests based on that distribution may be more robust than others (see Chapter 5) and large changes in the effective sample size due to spatial autocorrelation may have little effect on the interpretation of the data, particularly if $n$ is large to begin with. In contrast, the $\chi^2$ distribution changes markedly with the number of degrees of freedom and some tests that use it are also sensitive to the total sample size. This difference may lead to a different selection of statistical tests when there is a choice to be made.

We must remember that the critical probability levels used by ecologists, such as $\alpha = 0.05$, are only there to use as guidance for making decisions. In many cases, the fact that the nominal significance level of 5% is actually 9% or 2% because of spatial autocorrelation may not have a big effect on our interpretation of the data. Again, larger sample sizes can help. In addition, where testing and interpretation are sensitive to the distribution of the variable of interest (e.g. normal), larger sample sizes may allow us to be more confident in the analytic distribution of the variables we are using. Unfortunately, most of the time we do not know for sure that the variable actually follows the distribution we assume and we have to rely on the robustness (to departures from the assumptions) of the tests we use.

## 7.6 Randomization and restricted randomization tests

As mentioned in Section 7.3, an understanding of how ecological hypotheses translate into statistical hypotheses is essential in any study, but even more so when formulating randomization tests. This conversion has implications for the design and computation of the randomization tests. We need to understand each step involved in the analysis of ecological data and their interactions:

(1) the definition of the hypothesis,
(2) its translation into a statistical hypothesis, and
(3) the selection of the appropriate statistic and subsequent significance testing procedure (as illustrated in the example of Section 7.3).

In using parametric tests, ecologists rely on predefined statistical hypotheses, statistics and significance procedures that require the independence of the data and this may limit the scope of questions that can be asked.

Randomization tests provide an attractive alternative to parametric tests. One interesting feature of randomization tests is that significance is evaluated based on empirical distributions generated from the observed sample. This property is quite appealing to ecologists faced with small data sets that do not meet the assumed parametric distribution. Furthermore, although randomization tests do not offer the security of predefined methods, their flexibility provides the means to analyse complex ecological data using custom experimental designs for which classical tests have not been developed. Ecologists can also develop their own statistics, opening up the possibility to test novel questions. For example, the boundary statistics presented in Chapter 4 were developed to investigate and to test the properties of coherent boundaries.

While randomization tests may involve fewer assumptions, this does not mean that they have no assumptions. In fact, randomization tests are based on the premise that the data are independent such that re-arrangements (i.e. re-orders, exchanges, shuffles) of the data are equally likely. In a spatial context, this assumption corresponds to a statistical null hypothesis of complete spatial randomness (CSR) of the data. In the presence of spatially autocorrelated data, this assumption is invalid. Therefore, restricted randomization tests that consider the spatial structure of the data have been proposed (Legendre *et al.* 1990; Sokal *et al.* 1993; Manly 1997).

There are different ways to perform restricted randomization tests that keep the spatial pattern of the data (Manly 1997; Fortin *et al.* 2002). One of the earliest proposals to preserve the spatial structure of the data was the toroidal shift (as presented in Chapters 1 and 5). In this torus procedure, the restriction is that randomization is not performed at the level of the sampling locations but rather at the level of the entire study area. Indeed, a two-dimensional torus is constructed by connecting the study area margins and then sliding randomly the torus map as many times as is needed to generate a reference distribution. Such a torus procedure maintains most of the spatial structure of the data within the study area and assumes that the spatial process is stationary inside and outside the study area. If this is not the case, the torus procedure can produce a test that is too liberal.

As another example of potential problems with randomization procedures, consider the following situation: in a transect of 100 contiguous sampling units, we have recorded the presence or absence of tree canopy and the presence or absence of a shrub layer. By amazing coincidence, only 10 sampling units have no tree canopy (all in a row) and only 10 have a shrub layer (also all in a row). Interestingly, the canopy gap and the shrub patch are offset somewhat, so that only 7 sampling units have shrubs and no canopy. By a restricted randomization test (the one-dimensional torus method, also known as caterpillar randomization), this is not a significant match (with $\alpha = 0.05$) because there are 7 of 100 random relative positions that have an overlap as great, with overlaps of 7, 8, 9, 10, 9 and 8, and 7 sampling units. If the transect were just a little longer, and the pattern was maintained, the result would be significant!

Other restricted randomization procedures are more appropriate to keep the spatial structure of ecological data in generating the reference distribution using Monte Carlo procedures or stochastic spatial models (see Chapter 5 and Manly 1997; Fortin & Jacquez 2000; Fortin *et al.* 2003). Such restricted randomization tests assume that the underlying process is stationary within the study area. This is not the case when we are studying the significance of boundaries, where, by definition, ecological boundaries are at the interface between two patches, systems or ecosystems and which can be the result of a single process but more likely from the spatial interaction of two or more processes. This is the case where even restricted randomization tests do not seem appropriate and one would need to rely on modelling ecological processes.

We must exercise caution, therefore, to ensure that the null hypothesis implied by the restricted randomization is ecologically tenable. Randomization and restricted randomization tests in ecology are particularly prone to mis-specification of the null hypothesis, primarily because the null hypothesis is embedded in the randomization procedure and is not self-evident. A clear understanding of the null and alternative hypotheses of the chosen randomization test is required in order to ensure the biological and ecological questions under study are correctly addressed.

Furthermore, the null hypotheses, and subsequent randomization procedures, need to leave out the key process tested. Indeed, if all the processes are included in the null hypothesis, there is nothing left to test! For example, in Chapter 4, the overlap statistics have been developed to test whether or not the spatial locations of two boundaries, based on two different data sets (e.g. plant species and animal species), spatially overlap. In such a case, the null hypothesis is that there is not a spatial relationship between the plant and animal boundaries: $H_0$ = no spatial association between the boundaries. The alternatives are: $H_{1a}$ = the boundaries are spatially positively associated, and $H_{1b}$ = the boundaries are spatially negatively associated (i.e. spatially repulsing one another). So the null hypotheses are at the boundary level. Consequently, the randomization procedure should also be at the boundary level. There are several ways to do so, for example by using the torus procedure (Fortin *et al.* 1996) or by randomly placing the boundaries (location and orientation) within the study area (Sokal *et al.* 1988). We might be interested, however, in testing which ecological processes are involved in the actual location of the boundaries. In that case, the randomization should be at the species level. Then we could examine two ecological processes:

(1) the spatial structure (spatial dependence and spatial autocorrelation) of each species, and
(2) the spatial interaction among plant or animal species in the structure of the community.

If only the spatial structure of the species is of interest, then each species can be spatially randomized separately; if both the species and the community spatial structures are of interest then the spatial randomization of each species needs to be linked to the randomization procedure of the other species. These randomizations can be trickier to realize and require a clear understanding of the ecological processes, or concepts, that are involved.

In conclusion, randomization and restricted randomization tests free ecologists from parametric tests that were not designed to accommodate their novel questions and the inherent spatial structure of ecological data.

## 7.7 Complementarity of methods

Dale *et al.* (2002) described how many of the methods used for spatial analysis are closely related to one another, either conceptually or mathematically. This is true, in an even more general way, of the broader range of methods described in this book. Knowing the relationships enables us to choose and use sets of methods that provide complementary insights, as with descriptive methods and inferential statistics. The methods chosen can be complementary in the characteristics they detect or in the range of their treatment, global vs. local. The methods may also complement each other by the interconversion of data type (e.g. points vs. sampling units) or by being cumulative (blocking) as in TTLQV or Ripley's *K*-function vs.

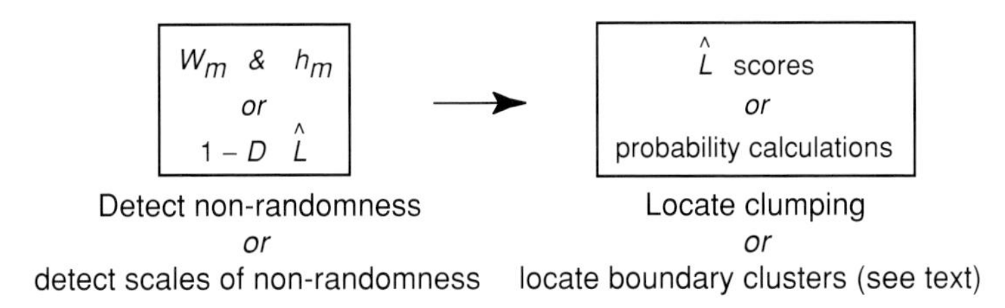


Figure 7.4 Complementary methods in one dimension: $W_m$ and $h_m$ or one-dimensional Ripley's $K$-function to characterize non-randomness and local scores or probability calculations to detect regions of event clumping.

decumulative (e.g. using individual units) as in PQV or Condit's Ω-function. Finally, methods may complement each other by having one method that provides evaluations at individual points in time (such as boundary detection) and another that provides evaluations at changes through time (such as polygon change analysis). It would be impossible to give an exhaustive list of all the combinations of methods that researchers might use to answer sets of related questions of their data, but we will give some examples to illustrate the concept and to provide some guidance.

The simplest spatial data may be a series of events in a single spatial dimension, like waterfalls along a river or termite nests along a line transect. To analyse such data, we can use the statistic $W_m$ to detect non-randomness and then $h_m$ to detect clumping of events (see Chapter 2), or we can use the one-dimensional Ripley's $K$-function analysis to detect scales of over- or underdispersion. If clumping is detected, we can plot the scores of the Ripley's analyses to find the locations of greatest clumping or those locations can be found using the probability calculations described in Dale (1999). That approach finds the sections of the transect where the probability of finding as many events as are observed in it has the lowest probability based on the null hypothesis of randomness. Whichever approach is used, the process of analysis involves the use of several complementary methods that detect different characteristics of the pattern (see Figure 7.4).

As another example, consider a forest plot that has been mapped, recording the positions, species and diameters of all the tree stems. A first analysis could be the modified Ripley's $K$-function analysis to determine the scales at which the stems (of any species and any size) are aggregated or overdispersed. This could be followed by a univariate version of Condit's Ω analysis, based on rings rather than circles, to determine whether there are any distance classes of particular interest. If the overall pattern of the stems is patchy, Getis' score mapping or circumcircle score mapping could be used to examine the data for non-stationarity and to identify the positions of patches and gaps. The next analysis might be a mark correlation analysis to determine the aggregation or segregation of tree sizes as a function of distance.

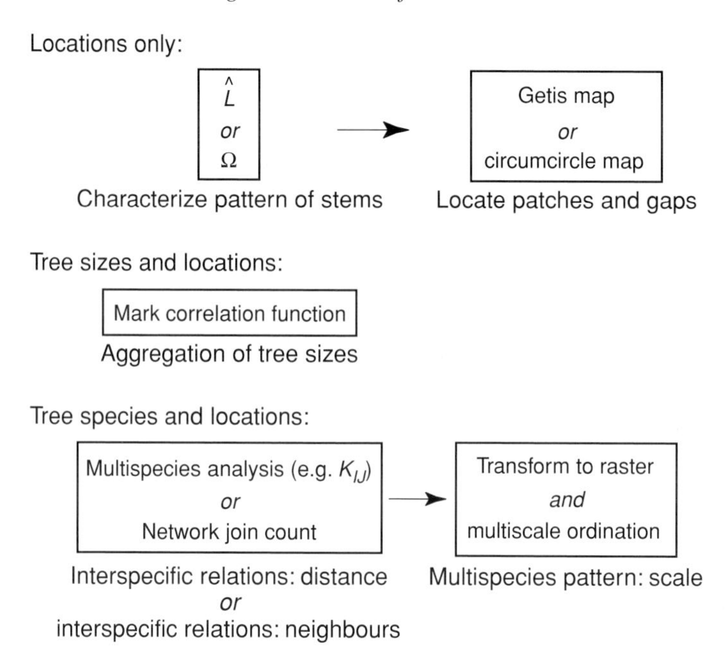


Figure 7.5 Complementary approaches to the analysis of marked event data (stems) in two dimensions: Ripley's $K$-function or Condit's Ω to characterize the non-randomness of the events' positions, with the Getis method or circumcircle scores to plot the locations of the centres of patches or of gaps for a given scale.

Any of a number of the multispecies analysis approaches described in Chapter 2 could be used to examine the interspecific associations. Having already used distance-based analysis on the data, however, a complementary approach would be to use the neighbour networks to look at the join counts for species pairs as in the Dixon method, which compares observed and expected counts. Finally, the data could be converted to raster format of counts or other quantitative data, using square units of a size chosen by the results of the previous analyses, and subjected to multi-scale ordination (MSO, also Chapter 2) to investigate the existence of multispecies pattern. This analysis scheme is illustrated in Figure 7.5.

Given abundance data for a single species from a transect of contiguous quadrats, 3TLQV or the Mexican hat wavelet could be used to determine the scales of pattern in the data, followed by NQV analysis to determine the sizes of the smaller phase in those scales of pattern. If the abundances are patchy, wavelets or a moving-split window (MSW) could be used to find the edges between the regions of high density and those of low density. If an environmental factor such as altitude was recorded for the same sampling units, any of the covariance methods described in Chapter 2 (3TLQC or wavelet covariance) could then be used to determine the scales at which

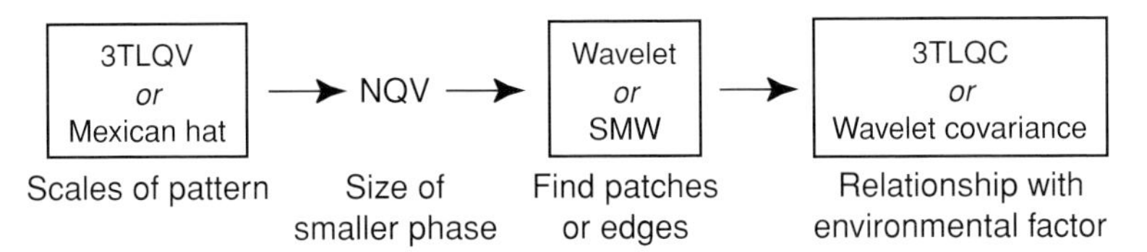


Figure 7.6 Complementary methods for analysing density data in a string of contiguous quadrats: 3TLQV or the Mexican hat wavelet analysis to detect the scales of pattern in the data, with Galiano's NQV to detect the size of the smaller phase, and local wavelet analysis or a split-moving window to detect patches or edges. If an environmental factor is also recorded, 3TLQC or wavelet covariance analysis can be used to detect the scales of positive or negative association of density with that factor.

the species abundance covaries (positively or negatively) with the environmental variable (see Figure 7.6).

For the quantitative single-variable data collected at spaced locations, there is also a range of methods that can be used to evaluate different spatial characteristics in the data (Chapter 3). For example, an omnidirectional correlogram or variogram can be used to evaluate the (isotropic) autocorrelation structure of the data as a function of distance. LISAs may then be used to plot localized areas of high and low spatial association. A complementary approach, not using actual distance would be to look at the correlation of first-order neighbours in a neighbour network, then second-order neighbours and so on. Then, spatially constrained clustering can be used to identify aggregations of similar values or triangulation-wombling can be used to detect boundaries (Chapter 4). Depending on the purpose of the study, having evaluated the nature of the spatial autocorrelation as a function of distance, interpolation techniques like kriging can be used to provide an estimate of the variable over the whole study area. Lastly, if that analysis indicates areas of high variance, and thus poor quality prediction, subsidiary sampling may be indicated to improve the quality of the interpolation. This sequence of analysis steps is shown in Figure 7.7.

As a last example, consider the hourly position records for a radio-collared animal and a habitat map on which those positions can be located. A first analysis would be to quantify the radial (distance) and angular autocorrelation as a function of lag, as described in Chapter 6. Local autocorrelation scores or local tortuosity measures would allow us to detect non-stationarity in the data. If the stationarity is a reasonable assumption (or piecewise stationarity), we might then model the data to achieve a reasonable basis for Monte Carlo generation of artificial data for comparison (cf. Chapter 5). We could then compare the actual habitat use with either these Monte Carlo 'data' or with randomized positions of the original path

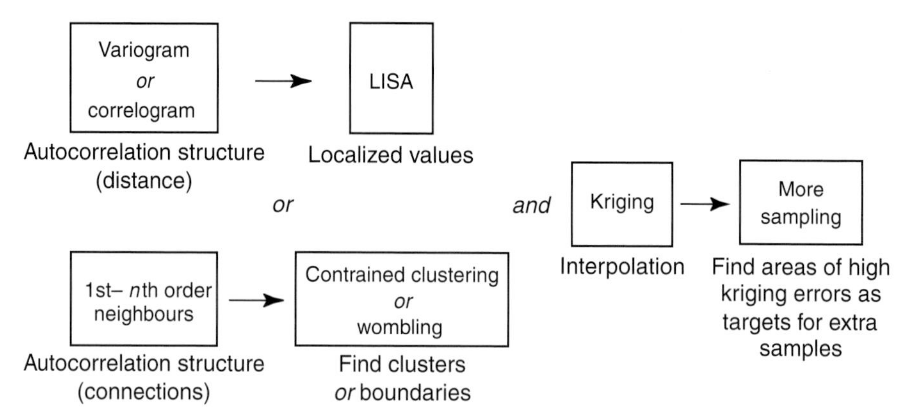


Figure 7.7 Complementary analysis of irregularly spaced records of a quantitative variable: variogram or correlogram analysis to characterize the overall autocorrelation structure as a function of distance, with LISA methods to detect local characteristics. Neighbour networks could be used to characterize the autocorrelation structure, based on connections rather than distance, with constrained clustering to find local clusters of similar values or wombling to detect boundaries between regions of different values. Kriging can be used to locate areas with high estimated variances, indicating a need for greater sampling intensity.

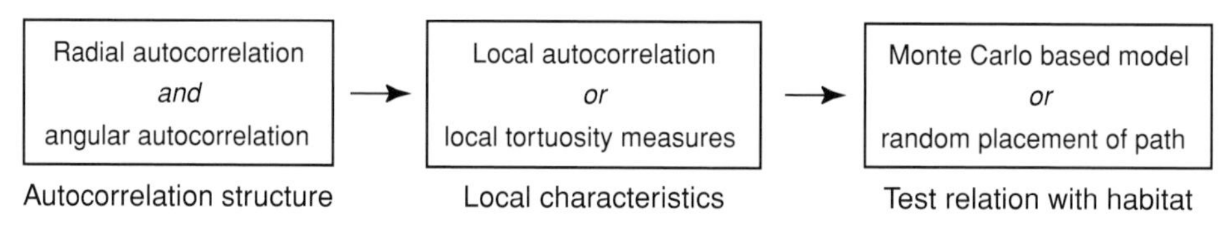


Figure 7.8 Complementary methods to analysis of radio-collar position data referenced to a map of habitat types: radial and angular autocorrelation analysis will characterize the autocorrelation structure, which might then be modelled. Local scores of measures such as tortuosity can then detect areas of behavioural intensity. The relationship between the path characteristics and the habitat can be tested by comparison with paths generated by a Monte Carlo method based on the model generated in the earlier steps, or by randomization of the path's position relative to the habitat map.

of movement on the habitat map to evaluate non-randomness in habitat use. This scheme for analysis is illustrated in Figure 7.8.

While clearly we cannot go through all possible combinations of analysis techniques that might be used, these few examples can at least provide an idea of what we mean when we talk about complementary techniques. Figure 7.9, which is based on the 'relationship' diagrams in Dale *et al.* (2002), shows the relationships among the groupings of methods we described in our examples. It is clear from the figure that there are many other combinations of complementary analyses that could be

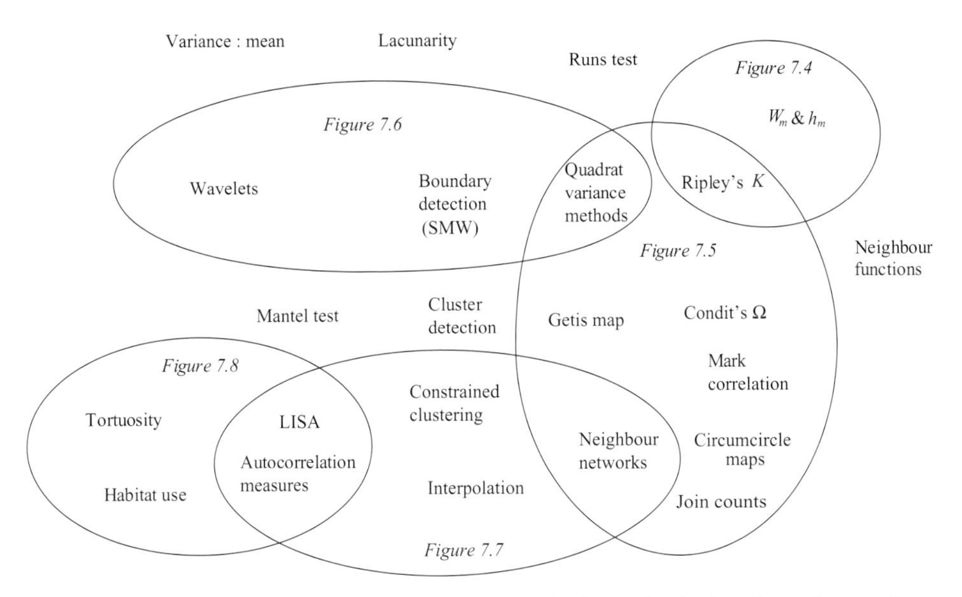


Figure 7.9 Relationships among the spatial analysis methods showing subsets of previous figures. Relative positions reflect the degree of similarities among the methods.

pursued. In some cases, the choice of subsequent methods will depend on the results of the preceding step.

## 7.8 Future work

Throughout this book, we have pointed out areas of methodological research that seem to need and deserve further efforts. A few of these are polygon change analysis, spatio-temporal analysis, dealing with autocorrelation in statistical tests, entropy approaches to categorical data and the analysis of multispecies point patterns. This list is not exhaustive, but gives some idea of the range of topics that are worth further investigation. For ecologists, the 'bottom line' is still to clarify the relationship between process and pattern. In many instances, our concerns about methods, and their possible weaknesses, would be solved by knowing more about the biology of the system being studied. In particular, very detailed knowledge about spatial processes would enable us to refine the methods we use. This parallels the suggestion that pilot studies (any prior information in fact) can help us make better decisions when we design surveys or experiments to investigate ecological phenomena.

We still face problems related to 'detrending' and 'pre-whitening' data prior to analysis. When all the processes and species' responses are linear and additive, these procedures seem to be straightforward. When the processes and species' responses are non-linear (quadratic, cubic, unimodal, multimodal), by removing the trends we may also remove embedded patterns so that the residuals may contain distorted and false spatial structure. The relationship between detrending and pre-whitening, on the one hand, and dealing with the effects of spatial autocorrelation, on the other, is not direct, and may need careful clarification when used together. As Haining (2003) pointed out, 'in general, pre-whitening alone is rarely sufficient to cope with the problem raised by spatial dependence'.

Another issue that is yet to be fully resolved is the testing of spatial indices and measures for statistical significance. As discussed in Chapter 2, the search for significance tests is greatly complicated by the various forms of lack of independence, both in the data themselves and the calculations on which our statistics are based (TTLQV, Moran's $I$, Geary's $c$, and semi-variance being extreme examples of the re-use of data). The sensible use of restricted randomizations seems to be the solution, again going beyond (we hope) the obvious and simple null hypothesis of 'there is no pattern'. Similarly, it is becoming more and more of interest to be able to compare the spatial pattern of two different areas or of the same area at different moments. When it is the case of the same area at different times, it is more likely that the same underlying process is at play; but in the case of two different areas, it is not so. Moreover, a spatial pattern is only one realization of a process and so the pattern may be 'significantly' different at two locations, despite having resulted from the same process. These issues related to the comparison of spatial pattern can be only partially addressed by means of stochastic spatial modelling (Chapter 3, Fortin *et al.* 2003).

In Chapter 5, we spent a considerable amount of effort discussing models of spatial autocorrelation, particularly as a learning tool. While the patchiness, which we determined is a common characteristic of ecological data, can be modelled in the AR–MA structure, it is not clear how biologically realistic such models would be. We have to ask what the ecological processes are that would give rise to the kinds of spatial autocorrelation we described. Dieckmann *et al.* (2000) provided some helpful discussion of this issue, but we will provide a couple of examples. In most statistics text books, the analysis of variance is often presented using a model such as:

$$X_{ij} = B_i + T_j + \varepsilon_{ij}. \tag{7.1}$$

The observed value (suppose it is crop yield) is interpreted as the sum of a block effect, $B_i$, a treatment effect, $T_j$, and an error term, $\varepsilon_{ij}$. When the error term is attributed to variation in soil nutrients, soil moisture and light availability, the autocorrelation in yield is very similar to the induced structure (Model 3) described in Chapter 5. In a well-known plant competition experiment, Franco & Harper (1988) found that the sizes of first neighbours were negatively correlated while those of second neighbours were positively autocorrelated. That is, large plants had smaller neighbours and small plants had larger neighbours, probably as a result of competition. In this case, a first-order autoregressive model (Model 2 in Chapter 5) with $\rho$ negative, would be a biologically realistic model of the spatial structure. As we suggested above, in many ecological examples, we expect the variable of interest to have both inherent and induced autocorrelation, but the biological and physical processes that give rise to it may not always be clear, and so realistic models may remain a challenge for us all.

Although we have tried, in this book, to cover a wide range of issues related to the spatial analysis of ecological data, we have left out several related areas of research such as the spatial aspects of species diversity and the spatial association of different species (cf. Dale & John 1999; Plotkin *et al.* 2000; Shimatani 2001).

Finally, the new challenging area of research is in the merging of spatial and temporal data, patterns and models to create a better understanding of spatial dynamics of ecological processes.

# Appendices

## Appendix 1 *Classification of spatial statistics according to data types*

| | Data[a] | | |
|---|---|---|---|
| | Population | | Sample |
| Data types | Point ($x$–$y$) | Lattice ($x$–$y$, $v$) | Sparse ($x$–$y$, $v$) |
| $x$–$y$ | Aggregation indices<br>$k$-nearest-neighbours [i, r, t]<br>Ripley's $K$ (uni-, bi-multivariate [i, r, p]<br>Circumcircle [r, p]<br>Fractal dimension [i, r]<br>Spectral analysis | | |
| Qualitative | Join count [i, r, t]<br>Mantel correlogram [i, r, t]<br>Mantel and partial Mantel tests [i, t]<br>Mark correlation | Join count [i, r, t]<br>Mantel correlogram [i, r, t]<br>Mantel and partial Mantel tests [i, t] | Mantel correlogram [i, r, t]<br>Mantel and partial Mantel tests [i, t] |
| Quantitative | Global Moran's $I$, Geary's $c$, [i, r, d, t], semi-variance [i, r, d, p]<br>Local Moran's $I$, Getis' $G^*$, Ord's $O$ [i, r, t]<br>Mantel correlogram [i, r, t]<br>Mantel and partial Mantel tests [i, t]<br>Fractal dimension [i, r]<br>Mark correlation | Global Moran's $I$, Geary's $c$ [i, r, d, t], semi-variance [i, r, d, p]<br>Local Moran's $I$, Getis' $G^*$, Ord's $O$ [i, r, t]<br>Mantel correlogram [i, r, t]<br>Mantel and partial Mantel tests [i, t]<br>Fractal dimension [i, r]<br>Lacunarity [i, r]<br>Block variance [i, r]<br>Spectral analysis [i, r]<br>Wavelets [i, r] | Global Moran's $I$, Geary's $c$ [i, r, d, t], semi-variance [i, r, d, p]<br>Local Moran's $I$, Getis' $G^*$, Ord's $O$ [i, r, t]<br>Mantel correlogram [i, r, t]<br>Mantel and partial Mantel tests [i, t]<br>Fractal dimension [i, r] |
| Ordinal/ranked | Join count [i, r, t]<br>Mark correlation | Join count [i, r, t]<br>Spearman Mantel tests [i, t] | Spearman Mantel tests [i, t] |

[a] v: 'value' of a given variable (either qualitative or quantitative) i: the method can estimate the intensity of the spatial structure; r: the method can estimate the spatial range (zone of influence) of the spatial pattern; d: the method can estimate the intensity of spatial pattern according to orientation/directionality (so it can differentiate isotropic from anisotropic patterns); t: significance tests (either analytic or randomization tests); p: significance tests (based on randomization tests).

Appendix 2 *Classification of spatial statistics according to the goals/algorithms of the statistics themselves*

| Algorithm family | Data: Population | | Data: Sample |
|---|---|---|---|
| | Point ($x$–$y$) | Lattice ($x$–$y$, $v$) | Sparse ($x$–$y$, $v$) |
| Topology | Networks (nearest-neighbour, relative neighbour graph, minimum spanning tree, Gabriel, Delaunay) | Rook<br>Bishop<br>Queen | Networks (nearest-neighbour, relative neighbour graph, minimum spanning tree, Gabriel, Delaunay) |
| First-order | | | |
| Aggregation indices | Variance-to-mean<br>Clumping index<br>Green's index<br>Lloyd's index | | |
| Distance | Morisita's index<br>$k$-nearest-neighbour | | |
| Second-order | | | |
| Distance | Ripley's $K$<br>circumcircle | | |
| Autocorrelation | | Moran's $I$<br>Geary's $c$<br>Semi-variance<br>Mantel and partial Mantel tests | Moran's $I$<br>Geary's $c$<br>Semi-variance<br>Mantel and partial Mantel tests |
| Others | | | |
| Interpolation | Voronoi polygons | Trend surface analysis<br>Kriging | Trend surface analysis<br>Kriging |
| Contiguous sampling units | | Block-variance methods<br>Lacunarity<br>Spectral analysis<br>Wavelets | |
| Spatial geometry | Fractal dimension | Fractal dimension | Fractal dimension |
| Boundary detection | | Moving-window<br>Lattice-wombling<br>Wavelet<br>Spatial clustering | Triangular-wombling<br>Categorical-wombling<br>Spatial clustering |

# References


Allain, C. & Cloitre, M. (1991). Characterizing the lacunarity of random and deterministic fractal sets. *Physical Review A*, **44**, 3552–8. (Cited on page 79).

Allen, T. F. H. & Hoekstra, T. W. (1992). *Toward a Unified Ecology*. New York: Columbia University Press. (Cited on page 140).

Alt, W. (1990). Correlation analysis of two-dimensional locomotion paths. In *Biological Motion*, eds. W. Alt & G. Hoffman, pp. 254–68. New York: Springer-Verlag. (Cited on page 276).

Andersen, M. (1992). Spatial analysis of two species interactions. *Oecologia*, **91**, 134–40. (Cited on pages 41, 44).

Anselin, L. (1988). *Spatial Econometrics: Methods and Models*. Dordrecht: Kluwer Academic Publishers. (Cited on page 1).

(1995). Local indicators of spatial association: LISA. *Geographical Analysis*, **27**, 93–115. (Cited on page 154).

Arbia, G., Benedetti, R. & Espa, G. (1996). Effects of the MAUP on image classification. *Geographical Systems*, **3**, 123–41. (Cited on page 146).

Armstrong, R. A. (1983). Growth curve of the lichen *Rhizocarpon geographicum*. *New Phytologist*, **94**, 619–22. (Cited on page 295).

(1992). A comparison of the growth curves of the foliose lichen *Parmelia conspersa* determined by a cross-sectional study and by direct measurement. *Environmental and Experimental Botany*, **32**, 221–7. (Cited on page 295).

Armstrong, R. A. & Smith, S. N. (1996). Experimental studies of hypothallus growth in the lichen *Rhizocarpon geographicum*. *New Phytologist*, **132**, 123–6. (Cited on page 295).

Baddeley, A. J., Howard, C. V., Boyde, A. & Reid, S. (1987). Three-dimensional analysis of the spatial distribution of particles using the tandem-scanning reflected light microscope. *Acta Stereologica*, **6**, 87–100. (Cited on page 81).

Bailey, T. C. & Gatrell, A. C. (1995). *Interactive Spatial Data Analysis*. Harlow: Longman Scientific & Technical. (Cited on pages 1, 9, 39, 230, 231, 233, 268).

Baker, W. L. & Cai, Y. (1992). The r.le programs for multiscale analysis of landscape structure using the GRASS geographical information system. *Landscape Ecology*, **7**, 291–302. (Cited on page 13).

Barbujani, G., Oden, N. N. & Sokal, R. R. (1989). Detecting regions of abrupt change in maps of biological variables. *Systematic Zoology*, **38**, 376–89. (Cited on pages 192, 195).

Barbujani, G. & Sokal, R. R. (1991). Zones of sharp genetic change in Europe are also linguistic boundaries. *Proceedings of the National Academy of Sciences of the United States of America*, **87**, 1816–19. (Cited on page 195).

Barot, S., Gignoux, J. & Menaut, J.-C. (1999). Demography of a savanna palm tree: predictions from comprehensive spatial pattern analysis. *Ecology*, **80**, 1987–2005. (Cited on pages 43, 44).

Bartlett, M. S. (1935). Some aspects of the time correlation problem in regard to tests of significance. *Journal of the Royal Statistical Society*, **98**, 536–43. (Cited on page 238).

Bascompte, J. & Solé, R. V. (1998). Spatiotemporal patterns in nature. *Trends in Ecology and Evolution*, **13**, 173–4. (Cited on page 304).

Batschelet, E. (1981). *Circular Statistics in Biology*. London: Academic Press. (Cited on page 276).

Blundell, G. M., Maier, J. A. K. & Debevec, E. M. (2001). Linear home ranges: effects of smoothing, sample size and autocorrelation on kernel estimates. *Ecological Monographs*, **71**, 469–89. (Cited on page 199).

Bellehumeur, C. & Legendre, P. (1997). Aggregation of sampling units: an analytical solution to predict variance. *Geographical Analysis*, **29**, 258–66. (Cited on page 146).

Bellehumeur, C., Legendre, P. & Marcotte, D. (1997). Variance and spatial scales in a tropical rain forest: changing the size of sampling units. *Plant Ecology*, **130**, 89–98. (Cited on pages 112, 146).

Benedict, J. B. (1967). Recent glacial history of an alpine area in the Colorado front range USA. I. Establishing a lichen-growth curve. *Journal of Glaciology*, **6**, 817–32. (Cited on page 295).

Bergman, C. M., Schaefer, J. A. & Luttich, S. N. (2000). Caribou movement as a correlated random walk. *Oecologia*, **123**, 364–74. (Cited on page 284).

Bishop, Y. M. M., Fienberg, S. E. & Holland, P. W. (1975). *Discrete Multivariate Analysis*. Cambridge, MA: MIT Press. (Cited on page 286).

Bivand, R. (1980). A Monte Carlo study of correlation coefficient estimation with spatially autocorrelated observations. *Quaestiones Geographicae*, **6**, 5–10. (Cited on page 238).

Bjørnstad, O. N. & Falck, W. (2001). Nonparametric spatial covariance functions: estimation and testing. *Environmental and Ecological Statistics*, **8**, 53–70. (Cited on pages 132, 241).

Bjørnstad, O. N., Ims, R. A. & Lambin, X. (1999a). Spatial population dynamics: analyzing patterns and processes of population synchrony. *Trends in Ecology and Evolution*, **14**, 427–32. (Cited on page 304).

Bjørnstad, O. N., Stenseth, N. C. & Saitoh, T. (1999b). Synchrony and scaling in dynamics of voles and mice in northern Japan. *Ecology*, **80**, 622–37. (Cited on page 304).

Boots, B. N. (1980). Weighting Thiessen polygons. *Economic Geography*, **56**, 248–59. (Cited on page 298).

(2002). Local measures of spatial association. *Écoscience*, **9**, 168–76. (Cited on pages 11, 13, 154, 155, 157, 158).

(2003). Developing local measures of spatial association for categorical data. *Journal of Geographical Systems*, **5**, 139–60. (Cited on pages 154, 159).

Borcard, D. & Legendre, P. (2002). All-scale spatial analysis of ecological data by means of principal coordinates of neighbour matrices. *Ecological Modelling*, **153**, 51–68. (Cited on page 153).

Borcard, D., Legendre, P. & Drapeau, P. (1992). Partialling out the spatial component of ecological variation. *Ecology*, **73**, 1045–55. (Cited on page 153).

Bowersox, M. A. & Brown, D. G. (2001). Measuring the abruptness of patchy ecotones. *Plant Ecology*, **156**, 89–103. (Cited on page 199).

Bradshaw, G. A. & Fortin, M.-J. (2000). Landscape heterogeneity effects on scaling and monitoring large areas using remote sensing data. *Geographic Information Sciences*, **6**, 61–8. (Cited on pages 6, 14, 17, 20, 146).

Bradshaw, G. A. & Spies, T. A. (1992). Characterizing canopy gap structure in forests using wavelet analysis. *Journal of Ecology*, **80**, 205–15. (Cited on page 97).

Brandtberg, T. (1999). Automatic individual tree based analysis of high spatial resolution aerial images on naturally regenerated boreal forests. *Canadian Journal of Forest Research*, **29**, 1464–78. (Cited on page 209).

Brodie, C., Houle, G. & Fortin, M.-J. (1995). Development of a *Populus balsamifera* clone in subarctic Québec reconstructed from spatial analyses. *Journal of Ecology*, **83**, 309–20. (Cited on page 292).

Brodo, I. M., Sharnoff, S. D. & Sharnoff, S. (2001). *Lichens of North America*. New Haven, CT: Yale University Press. (Cited on page 293).

Brown, D. G. (1998). Classification and boundary vagueness in mapping presettlement forest types. *International Journal of Geographical Information Science*, **12**, 105–29. (Cited on page 183).

Brunt, J. W. & Conley, W. (1990). Behavior of a multivariate algorithm for ecological edge detection. *Ecological Modelling*, **49**, 179–203. (Cited on page 188).

Burrough, P. A. (1981). Fractal dimensions of landscapes and other environmental data. *Nature*, **294**, 240–2. (Cited on pages 140, 186).

(1986). Principles of geographical information systems for land resources assessment. *Monographs on Soil and Resources Survey No 12*. Oxford: Clarendon Press. (Cited on page 186).

(1987). Spatial aspects of ecological data. In *Community and Landscape Ecology*, eds. R. H. G. Jongman, C. J. F. ter Braak & O. F. R. van Tongeren, pp. 213–51. Nertherlands: Pudoc Wageningen. (Cited on page 11).

Burrough, P. A. & Frank, A. eds. (1996). *Geographic Objects with Indeterminate Boundaries*. London: Taylor and Francis. (Cited on page 176).

Burrough, P. A. & McDonnell, R. A. (1998). *Principles of Geographical Systems*. Oxford: Oxford University Press. (Cited on pages 17, 181, 182, 184, 320).

Cain, M. L. (1989). The analysis of angular data in ecological field studies. *Ecology*, **70**, 1540–3. (Cited on page 274).

(1990). Models of clonal growth in *Solidago altissima*. *Journal of Ecology*, **78**, 27–46. (Cited on page 284).

Cain, M. L. & Damman, H. (1997). Clonal growth and ramet performance in the woodland herb, *Asarum canadense*. *Journal of Ecology*, **85**, 883–97. (Cited on page 293).

Cain, M. L., Pacala, S. W., Silander, J. A. & Fortin, M.-J. (1995). Neighborhood models of clonal growth in the white clover, *Trifolium repens*. *American Naturalist*, **145**, 888–917. (Cited on page 284).

Campbell, J. E., Franklin, S. B., Gibson, D. J. & Newman, J. A. (1998). Permutation of two-term local quadrat variance analysis: general concepts for interpretation of peaks. *Journal of Vegetation Science*, **9**, 41–4. (Cited on page 87).

Candeau, J.-N., Fleming, R. A. & Hopkin, A. (1998). Spatiotemporal patterns of large-scale defoliation caused by the spruce budworm in Ontario since 1941. *Canadian Journal of Forest Research*, **28**, 1733–41. (Cited on page 304).

Canny, J. (1986). A computational approach to edge detection. *IEEE Transitional Pattern Analysis of Machine Intelligence*, **8**, 679–98. (Cited on page 208).

Cantwell, M. D. & Forman, T. T. (1993). Landscape graphs: ecological modeling with graph theory to detect configurations common to diverse landscapes. *Landscape Ecology*, **8**, 239–55. (Cited on page 64).

Carrer, M. & Urbinati, C. (2001). Spatial analysis of structural and tree-ring related parameters in a timberline forest in the Italian Alps. *Journal of Vegetation Science*, **12**, 643–52. (Cited on page 292).

Cerioli, A. (1997). Modified tests of independence in 2 $\times$ 2 tables with spatial data. *Biometrics*, **53**, 619–28. (Cited on page 235).

Chadoeuf, J., Brix, A., Pierret, A. & Allard, D. (2000). Testing local dependence of spatial structures on images. *Journal of Microscopy*, **200**, 32–41. (Cited on page 105).

Chatfield, C. (1975). *The Analysis of Time Series: Theory and Practice*. London: Chapman & Hall. (Cited on page 225).

Chilès, J.-P. & Delfiner, P. (1999). *Geostatistics: Modeling Spatial Uncertainty*. New York: Wiley. (Cited on pages 132, 137, 138, 160, 169, 170).

Choesin, D. & Boerner, R. E. J. (2002). Vegetation boundary detection: a comparison of two approaches applied to field data. *Plant Ecology*, **158**, 85–96. (Cited on page 187).

Civerolo, K. & Rao, S. T. (2001). Space-time analysis of precipitation-weighted sulfate concentrations over the eastern US. *Atmospheric Environment*, **35**, 5657–61. (Cited on page 257).

Claramunt, C. & Thériault, M. (1997). Towards semantics for modeling spatio-temporal processes within GIS. In *Advances in GIS Research II*, eds. M. J. Kraak & M. Molenaar, pp. 47–63. London: Taylor & Francis. (Cited on page 271).

Claussen, D. L., Finkler, M. S. & Smith, M. M. (1997). Thread trailing of turtles: methods for evaluating spatial movements and pathway structure. *Canadian Journal of Zoology*, **75**, 2120–28. (Cited on page 289).

Cliff, A. D. & Ord, J. K. (1973). *Spatial Autocorrelation*. London: Pion. (Cited on pages 1, 119, 120, 124, 126).

(1981). *Spatial Processes: Models and Applications*. London: Pion. (Cited on pages 1, 29, 118, 120, 124, 126, 131, 220, 264).

Clifford, P., Richardson S. & Hémon, D. (1989). Assessing the significance of correlation between two spatial processes. *Biometrics*, **45**, 123–34. (Cited on pages 222, 235, 238).

Cohn, R. D. (1999). Comparisons of multivariate relational structures in serially correlated data. *Journal of Agricultural, Biological & Environmental Statistics*, **4**, 238–57. (Cited on page 241).

Condit, R., Ashton, P. S., Baker, P., Bunyavejchewin, S., Gunatilleke, S., Gunatilleke, N., Hubbell, S., Foster, R. B., Itoh, A., Lafrankie, J. V., Lee, H. S., Losos, E., Manokaran, N., Sukumar, R. & Yamakura, T. (2000). Spatial patterns in the distribution of tropical tree species. *Science*, **288**, 1414–17. (Cited on page 53).

Conover, W. J. (1980). *Practical Nonparametric Statistics*, 2nd edn. New York: Wiley. (Cited on pages 288, 304).

Cressie, N. A. C. (1991). *Statistics for Spatial Data*. New York: Wiley. (Cited on pages 1, 216, 218, 222, 233).

(1993). *Statistics for Spatial Data*, revised edn. New York: Wiley. (Cited on pages xi, 1, 2, 9, 13, 22, 29, 39, 132, 138, 160).

(1996). Change of support and the modifiable areal unit problem. *Geographical Systems*, **3**, 159–80. (Cited on page 146).

Csillag, F., Boots, B., Fortin, M.-J., Lowell, K. & Potvin, F. (2001). Multiscale characterization of boundaries and landscape ecological patterns. *Geomatica*, **55**, 291–307. (Cited on pages 176, 186, 196).

Csillag, F., Fortin, M.-J. & Dungan, J. (2000). On the limits and extensions of the definition of scale. *Bulletin of the ESA*, **81**, 230–2. (Cited on pages 5, 6).
Csillag, F. & Kabos, S. (1996). Hierarchical decomposition of variance with applications in environmental mapping based on satellite images. *Mathematical Geology*, **28**, 385–405. (Cited on pages 98, 206).
(2002). Wavelets, boundaries, and the spatial analysis of landscape pattern. *Écoscience*, **9**, 177–90. (Cited on page 206).
Dale, M. R. T. (1977). Graph theoretical analysis of the phytosociological structure of plant communities: the theoretical basis. *Vegetatio*, **34**, 137–54. (Cited on page 64).
(1985). A geometric technique for evaluating lichen growth models using the boundaries of competing thalli. *Lichenologist*, **17**, 141–8. (Cited on pages 294, 295).
(1986). Overlap and spacing of species' ranges on an environmental gradient. *Oikos*, **47**, 303–8. (Cited on page 189).
(1988). The spacing and intermingling of species boundaries on an environmental gradient. *Oikos*, **53**, 351–6. (Cited on page 189).
(1995). Spatial pattern in communities of crustose saxicolous lichens. *Lichenologist*, **27**, 495–503. (Cited on pages 91, 266, 296).
(1999). *Spatial Pattern Analysis in Plant Ecology*. Cambridge: Cambridge University Press. (Cited on pages xi, 1, 25, 33, 75, 76, 82, 84, 85, 88, 90, 91, 97, 247, 266, 329).
(2000). Lacunarity analysis of spatial pattern: a comparison. *Landscape Ecology*, **15**, 467–78. (Cited on pages 80, 84).
Dale, M. R. T. & Blundon, D. J. (1991). Quadrat covariance analysis and the scales of interspecific association during primary succession. *Journal of Vegetation Science*, **2**, 103–12. (Cited on page 88).
Dale, M. R. T., Blundon, D. J., MacIsaac, D. A. & Thomas, A. G. (1991). Multiple species effects and spatial autocorrelation in detecting species associations. *Journal of Vegetation Science*, **2**, 635–42. (Cited on pages 50, 235).
Dale, M. R. T., Dixon, P., Fortin, M.-J., Legendre, P., Myers, D. E. & Rosenberg, M. (2002). The conceptual and mathematical relationships among methods for spatial analysis. *Ecography*, **25**, 558–77. (Cited on pages 33, 80, 97, 105, 328, 332).
Dale, M. R. T. & Fortin, M.-J. (2002). Spatial autocorrelation and statistical tests in ecology. *Écoscience*, **9**, 162–7. (Cited on pages 11, 29, 223, 225).
Dale, M. R. T., Henry, G. H. R. & Young, C. (1993). Markov models of spatial dependence in vegetation. *Coenoses*, **8**, 21–4. (Cited on page 236).
Dale, M. R. T. & John, E. A. (1999). Neighbour diversity in lichen-dominated communities. *Journal of Vegetation Science*, **10**, 571–8. (Cited on page 335).
Dale, M. R. T. & Mah, M. (1998). The use of wavelets for spatial pattern analysis in ecology. *Journal of Vegetation Science*, **9**, 805–14. (Cited on page 97).
Dale, M. R. T. & Powell, R. D. (1994). Scales of segregation and aggregation of plants of different kinds. *Canadian Journal of Botany*, **72**, 448–53. (Cited on pages 46, 63, 102, 266).
(2001). A new method for characterizing point patterns in plant ecology. *Journal of Vegetation Science*, **12**, 597–608. (Cited on pages 44, 98, 99, 100).
Dale, M. R. T. & Zbigniewicz, M. W. (1995). The evaluation of multi-species pattern. *Journal of Vegetation Science*, **6**, 391–8. (Cited on pages 89, 90).
(1997). Spatial pattern in boreal shrub communities: effects of a peak in herbivore density. *Canadian Journal of Botany*, **75**, 1342–8. (Cited on pages 221, 261).
Daubechies, I. (1993). *Different Perspectives on Wavelets*. Providence: American Mathematical Society. (Cited on pages 97, 206).

de Jong, P., Aarssen, L. W. & Turkington, R. (1980). The analysis of contact sampling data. *Oecologia*, **45**, 322–4. (Cited on page 50).
Dennis, B., Desharnais, R. A., Cushing, J. M., Henson, S. M. & Costantino, R. F. (2001). Estimating chaos and complex dynamics in an insect population. *Ecological Monographs*, **71**, 277–303. (Cited on page 310).
Deutsch, C. V. & Journel, A. G. (1992). GSLIB. *Geostatistical Software Library and User's Guide*. New York: Oxford University Press. (Cited on pages 138, 169, 170).
de Solla, S. R., Bonduriansky, R. & Brooks, R. J. (1999). Eliminating autocorrelation reduces biological relevance of home range estimates. *Journal of Animal Ecology*, **68**, 221–34. (Cited on page 289).
Dieckmann, U., Law, R. & Metz, J. A. J., eds. (2000). *The Geometry of Ecological Interactions*. Cambridge: Cambridge University Press. (Cited on pages 257, 334).
Dietz, E. J. (1983). Permutation tests for association between two distance matrices. *Systematic Zoology*, **32**, 21–6. (Cited on page 149).
Diggle, P. J. (1979). Statistical methods for spatial point patterns in ecology. In *Spatial and Temporal Analysis in Ecology*, eds. R. M. Cormack & J. K. Ord, pp. 95–150. Fairland, MD: International Cooperative Publishing House. (Cited on pages 35, 36).
(1983). *Statistical Analysis of Spatial Point Patterns*. London: Academic Press. (Cited on pages 38, 103).
Diggle, P. J. & Chetwynd A. G. (1991). Second-order analysis of spatial clustering for inhomogeneous populations. *Biometrics*, **47**, 1155–63. (Cited on pages 44, 266).
Diks, C., Takens, F. & DeGoede, J. (1997). Spatio-temporal chaos: a solvable model. *Physica D*, **104**, 269–85. (Cited on page 314).
Dixon, P. M. (2002). Nearest-neighbor contingency table analysis of spatial segregation for several species. *Écoscience*, **9**, 142–51. (Cited on pages 49, 50).
Doak, P. (2000). Population consequences of restricted dispersal for an insect herbivore in a subdivided habitat. *Ecology*, **81**, 1828–41. (Cited on page 284).
Doebeli, M. & Ruxton, G. D. (1998). Stabilization through spatial pattern formation in metapopulations with long-range dispersal. *Proceedings of the Royal Society of London Series B*, **265**, 1325–32. (Cited on page 310).
Dungan, J. L. (2001). Scaling up and scaling down: relevance of the support effect on remote sensing of vegetation. In *Modelling Scale in Geographical Information Science*, eds. N. J. Tate & P. M. Atkinson, pp. 231–5. New York: Wiley. (Cited on page 146).
Dungan, J. L., Perry, J. N., Dale, M. R. T., Legendre, P., Citron-Pousty, S., Fortin, M.-J., Jakomulska, A., Miriti, M. & Rosenberg, M. S. (2002). A balanced view of scale in spatial statistical analysis. *Ecography*, **25**, 626–40. (Cited on pages 5, 6, 14, 18, 21, 140, 142, 146, 174, 184).
Dutilleul, P. (1993a). Spatial heterogeneity and the design of ecological field experiments. *Ecology*, **74**, 1646–58 (Cited on page 249).
(1993b). Modifying the *t* test for assessing the correlation between two spatial processes. *Biometrics*, **49**, 305–14. (Cited on pages 222, 235, 238).
Dutilleul, P., Stockwell, J. D., Frigon, D. & Legendre, P. (2000). The Mantel–Pearson paradox: statistical considerations and ecological implications. *Journal of Agricultural Biology and Environmental Statistics*, **5**, 131–50. (Cited on pages 149, 152).
Edgington, E. S. (1995). *Randomization Tests*, 3rd edn. New York: Marcel Dekker. (Cited on page 26).
Edwards, G. & Fortin, M.-J. (2001). Cognitive view of spatial uncertainty. In *Spatial Uncertainty in Ecology: Implications for Remote Sensing and GIS Applications*, eds.

C. Hunsaker, M. Goodchild, M. Friedl & T. Case, pp. 133–57. New York: Springer-Verlag. (Cited on page 176).

Edwards, G. & Lowell, K. E. (1996). Modeling uncertainty in photointerpreted boundaries. *Photogrammetric Engineering and Remote Sensing*, **62**, 337–91. (Cited on page 183).

Efron, B. & Tibshirani, R. J. (1993). *An Introduction to the Bootstrap*. New York: Chapman & Hall. (Cited on page 26).

Ellner, S. & Turchin, P. (1995). Chaos in a noisy world: new methods and evidence from time-series analysis. *American Naturalist*, **145**, 343–75. (Cited on page 310).

Englund, E. J. (1990). A variance of geostatisticians. *Mathematical Geology*, **22**, 417–55. (Cited on page 325).

Epperson, B. K. (2003). Covariances among join-count spatial autocorrelation measures. *Theoretical Population Biology*, **64**, 81–7. (Cited on page 122).

Evans, J. P. & Cain, M. L. (1995). A spatially explicit test of foraging behaviour in a clonal plant. *Ecology*, **76**, 1147–55. (Cited on page 292).

Fagan, W. F., Fortin, M.-J. & Soykan, C. (2003). Integrating edge detection and dynamic modeling in quantitative analyses of ecological boundaries. *BioScience*, **53**, 730–8. (Cited on page 195).

Faghih, F. & Smith, M. (2002). Combining spatial and scale-space techniques for edge detection to provide a spatially adaptive wavelet-based noise filtering algorithm. *IEEE Transactions on Image Processing*, **11**, 1069–71. (Cited on page 208).

Fehmi, J. S. & Bartolome, J. W. (2001). A grid-based method for sampling and analysing spatially ambiguous plants. *Journal of Vegetation Science*, **12**, 467–72. (Cited on page 105).

Fisher, R. A. (1932). *Statistical Methods for Research Workers*, 4th edn. London: Oliver & Boyd. (Cited on pages 242, 247).

Fortin, M.-J. (1992). Detection of ecotones: definition and scaling factors. Ph.D. thesis, Department of Ecology and Evolution, State University of New York at Stony Brook, pp. 258. (Cited on pages 149, 154).

(1994). Edge detection algorithms for two-dimensional ecological data. *Ecology*, **75**, 956–65. (Cited on pages 64, 186, 192, 195, 196).

(1997). Effects of data types on vegetation boundary delineation. *Canadian Journal of Forest Research*, **27**, 1851–8. (Cited on pages 178, 184, 192).

(1999a). Effects of sampling unit resolution on the estimation of the spatial autocorrelation. *Écoscience*, **6**, 636–41. (Cited on pages 10, 17, 18, 20, 112, 117, 142, 144).

(1999b). The effects of quadrat size and data measurement on the detection of boundaries. *Journal of Vegetation Science*, **10**, 43–50. (Cited on pages 184, 186, 192, 196, 208).

Fortin, M.-J., Boots, B., Csillag, F. & Remmel, T. K. (2003). On the role of spatial stochastic models in understanding landscape indices in ecology. *Oikos*, **102**, 203–12. (Cited on pages 5, 11, 29, 170, 176, 327, 334).

Fortin, M.-J. & Drapeau, P. (1995). Delineation of ecological boundaries: comparisons of approaches and significance tests. *Oikos*, **72**, 323–32. (Cited on pages 184, 193, 197, 198, 199).

Fortin, M.-J., Drapeau, P. & Jacquez, G. M. (1996). Quantification of the spatial co-occurrences of ecological boundaries. *Oikos*, **77**, 51–60. (Cited on pages 29, 202, 204, 328).

Fortin, M.-J., Drapeau, P. & Legendre, P. (1989). Spatial autocorrelation and sampling design in plant ecology. *Vegetatio*, **83**, 209–22. (Cited on pages 18, 21, 22, 144).

Fortin, M.-J. & Gurevitch, J. (2001). Mantel tests: spatial structure in field experiments. In *Design and Analysis of Ecological Experiments*, 2nd edn, eds. S. M. Scheiner & J. Gurevitch, pp. 308–26. Oxford: Oxford University Press. (Cited on pages 149, 150, 151).

Fortin, M.-J. & Jacquez, G. M. (2000). Randomization tests and spatially autocorrelated data. *Bulletin of the Ecological Society of America*, **81**, 201–5. (Cited on pages 195, 239, 327).

Fortin, M.-J., G. M. Jacquez, G. M. & Shipley, B. (2002). Computer-intense methods. In *Encyclopedia of Environmetrics*, eds. A. El-Shaarawi & W. W. Piegorsch, pp. 399–402. Chichester: Wiley. (Cited on pages 149, 326).

Fortin, M.-J., Olson, R. J., Ferson, S., Iverson, L., Hunsaker, C., Edwards, G., Levine, D., Butera, K. & Klemas, V. (2000). Issues related to the detection of boundaries. *Landscape Ecology*, **15**, 453–66. (Cited on page 184).

Fortin, M.-J. & Payette, S. (2002). How to test the significance of the relation between spatially autocorrelated data at the landscape scale: a case study using fire and forest maps. *Écoscience*, **9**, 213–18. (Cited on page 151).

Fotheringham, A. S., Brunsdon, C. & Charlton, M. (2000). *Quantitative Geography: Perspectives on Spatial Data Analysis*. London: Sage Publications. (Cited on pages 1, 29, 154).

Fotheringham, A. S. & Zhan, F. B. (1996). A comparison of three exploratory methods for cluster detection in spatial point patterns. *Geographical Analysis*, **28**, 200–18. (Cited on page 266).

Fowler, H. W. & Fowler, F. G. (1976). *The Concise Oxford Dictionary of Current English*, 6th edn, ed. J. B. Sykes. Oxford: Oxford University Press. (Cited on page 5).

Franco, M. & Harper, J. (1988). Competition and the formation of spatial pattern in spacing gradients: an example using *Kochia scoparia*. *Journal of Ecology*, **76**, 959–74. (Cited on page 334).

Friedman, M. (1937). The use of ranks to avoid the assumption of normality implicit in the analysis of variance. *Journal of the American Statistical Association*, **32**, 675–701. (Cited on page 288).

Frost, H. J. & Thompson, C. V. (1988). Development of microstructure in thin films. *Proceedings of SPIE, The International Society for Optical Engineering*, **821**, 77–87. (Cited on page 298).

Fukushima, Y., Hiura, T. & Tanabe, S. I. (1998). Accuracy of the MacArthur–Horn method for estimating a foliage profile. *Agricultural and Forest Methodology*, **92**, 203–10. (Cited on page 95).

Gabriel, K. R. & Sokal, R. R. (1969). A new statistical approach to geographic variation analysis. *Systematic Ecology*, **18**, 259–70. (Cited on page 59).

Galiano, E. F. (1982). Pattern detection in plant populations through the analysis of plant-to-all-plants distances. *Vegetatio*, **49**, 39–43. (Cited on page 84).

(1983). Detection of multi-species patterns in plant populations. *Vegetatio*, **53**, 129–38. (Cited on page 262).

Gavrikov, V. & Stoyan, D. (1995). The use of marked point processes in ecological and environmental forest studies. *Environmental and Ecological Statistics*, **2**, 331–44. (Cited on page 55).

Geary, R. C. (1954). The contiguity ratio and statistical mapping. *The Incorporated Statistician*, **5**, 115–45. (Cited on page 126).

Gerrard, D. J. (1969). Competition quotient: a new measure of the competition affecting individual forest trees. *Research Bulletin 20*. Agricultural Experiment Station, Michigan State University. (Cited on page 103).

Getis, A. (1991). Spatial interaction and spatial autocorrelation: a cross product approach. *Environment and Planning A*, **23**, 1269–77. (Cited on pages 105, 147).

Getis, A. & Boots, B. (1978). *Models of Spatial Processes: An Approach to the Study of Point, Line and Area Patterns*. Cambridge: Cambridge University Press. (Cited on pages 1, 29).

Getis, A. & Franklin, J. (1987). Second-order neighborhood analysis of mapped point patterns. *Ecology*, **68**, 473–7. (Cited on pages 41, 102).

Getis A. & Ord, J. K. (1992). The analysis of spatial association by use of distance statistics. *Geographical Analysis*, **24**, 189–206. (Cited on pages 154, 157).

(1996). Local spatial statistics: an overview. In *Spatial Analysis: Modelling in a GIS Environment*, eds. P. Longley & M. Batty, pp. 261–77. Cambridge: GeoInformation International. (Cited on page 158).

Gignoux, J., Camille, D. & Sebastien, B. (1999). Comparing the performances of Diggle's tests of spatial randomness for small samples with and without edge-effect correction: application to ecological data. *Biometrics*, **55**, 156–64. (Cited on page 39).

Gonzalez-Andujar, J. L. & Perry, J. N. (1993). Chaos, metapopulations and dispersal. *Ecological Modelling*, **65,** 255–63. (Cited on page 310).

Good, P. (1993). *Permutation Tests: A Practical Guide to Resampling Methods for Hypothesis Testing*. New York: Springer-Verlag. (Cited on pages 26, 28).

Goovaerts, P. (1997). *Geostatistics for Natural Resources Evaluation*. New York: Oxford University Press. (Cited on pages 132, 137, 138, 160, 169, 170).

Gordon, A. D. (1999). Classification. *Monographs on Statistics and Applied Probability 82*, 2nd edn. London: Chapman & Hall/CRC. (Cited on page 178).

Goreaud, F. & Pélissier, R. (1999). On explicit formulas of edge effect correction for Ripley's *K*-function. *Journal of Vegetation Science*, **10**, 433–8. (Cited on page 38).

Gosz, J. R. (1993). Ecotone hierarchies. *Ecological Applications*, **3**, 368–76. (Cited on pages 184, 186).

Goulard, M., Pagès, L. & Cabanettes, A. (1995). Marked point process: using correlation functions to explore a spatial data set. *Biometrical Journal*, **37**, 837–53. (Cited on page 57).

Greig-Smith, P. (1961). Data on pattern within plant communities. I. The analysis of pattern. *Journal of Ecology*, **49**, 695–702. (Cited on pages 88, 146).

(1983). *Quantitative Plant Ecology*, 3rd edn. Berkeley: University of California Press. (Cited on page 88).

Griffith, D. A. (1981). Interdependence in space and time: numerical and interpretative considerations. In *Dynamic Spatial Models*, eds. D. A. Griffith & R. D. MacKinnon, pp. 258–87. Netherlands: Sijthoff & Noordhoff. (Cited on pages 258, 264).

(1988). Estimating spatial autoregressive model parameters with commercial statistical packages. *Geographical Analysis*, **20**, 176–86. (Cited on page 233).

(1992). What is spatial autocorrelation? Reflection on the past 25 years of spatial statistics. *L'Espace géographique*, **3**, 265–80. (Cited on page 30).

Gustafson, E. J. (1998). Quantifying landscape spatial pattern: what is the state of the art? *Ecosystems*, **1**, 143–56. (Cited on pages 13, 139, 175).

Haase, P. (1995). Spatial pattern analysis in ecology based on Ripley's K-function: introduction and methods of edge correction. *Journal of Vegetation Science*, **6**, 575–82. (Cited on pages 22, 38, 39, 41).

Haining, R. P. (1990). *Spatial Data Analysis in the Social and Environmental Sciences.* Cambridge: Cambridge University Press. (Cited on pages 1, 11, 13, 22, 24, 29, 132, 160).

(1991). Bivariate correlation with spatial data. *Geographical Analysis*, **23**, 210–27. (Cited on page 221).

(2003). *Spatial Data Analysis: Theory and Practice*. Cambridge: Cambridge University Press. (Cited on pages xii, 1, 9, 22, 132, 160, 334).

Haining, R. P. (1978). The moving average model for spatial interaction. *Transactions of the Institute for British Geographers*, **NS3**, 202–25. (Cited on page 233).

Hall, K. R. & Maruca, S. L. (2001). Mapping a forest mosaic: a comparison of vegetation and songbird distributions using geographic boundary analysis. *Plant Ecology*, **156**, 105–20. (Cited on page 204).

Handcock, R. & Csillag, F. (2002). Ecoregionalization assessment: spatio-temporal analysis of net primary production across Ontario. *Écoscience*, **9**, 219–230. (Cited on pages 186, 196).

Hansen, A. & di Castri, F. (1992). *Landscape Boundaries: Consequences for Biotic Diversity and Ecological Flows*. New York: Springer-Verlag. (Cited on page 184).

Hanski, I. & Woiwod, I. P. (1993). Spatial synchrony in the dynamics of moth and aphid populations. *Journal of Animal Ecology*, **62**, 656–68. (Cited on pages 301, 302).

Harary, F. (1967). Topological concepts in graph theory. In *A Seminar on Graph Theory*, ed. F. Harary, pp. 13–24. New York: Holt, Rinehart & Winston. (Cited on page 57).

Hargrove, W. W., Hoffman, F. M. & Schwartz, P. M. (2002). A fractal landscape realizer for generating synthetic maps. *Conservation Ecology*, **6**, 2. Online www.consecol.org/vol6/iss1/art2. (Cited on page 139).

Heagerty, P. J. & Lumley, T. (2000). Window subsampling of estimating functions with application to regression models. *Journal of the American Statistical Society*, **95**, 197–211. (Cited on page 241).

Henebry, G. M. (1995). Spatial model error analysis using autocorrelation indices. *Ecological Modelling*, **82**, 75–91. (Cited on pages 258, 264).

Hill, D. J. (1981). The growth of lichens with special reference to the modeling of circular thalli. *Lichenologist*, **13**, 265–87. (Cited on page 295).

Hill, M. O. (1973). The intensity of spatial pattern in plant communities. *Journal of Ecology*, **61**, 225–35. (Cited on pages 82, 84).

Holland, M. M., Risser, P. G. & Naiman, R. J., eds. (1991). *Ecotones*. New York: Chapman & Hall. (Cited on page 184).

Hope, A. C. A. (1968). A simplified Monte Carlo significance test procedure. *Journal of the Royal Statistical Society B*, **30**, 582–98. (Cited on page 149).

Hudak, A. T., Lefsky, M. A., Cohen, W. B. & Berterretche, M. (2002). Integration of lidar and Landsat ETM plus data for estimating and mapping forest canopy height. *Remote Sensing of Environment*, **82**, 397–416 (Cited on page 170).

Hunsaker C., Goodchild, M., Friedl, M. & Case, T., eds. (2001). *Spatial Uncertainty in Ecology. Implications for Remote Sensing and GIS Applications*. New York: Springer-Verlag. (Cited on page 17).

Hurlbert, S. H. (1984). Pseudoreplication and the design of ecological field experiments. *Ecological Monographs*, **54**, 187–211. (Cited on pages 285, 287).

Isaaks, E. H. & Srivastava, R. M. (1989). *An Introduction to Applied Geostatistics*. Oxford: Oxford University Press. (Cited on pages 29, 132, 135, 160).

Jacquez, G. M. (1995). The map comparison problem: tests for the overlap of geographical boundaries. *Statistical Medicine*, **14**, 2343–61. (Cited on page 202).

(1996). A *k* nearest neighbour test for space–time interaction. *Statistical Medicine*, **15**, 1935–49. (Cited on page 266).

Jacquez, G. M., Maruca, S. L. & Fortin, M.-J. (2000). From fields to objects: a review of geographic boundary analysis. *Journal of Geographical Systems*, **2**, 221–41. (Cited on page 181).

Jean, R. (1994). *Phyllotaxis: A Systemic Study of Plant Pattern Morphogenesis*. Cambridge: Cambridge University Press. (Cited on page 250).

Jelinski, D. E. & Wu, J. (1996). The modifiable areal unit problem and implications for landscape ecology. *Landscape Ecology*, **3**, 129–40. (Cited on pages 18, 112).

Jenkins, G. M. & Watts, D. G. (1968). *Spectral Analysis and Its Applications*. San Francisco: Holden-Day. (Cited on page 219).

Johnston, C. A., Pastor, J. & Pinay, G. (1992). Quantitative methods for studying landscape boundaries, In *Landscape Boundaries*, eds. A. J. Hansen & F. di Castri, pp. 107–25. New York: Springer-Verlag. (Cited on page 187).

Jordan, G. J. (2002). Space, time and uncertainty: detecting and characterizing boundaries in forest fire disturbances. Ph.D. thesis, School of Resource and Environmental Management, Simon Fraser University. (Cited on page 184).

Journel, A. G. & Huijbregts, C. (1978). *Mining Geostatistics*. London: Academia Press. (Cited on pages 29, 132, 137, 146, 160, 165, 166).

Kabos, S. & Csillag, F. (2002). The analysis of spatial association of nominal data on regular lattice by join-count-statistic without first-order homogeneity. *Computers and Geosciences*, **28**, 901–10. (Cited on pages 122, 154, 159).

Kareiva, P. & Shigesada, N. (1983). Analyzing insect movement as a correlated random walk. *Oecologia*, **56**, 234–8. (Cited on page 284).

Kenkel, N. C. (1991). Spatial competition models for plant populations. In *Computer Assisted Vegetation Analysis*, eds. E. Feoli & L. Orlóci, pp. 387–97. Netherlands: Kluwer Academic Publishers. (Cited on pages 298, 299).

(1993). Modeling Markovian dependence in populations of *Aralia nudicaulis*. *Ecology*, **74**, 1700–6. (Cited on page 41).

Kenkel, N. C., Hendrie, M. L. & Bella, I. E. (1997). A long-term study of *Pinus banksiana* population dynamics. *Journal of Vegetation Science*, **8**, 241–54. (Cited on pages 263, 291).

Kenkel, N. C. & Walker, D. J. (1993). Fractals and ecology. *Abstracta Botanica*, **17**, 53–70. (Cited on page 139).

Knapp, P. (1998). Spatio-temporal patterns of large grassland fires in the Intermountain West, USA. *Global Ecology and Biogeography Letters*, **7**, 259–72. (Cited on page 257).

Knox, E. G. (1964). The detection of space-time interactions. *Applied Statistics*, **13**, 25–9. (Cited on page 267).

Koenig, W. D. & Knops, J. M. (1998). Testing for spatial autocorrelation in ecological time series. *Ecography*, **21**, 423–9. (Cited on page 302).

König, D., Carvajal-Gonzalez, S., Downs, A. M., Vassy, J. & Rigaut, J. P. (1991). Modeling and analysis of 3-D arrangements of particles by point processes with examples of application to biological data obtained by confocal scanning light microscopy. *Journal of Microscopy*, **161**, 405–33. (Cited on page 81).

Krebs, C. J. (2002). *Ecology: The Experimental Analysis of Distribution and Abundance: Hands-On Field Package*, 5th edn. San Francisco, CA: Prentice Hall. (Cited on pages 25, 300).

Krebs, C. J., Boutin, S. & Boonstra, R. (2001). *Ecosystem Dynamics of the Boreal Forest: The Kluane Project*. Oxford: Oxford University Press. (Cited on page 262).

Krige, D. G. (1966). Two-dimensional weighted moving average trend surfaces for ore-evaluation. *Journal of the South Africa Institute of Mining and Metallurgy*, **66**, 13–38. (Cited on page 165).

Lambin, X., Elston, D. A., Petty, S. J. & MacKinnon, J. L. (1998). Spatial asynchrony and periodic traveling waves in cyclic populations of field voles. *Proceedings of the Royal Society of London Series B*, **265**, 1491–6. (Cited on page 304).

Law, R., Herben, T. & Dieckmann, U. (1997). Non-manipulative estimates of competition coefficients in a montane grassland community. *Journal of Ecology*, **85**, 505–17. (Cited on pages 259, 261).

Leduc, A., Drapeau, P., Bergeron, Y. & Legendre, P. (1992). Study of spatial components of forest cover using partial Mantel tests and path analysis. *Journal of Vegetation Science*, **3**, 69–78. (Cited on page 151).

Legendre, P. (1993). Spatial autocorrelation: trouble or new paradigm? *Ecology*, **74**, 1659–73. (Cited on pages 9, 11, 29, 213, 255).

(2000). Comparison of permutation methods for the partial correlation and partial Mantel tests. *Journal of Statistical Computer Simulation*, **67**, 37–73. (Cited on page 152).

Legendre, P., Dale, M. R. T., Fortin, M.-J., Casgrain, P. & Gurevitch, J. (2004). Effects of spatial structures on the results of field experiments. *Ecology*, **85**, 3202–14. (Cited on pages 249, 251).

Legendre, P., Dale, M. R. T., Fortin, M.-J., Gurevitch, J., Hohn, M. & Myers, D. E. (2002). The consequences of spatial structure for the design and analysis of ecological field surveys. *Ecography*, **25**, 601–15. (Cited on pages 212, 239, 251, 319, 324).

Legendre, P. & Fortin, M.-J. (1989). Spatial pattern and ecological analysis. *Vegetatio*, **80**, 107–38. (Cited on pages 18, 150, 151, 165, 177).

Legendre, P. & Legendre, L. (1998). *Numerical Ecology*, 2nd English edn. Amsterdam: Elsevier. (Cited on pages 1, 9, 21, 28, 29, 95, 116, 127, 148, 149, 150, 151, 153, 176, 177, 187, 219, 222, 239).

Legendre, P., Sokal, R. R., Oden, N. L. Vaudor, A. & Kim, J. (1990). Analysis of variance with spatial autocorrelation in both the variable and the classification criterion. *Journal of Classification*, **7**, 53–75. (Cited on pages 241, 326).

Lejeune, O. & Tlidi, M. (1999). A model for the explanation of vegetation stripes (tiger bush). *Journal of Vegetation Science*, **10**, 201–8. (Cited on page 10).

Lele, S. (1991). Jackknifing linear estimating equations: asymptotic theory and applications in stochastic processes. *Journal of the Royal Statistical Society B*, **53**, 253–67. (Cited on page 241).

Leung, Y. (1987). On the imprecision of boundaries. *Geographical Analysis*, **19**, 125–51. (Cited on page 183).

Levin, S. A. (1992). The problem of pattern and scale in ecology. *Ecology*, **73**, 1943–67. (Cited on page 1).

Li, H. & Reynolds, J. F. (1995). On definition and quantification of heterogeneity. *Oikos*, **73**, 280–4. (Cited on page 175).

Lindeberg, T. (1994). *Scale-Space Theory in Computer Vision*. Dordrecht: Kluwer. (Cited on page 208).

Little, L. R. & Dale, M. R. T. (1999). A method for analyzing spatio-temporal pattern in plant establishment, tested on a *Populus balsamifera* clone. *Journal of Ecology*, **87**, 620–7. (Cited on page 264).

Liu, C. (2001). A comparison of five distance-based methods for pattern analysis. *Journal of Vegetation Science*, **12**, 411–16. (Cited on page 35).

Lotwick, H. W. & Silverman, B. W. (1982). Methods for analysing spatial processes of several types of points. *Journal of the Royal Statistical Society B*, **44**, 406–13. (Cited on pages 47, 52).

Lovett Doust, J. (1981). Population dynamics and local specialization in a clonal perennial (*Ranunculus repens*): I. The dynamics of ramets in contrasting habitats. *Journal of Ecology*, **69**, 743–55. (Cited on page 264).

Lowell, K. (1997). Effect(s) of the "no-same-color-touching" constraint on the join-count statistic: a simulation study. *Geographical Analysis*, **29**, 339–53. (Cited on page 122).

Ludwig, J. A. & Cornelius, J. M. (1987). Locating discontinuities along ecological gradients. *Ecology*, **68**, 448–50. (Cited on page 188).

Ludwig, J. A. & Goodall, D. W. (1978). A comparison of paired with blocked-quadrat variance methods for the analysis of spatial pattern. *Vegetatio*, **38**, 49–59. (Cited on page 83).

Ludwig, J. A. & Reynolds, J. F. (1988). *Statistical Ecology*. New York: Wiley. (Cited on page 83).

MacKinnon, J. L., Petty, S. J., Elston, D. A., Thomas, C. J., Sherratt, T. N. & Lambin, X. (2001). Scale invariant spatio-temporal patterns of field vole density. *Journal of Animal Ecology*, **70**, 101–11. (Cited on page 304).

Mandelbrot, B. B. (1983). *The Fractal Geometry of Nature*. New York: Freeman. (Cited on pages 139, 186).

Manly, B. F. J. (1997). *Randomization, Bootstrap, and Monte Carlo Methods in Biology*, 2nd edn. London: Chapman & Hall. (Cited on pages 1, 26, 28, 29, 35, 41, 149, 239, 242, 285, 326, 327).

Manly, B. F. J., McDonald, L. L., Thomas, D. L., McDonald, T. L. & Erickson, W. P. (2002). *Resource Selection by Animals: Statistical Design and Analysis for Field Studies*, 2nd edn. Dordrecht: Kluwer Academic.

Mantel, N. (1967). The detection of disease clustering and a generalized regression approach. *Cancer Research*, **27**, 209–20. (Cited on page 147).

Marr, D. & Hildreth, E. (1980). Theory of edge detection. *Proceedings of the Royal Society of London, Series B*, **207**, 187–217. (Cited on page 208).

Matheron, G. (1970). La théorie des variables régionalisées, et ses applications. *Les Cahiers du Centre de Morphologie Mathématique de Fontainebleau*. Fontainebleau: Fascicule 5. (Cited on page 132).

Matula, D. W. & Sokal, R. R. (1980). Properties of Gabriel graphs relevent to geographic variation research and the clustering of points in the plane. *Geographical Analysis*, **12**, 205–22. (Cited on page 60).

McBratney, A. B. & de Gruijter, J. J. (1992). A continuum approach to soil classification by modified fuzzy *k*-means with extra grades. *Journal of Soils Science*, **43**, 159–76. (Cited on page 182).

McCoy, E. D., Bell, S. S. & Walters, K. (1986). Identifying biotic boundaries along environmental gradients. *Ecology*, **67**, 749–59. (Cited on page 189).

McGarigal, K. & Marks, B. J. (1995). FRAGSTATS: spatial pattern analysis program for quantifying landscape structure. *US Forest Service General Technical Report PNW 351*. Corvallis, OR: US Forest Service. (Cited on page 13).

McIntosh, R. P. (1985). *The Background of Ecology: Concept and Theory*. Cambridge: Cambridge University Press. (Cited on page 1).

Mead, R. (1966). A relationship between individual plant spacing and yield. *Annals of Botany*, **30**, 301–9. (Cited on page 298).

Millspaugh, J. J., Skalski, J. R., Kernohan, B. J., Raedeke, K. J., Brundige, G. C. & Cooper, A. B. (1998). Some comments on spatial independence in studies of resource selection. *Wildlife Society Bulletin*, **26**, 232–6. (Cited on page 288).

Milne, B. T. (1992). Spatial aggregation and neutral models in fractal landscapes. *American Naturalist*, **139**, 32–57. (Cited on page 139).

Minta, S. C. (1992). Tests of spatial and temporal interaction among animals. *Ecological Applications*, **2**, 178–88. (Cited on page 288).

Mithen, R., Harper, J. L. & Weiner, J. (1984). Growth and mortality of individual plants as a function of "available area". *Oecologia*, **62**, 57–60. (Cited on pages 61, 298).

Mizon, G. E. (1995). A simple message for autocorrelation correctors: don't. *Journal of Econometrics*, **69**, 267–89. (Cited on page 242).

Moran, P. A. P. (1948). The interpretation of statistical maps. *Journal of the Royal Statistical Society B*, **10**, 243–51. (Cited on pages 118, 124).

Morisawa, M. (1985). *Rivers: Form and Process.* New York: Longman. (Cited on page 105).

Moss, R., Elston, D. A. & Watson, A. (2000). Spatial asynchrony and demographic traveling waves during red grouse population cycles. *Ecology*, **81**, 981–9. (Cited on page 304).

Mugglestone, M. A. (1996). The role of tessellation methods in the analysis of three-dimensional spatial point patterns. *Journal of Agricultural, Biological & Environmental Statistics*, **1**, 141–53. (Cited on page 81).

Mugglestone, M. A. & Renshaw, E. (1996). A practical guide to the spectral analysis of spatial point processes. *Computational Statistics & Data Analysis*, **21**, 43–65. (Cited on page 95).

Nestel, D. & Klein, M. (1995). Geostatistical analysis of leafhopper (Homoptera: Cicadellidae) colonization and spread of deciduous orchards. *Environmental Entomology*, **24**, 1032–9. (Cited on page 263).

Neu, C. W., Byers, C. R., Peek, J. M. & Boy, V. (1974). A technique for analysis of utilization-availability data. *Journal of Wildlife Management*, **38**, 541–5. (Cited on pages 286, 287).

Noy-Meir, I. & Anderson, D. (1971). Multiple pattern analysis or multiscale ordination: towards a vegetation hologram. In *Statistical Ecology*, vol. 3: *Populations, Ecosystems, and Systems Analysis*, eds. G. P. Patil, E. C. Pielou & W. E. Waters, pp. 207–32. University Park: Pennsylvania State University Press. (Cited on page 89).

Oden, N. L. (1984). Assessing the significance of a spatial correlogram. *Geographical Analysis*, **16**, 1–16. (Cited on page 126).

Oden, N. L. & Sokal, R. R. (1986). Directional autocorrelation: an extension of spatial correlograms to two dimensions. *Systematic Zoology*, **35**, 608–17. (Cited on pages 131, 150).

(1992). An investigation of 3-matrix permutation tests. *Journal of Classification*, **9**, 275–90. (Cited on page 152).

Oden, N. L., Sokal, R. R., Fortin, M.-J. & Goebl, H. (1993). Categorical wombling: detecting regions of significant change in spatially located categorical variables. *Geographical Analysis*, **25**, 315–36. (Cited on pages 195, 197, 198, 199, 202).

Ohman, M. E. (1990). The demographic effects of diel vertical migration by zooplankton. *Ecological Monographs*, **60**, 257–82. (Cited on page 258).

Okabe, A., Boots, B. & Sugihara, K. (1992). *Spatial Tesselations: Concepts and Applications of Voronoi Diagrams*. New York: Wiley. (Cited on pages 60, 160, 298).

Okabe, A. & Yamada, I. (2001). The *K*-function method on a network and its computational implementation. *Geographical Analysis*, **33**, 270–90. (Cited on page 79).

O'Neill, R. V., Hunsaker, C. T., Timmins, S. P., Jackson, B. L., Jones, K. B., Riitters, K. H. & Wickham, J. D. (1996). Scale problems in reporting landscape pattern at the regional scale. *Landscape Ecology*, **11**, 169–80. (Cited on pages 18, 20).

O'Neill, R. V., Krummel, J. R., Gardner, R. H., Sugihara, G., Jackson, B., DeAngelis, D. L., Milne, B. T., Turner, M. G., Zygmunt, B., Christenson, S. W., Dale, V. H. & Graham, R. L. (1988). Indices of landscape pattern. *Landscape Ecology*, **1**, 153–62. (Cited on page 13).

O'Neill, R. V., Riitters, K. H., Wickham, J. D. & Jones, K. B. (1999). Landscape pattern metrics and regional assessment. *Ecosystem Health*, **5**, 225–33. (Cited on pages 18, 20).

Openshaw, S. (1984). *The Modifiable Areal Unit Problem*. Norwich: Geo Books. (Cited on page 146).

Ostendorf, B. & Reynolds, J. F. (1998). A model of arctic tundra vegetation derived from topographic gradients. *Landscape Ecology*, **13**, 187–201. (Cited on page 222).

O'Sullivan, D. & Unwin, D. (2003). *Geographic Information Analysis*. New Jersey: Wiley. (Cited on pages 1, 146, 160).

Otis, D. L. (1997). Analysis of habitat selection studies with multiple patches within cover types. *Journal of Wildlife Management*, **61**, 1016–22. (Cited on page 288).

Otis, D. L. & White, G. C. (1999). Autocorrelation of location estimates and the analysis of radiotracking data. *Journal of Wildlife Management*, **63**, 1039–44. (Cited on page 288).

Owens, M. K. & Norton, B. E. (1989). The impact of 'available area' on *Artemisia tridentata* seedling dynamics. *Vegetatio*, **82**, 155–62. (Cited on page 298).

Palmer, M. W. (1992). The coexistence of species in fractal landscapes. *American Naturalist*, **139**, 375–97. (Cited on page 139).

Pelletier, B., Fyles, J. W. & Dutilleul, P. (1999). Tree species control and spatial structure of forest floor properties in a mixed-species stand. *Écoscience*, **6**, 79–91. (Cited on page 153).

Penttinen, A. K., Stoyan, D. & Henttonen, H. M. (1992). Marked point processes in forest statistics. *Forest Science*, **38**, 806–24. (Cited on page 55).

Peralta, P. & Mather, P. (2000). An analysis of deforestation patterns in the extractive reserves of Acre, Amazonia, from satellite imagery: a landscape ecological approach. *International Journal of Remote Sensing*, **21**, 2555–70. (Cited on pages 95, 269).

Perry, J. N. (1994). Chaotic dynamics can generate Taylor's power-law. *Proceedings of the Royal Society of London Series B, Biological Sciences*, **257**, 221–6. (Cited on page 94).

(1995). Spatial analysis of distance indices. *Journal of Animal Ecology*, **64**, 303–14. (Cited on page 94).

(1996). Simulating spatial patterns of counts in agriculture and ecology. *Computers and Electronics in Agriculture*, **15**, 93–109. (Cited on page 94).

(1999). Red–blue plots for detecting clusters in count data. *Ecology Letters*, **2**, 106–13. (Cited on page 94).

Perry, J. N., Smith, R. H., Woiwod, I. P. & Morse, D. R., eds. (2000). *Chaos in Real Data: The Analysis of Non-Linear Dynamics from Short Ecological Time Series*. 225pp. Dordrecht: Kluwer Academic. (Cited on pages 307, 310).

Perry, J. N., Woiwod, I. P. & Hanski, I. (1993). Using response-surface methodology to detect chaos in ecological time series. *Oikos*, **68**, 329–39. (Cited on page 310).

Peters, V. S. (2003). Keystone processes affect succession in boreal mixed woods: the relationship between masting in white spruce and fire history. Ph.D. Thesis, University of Alberta. (Cited on page 302).

Petrovskii, S. V. & Malchow, H. (2001). Wave of chaos: new mechanism of pattern formation in spatio-temporal population dynamics. *Theoretical Population Biology*, **59**, 157–74. (Cited on page 313).

Pielou, E. C. (1959). The use of point-to-plant distances in the study of the pattern of plant populations. *Journal of Ecology*, **47**, 607–13. (Cited on page 34).
(1977). *Mathematical Ecology*. New York: Wiley. (Cited on pages 41, 58, 239).
Pitas, I. (2000). *Digital Image Processing Algorithms and Applications*. New York: Wiley-Interscience. (Cited on pages 190, 207).
Plotkin, J. B., Potts, M. D., Leslie, N., Manokaran, N., LaFrankie, J. & Ashton, P. S. (2000). Species-area curves, spatial aggregation, and habitat specialization in tropical forests. *Journal of Theoretical Biology*, **207**, 81–99. (Cited on page 335).
Plotnick, R. E., Gardner, R. H., Hargrove, W. W., Pretegaard, K. & Perlmutter, M. (1996). Lacunarity analysis: a general technique for the analysis of spatial patterns. *Physical Review E*, **53**, 5461–8. (Cited on pages 79, 80).
Pollard, J. H. (1971). On distance estimators of density in randomly distributed forests. *Biometrics*, **27**, 991–1002. (Cited on page 35).
Porteus, B. T. (1987). The mutual independence hypothesis for categorical data in complex sampling schemes. *Biometrika*, **74**, 857–62. (Cited on page 235).
Qi, Y. & Wu, J. (1996). Effects of changing scale on the results of landscape pattern analysis using spatial autocorrelation indices. *Landscape Ecology*, **11**, 39–50. (Cited on pages 18, 112, 144).
Ranta, E., Kaitala, V., Lindström, J. & Helle, E. (1997). The Moran effect and synchrony in population dynamics. *Oikos*, **78**, 136–42. (Cited on page 302).
Redding, T. E., Hope, G. D., Fortin, M.-J., Schmidt, M. G. & Bailey, W. G. (2003). Spatial patterns of soil temperature and moisture across subalpine forest-clearcut edges in the southern interior of British Columbia. *Canadian Journal of Soil Science*, **83**, 121–30. (Cited on page 206).
Reich, R. M., Metzger, K. L. & Bonham, C. D. (1997). Application of permutation procedures for comparing multi-species point patterns of grassland plants. *Grassland Science*, **43**, 189–95. (Cited on pages 41, 50, 51, 52).
Renshaw, E. & Ford, E. D. (1984). The description of spatial pattern using two-dimensional spectral analysis. *Vegetatio*, **56**, 75–85. (Cited on page 95).
Ripley, B. D. (1976). The second-order analysis of stationary point processes. *Journal of Applied Probability*, **13**, 255–66. (Cited on pages 37, 41).
(1978). Spectral analysis and the analysis of pattern in plant communities. *Journal of Ecology*, **66**, 965–81. (Cited on pages 95, 98).
(1981). *Spatial Processes*. New York: Wiley. (Cited on pages 1, 2, 233).
(1988). *Statistical Inference for Spatial Processes*. Cambridge: Cambridge University Press. (Cited on page 39).
Rooney, S. M., Wolfe, A. & Hayden, T. J. (1998). Autocorrelated data in telemetry studies: time to independence and the problem of behavioural effects. *Mammal Review*, **29**, 89–98. (Cited on pages 288, 289).
Rossi, R. E., Mulla, D. J., Journel, A. J. & Franz, E. H. (1992). Geostatistical tools for modeling and interpreting ecological spatial dependence. *Ecological Monographs*, **62**, 277–314. (Cited on pages 135, 137, 165).
Ruxton, G. D. (1993). Linked populations can still be chaotic. *Oikos*, **68**, 347–8. (Cited on page 310).
Sadahiro, Y. & Umemura, M. (2002). A computational approach for the analysis of changes in polygon distributions. *Journal of Geographical Systems*, **3**, 137–54. (Cited on page 269).
Salvatori, V., Skidmore, A. K., Corsi, F. & Van der Meer, F. (1999). Estimating temporal independence of radio-telemetry data on animal activity. *Journal of Theoretical Biology*, **198**, 567–74. (Cited on page 288).

Sanuy, D. & Bovet, P. (1997). A comparative study on the paths of five anura species. *Behavioural Processes*, **41**, 193–9. (Cited on page 284).

Schroeder, M. (1991). *Fractals, Chaos, Power Laws*. New York: Freeman. (Cited on pages 305, 307).

Schultz, C. B. & Crone, E. E. (2001). Edge-mediated dispersal behavior in a prairie butterfly. *Ecology*, **82**, 1879–92. (Cited on pages 284, 285, 292).

Setzer, R. W. (1985). Spatio-temporal patterns of mortality in *Pemphigus populicaulis* and *P. populitransversus* on cottonwoods. *Oecologia*, **67**, 310–21. (Cited on page 258).

Sherratt, J. A. (2001). Periodic travelling waves in cyclic predator–prey systems. *Ecology Letters*, **4**, 30–7. (Cited on page 304).

Sherratt, T. N., Lambin, X., Petty, S. J., MacKinnon, J. L., Coles, C. F. & Thomas, C. J. (2000). Use of coupled oscillator models to understand synchrony and traveling waves in populations of the field vole *Microtus agrestis* in northern England. *Journal of Applied Ecology*, **37**, 148–58. (Cited on page 304).

Shimatani, K. (2001). Multivariate point processes and spatial variation of species diversity. *Forest Ecology and Management*, **142**, 215–29. (Cited on page 335).

Simard, Y., Marcotte, D. & Naraghi, K. (2003). Three-dimensional acoustic mapping and simulation of krill distribution in the Saguenay–St. Lawrence Marine Park whale feeding ground. *Aquatic Living Resources*, **16**, 137–44. (Cited on page 165).

Simberloff, D. (1979). Nearest neighbor assessments of spatial configurations of circles rather than points. *Ecology*, **60**, 679–85. (Cited on page 103).

Smouse, P. E., Long, J. C. & Sokal, R. R. (1986). Multiple regression and correlation extensions of the Mantel test of matrix correspondence. *Systematic Zoology*, **35**, 627–32. (Cited on page 150).

Sokal, R. R. & Oden, N. L. (1978). Spatial autocorrelation in biology. 1. Methodology. *Biological Journal of the Linnean Society*, **10**, 199–228. (Cited on page 120).

Sokal, R. R., Oden, N. L. & Thomson, B. A. (1988). Genetic changes across language boundaries in Europe. *American Journal of Physical Anthropology*, **76**, 337–61. (Cited on page 328).

(1998). Local spatial autocorrelation in biological variables. *Biological Journal of the Linnean Society*, **65**, 41–62. (Cited on page 154).

Sokal, R. R., Oden, N. L., Thomson, B. A. & Kim, J. (1993). Testing for regional differences in means: distinguishing inherent from spurious spatial autocorrelation by restricted randomization. *Geographical Analysis*, **25**, 199–210. (Cited on pages 151, 326).

Sokal, R. R. & Rohlf, F. J. (1981). *Biometry*. 2nd edn. San Fransisco: Freeman. (Cited on page 235).

(1995). *Biometry*. 3rd edn. San Fransisco: Freeman. (Cited on pages 227, 236, 243, 247, 277).

Sokal, R. R. & Wartenberg, D. E. (1983). A test of spatial autocorrelation using an isolation-by-distance model. *Genetics*, **105**, 219–37. (Cited on page 116).

Solé, R. V. & Bascompte, J. (1995). Measuring chaos from spatial information. *Journal of Theoretical Biology*, **175**, 139–47. (Cited on page 307).

Solow, A. R. (1989). Bootstrapping sparsely sampled spatial point patterns. *Ecology*, **70**, 379–82. (Cited on page 288).

St-Louis, V., Fortin, M.-J. & Desrochers, A. (2004). Association between microhabitat and territory boundaries of two forest songbirds. *Landscape Ecology*, **19**, 591–601. (Cited on page 204).

Stone, L. & Ezrati, S. (1996). Chaos, cycles and spatiotemporal dynamics in plant ecology. *Journal of Ecology*, **84**, 279–91. (Cited on pages 307, 310).

Stoyan, D., Kendall, W. S. & Mecke, J. (1995). *Stochastic Geometry and its Applications*. New York: Wiley. (Cited on page 105).
Stoyan, D. & Penttinen, A. (2000). Recent applications of point process methods in forestry statistics. *Statistical Science*, **15**, 61–78. (Cited on page 55).
Sutcliffe, O. L., Thomas, C. D. & Moss, D. (1996). Spatial synchrony and asynchrony in butterfly population dynamics. *Journal of Animal Ecology*, **65**, 85–95. (Cited on page 303).
Swihart, R. K. & Slade, N. A. (1985). Testing for independence of observations in animal movements. *Ecology*, **66**, 1176–84. (Cited on page 288).
(1986). The importance of statistical power when testing for independence in animal movements. *Ecology*, **67**, 255–8. (Cited on page 288).
Tavaré, S. (1983). Serial dependence in contingency tables. *Journal of the Royal Statistical Society B*, **45**, 100–6. (Cited on pages 234, 235).
Tavaré, S. & Altham, P. M. E. (1983). Serial dependence of observations leading to contingency tables, and corrections to chi-squared statistics. *Biometrika*, **70**, 139–44. (Cited on pages 234, 238).
TerraSeer (2001). *BoundarySeer*. [V1.0]. Online: www.terraseer.com. (Cited on page xii).
Thomas, D. L. & Taylor, E. J. (1990). Study designs and tests for comparing resource use and availability. *Journal of Wildlife Management*, **54**, 322–30. (Cited on page 287).
Tischendorf, L. (2001). Can landscape indices predict ecological processes consistently? *Landscape Ecology*, **16**, 235–54. (Cited on page 175).
Tobin, P. C. & Bjørnstad, O. N. (2003). Spatial dynamics and cross-correlation in a transient predator–prey system. *Journal of Animal Ecology*, **72**, 460–7. (Cited on page 301).
Tobler, W. F. (1970). A computer movie simulating urban growth in the Detroit region. *Economic Geography*, **46**, 234–40. (Cited on pages 7, 212).
Todd, K. W., Csillag, F. & Atkinson, P. M. (2003). Three-dimensional mapping of light transmittance and foliage distribution using lidar. *Canadian Journal of Remote Sensing*, **29**, 544–55. (Cited on page 170).
Toussaint, G. T. (1980). The relative neighbourhood graph of a finite planar set. *Pattern Recognition*, **12**, 261–8. (Cited on page 59).
Turchin, P. (1996). Fractal analyses of animal movement: a critique. *Ecology*, **77**, 2086–90. (Cited on page 285).
(1998). *Quantitative Analysis of Movement*. Sunderland: Sinauer. (Cited on pages 276, 277, 284, 289).
Turchin, P. & Taylor, A. D. (1992). Complex dynamics in ecological time series. *Ecology*, **73**, 289–305. (Cited on page 310).
Turner, M. G., Gardner, R. H. & O'Neill, R. V. (2003). *Landscape Ecology in Theory and Practice: Pattern and Process*. New York: Springer-Verlag. (Cited on page 176).
Upton, G. J. G. & Fingleton, B. (1985). *Spatial Data Analysis by Example*, vol. I: *Point Pattern and Quantitative Data*. New York: Wiley. (Cited on pages 1, 34, 36, 44, 62, 233, 241).
(1989). *Spatial Data Analysis by Example*, vol. II: *Categorical and Directional Data*. New York: Wiley. (Cited on pages 234, 274, 276).
Urban, D. & Keitt, T. (2001). Landscape connectivity: a graph-theoretic perspective. *Ecology*, **82**, 1205–18. (Cited on pages 64, 66, 67, 74).
Vacek, S. & Leps, J. (1996). Spatial dynamics of forest decline: the role of neighbouring trees. *Journal of Vegetation Science*, **7**, 789–98. (Cited on page 263).

van Es, H. M. & van Es, C. L. (1993). The spatial nature of randomization and its effects on the outcome of field experiments. *Agronomy Journal*, **85**, 420–8. (Cited on page 249).

van Lieshout, M. N. M. & Baddeley, A. J. (1999). Indices of dependence between types in multivariate point patterns. *Scandinavian Journal of Statistics*, **26**, 511–32. (Cited on page 48).

Venables, W. N. & Ripley, B. D. (2002). *Modern Applied Statistics with S*, 4th edn. New York: Springer-Verlag. (Cited on page 29).

Ver Hoef, J. M., Cressie, N. A. C. & Glenn-Lewin, D. C. (1993). Spatial models for spatial statistics: some unification. *Journal of Vegetation Science*, **4**, 441–52. (Cited on page 83).

Ver Hoef, J. M. & Glenn-Lewin, D. C. (1989). Multiscale ordination: a method for detecting pattern at several scales. *Vegetatio*, **82**, 59–67. (Cited on pages 89, 90).

Viljugrein, H., Lingjærde, O. C., Stenseth, N. C. & Boyce, M. S. (2001). Spatio-temporal patterns of mink and muskrat in Canada during a quarter century. *Journal of Animal Ecology*, **70**, 671–82. (Cited on page 304).

Wackernagel, H. (2003). *Multivariate Geostatistics: An Introduction with Applications*. New York: Springer-Verlag. (Cited on pages 147, 170).

Wakefield, J. C., Kelsall, J. E. & Morris, S. E. (2000). Clustering, cluster detection, and spatial variation in risk. In *Spatial Epidemiology: Methods and Applications*, eds. P. Elliott, J. C. Wakefield, N. G. Best & D. J. Briggs, pp. 128–52. Oxford: Oxford University Press. (Cited on pages 266, 267).

Wallerman, J., Joyce, S., Vencatasawmy, C. P. & Olsson, H. (2002). Prediction of forest stem volume using kriging adapted to detected edges. *Canadian Journal of Forest Research*, **32**, 509–18. (Cited on page 169).

Wartenberg, D. (1985). Multivariate spatial autocorrelation: a method for explanatory geographical analysis. *Geographical Analysis*, **17**, 263–83. (Cited on page 147).

Watkinson, A. R., Lonsdale, W. M. & Firbank, L. G. (1983). A neighbourhood approach to self-thinning. *Oecologia*, **56**, 381–4. (Cited on page 298).

Watt, A. S. (1947). Pattern and process in the plant community. *Journal of Ecology*, **35**, 1–22. (Cited on pages 256, 257, 263).

Weber, K. T., Burcham, M. & Marcum, C. L. (2001). Assessing independence of animal locations with association matrices. *Journal of Rangeland Management*, **54**, 21–4. (Cited on page 288).

Webster's (1989). *Webster's Encyclopaedic Unabridged Dictionary of the English Language*. New Jersey: Gramercy Books. (Cited on page 5).

Webster, R. (1973). Automatic soil-boundary location from transect data. *Mathematical Geology*, **5**, 27–37. (Cited on page 187).

Webster, R. & Oliver, M. (2001). *Geostatistics for Environmental Scientists*. Chichester: Wiley. (Cited on pages 22, 112, 132, 137, 144, 160, 169, 170).

Weishampel, J. F., Godin, J. R. & Henebry, G. M. (2001). Pantropical dynamics of "intact" rain forest canopy structure. *Global Ecology and Biogeography*, **10**, 389–97. (Cited on page 95).

White, G. C. & Garrott, R. A. (1990). *Analysis of Wildlife Radio-Tracking Data*. New York: Academic Press. (Cited on page 287).

Whittaker, R. H. (1972). Evolution and measurement of species diversity. *Taxon*, **21**, 213–51. (Cited on page 187).

Whittle, P. (1954). On stationary processes in the plane. *Biometrika*, **41**, 434–49. (Cited on page 219).

Wiens, J. A. (1989). Spatial scaling in ecology. *Functional Ecology*, **3**, 385–97. (Cited on page 5).
Wiens, J. A., Crist, T. O. & Milne, B. T. (1993). On quantifying insect movements. *Environmental Entomology*, **22**, 709–15. (Cited on pages 284, 285).
Williams, B. (1992). The measurement of sinuosity in correlated random walks. *Journal of Theoretical Biology*, **155**, 437–42. (Cited on page 284).
Wilson, M. V. & Mohler, C. L. (1983). Measuring compositional change along gradients. *Vegetatio*, **54** 129–41. (Cited on page 187).
Wolfram, S. (2002). *A New Kind of Science*. Champagne: Wolfram Media. (Cited on pages 312, 313).
Womble, W. H. (1951). Differential systematics. *Science*, **114**, 315–22. (Cited on page 190).
Woodcock, C. & Strahler, A. (1987). The factor of scale in remote sensing. *Remote Sensing of Environment*, **21**, 311–22. (Cited on page 146).
Woollons, R. C. (1998). Even-aged stand mortality estimation through a two-step regression process. *Forest Ecology and Management*, **105**, 189–95. (Cited on page 292).
Wu, J. & Qi, Y. (2000). Dealing with scale in landscape analysis: an overview. *Geographic Information Sciences*, **6**, 1–5. (Cited on page 146).
Wu, J. G., Shen, W. J. & Sun, W. Z. (2002). Empirical patterns of the effects of changing scale on landscape metrics. *Landscape Ecology*, **117**, 761–82.
Wu, X. B., Thuro, T. L. & Whisenant, S. G. (2000). Fragmentation and changes in hydrologic function of tiger bush landscapes, south-west Niger. *Journal of Ecology*, **88**, 790–800. (Cited on pages 10, 95, 261).
Wulder, M. & Boots, B. (1998). Local spatial autocorrelation characteristics of remotely sensed imagery assessed with the Getis statistic. *International Journal of Remote Sensing*, **19**, 2223–331. (Cited on pages 158, 159).
Yarranton, G. A. (1966). A plotless method of sampling vegetation. *Journal of Ecology*, **59**, 224–37. (Cited on page 50).
Young, C. G., Dale, M. R. T., & Henry, G. H. R (1999). Spatial pattern of vegetation in high Arctic sedge meadows. *Écoscience*, **6**, 556–64. (Cited on pages 89, 236).
Zadeh, L. A. (1965). Fuzzy sets. *Information and Control*, **8**, 338–53. (Cited on page 181).
Zar, J. H. (1984). *Biostatistical Analysis*, 2nd edn. Englewood Cliffs, NJ: Prentice-Hall. (Cited on page 276).

# Index